Immunopharmacology

KU-735-515

LIVERPOOL
JOHN MOORES UNIVERSITY
AVRIL ROBARTS LRC
TITHEBARN STREET
LIVERPOOL L2 2ER
TEL 0151 231 4022

WITHDRAWN

LIVERPOOL JMU LIBRARY

3 1111 01187 4946

Preface

As we entered the twenty-first century, major advances in the arena of recombinant DNA, hybridoma and transgenic technologies had not only revolutionized the understanding of the etiology and pathogenesis of a number of debilitating and life-threatening diseases, but also provided novel modes of treatment. Whether it is the clinical application of recombinant cytokines, their agonists or antagonists, monoclonal antibodies, regulatory T cells, gene therapy or the concept of T cell vaccines, all these require understanding of an evolving discipline that worked on the interface of immunology, pathology, pharmacology and genetics called Immunopharmacology. The initial emphasis of the discipline was the development of the drugs that suppressed immune response to prevent tissue rejection after organ transplantation. The field once considered restricted only to protect the host from invading organisms by mounting immune and inflammatory responses evolved exponentially as we gradually learned about the exciting and sometimes adverse role of the products of the immune response in a very wide range of physiological and pathological settings ranging from cardiovascular, pulmonary, gastrointestinal to neurological function. A number of these products and therapies based on their understanding not only continue to become symptomatic and curative therapeutic agents but have extensively contributed to the early diagnosis of a number of dreadful disorders.

This book is written for graduate students in pharmacology and professional students in pharmacy and medicine. The introductory chapter is aimed for the students who have not previously taken a course in Basic Immunology. The Chapters 2 and 3 focus on cytokines, their receptors, pharmacology and clinical applications. The next section is devoted to the pharmacology of immune regulatory agents, monoclonal antibodies and etiology and mechanisms of IgE-mediated responses and immunotherapy for allergic disease. The following section includes chapters on the mechanisms of allograft rejections with description of the requirements for different types of clinical tissue transplantation and immunological basis of acquired immune deficiency disease. In the chapter on AIDS, the emphasis has been on the life cycle of HIV, available therapeutic options and the difficulties associated with the development of a vaccine for AIDS and why an HIV vaccine does not fit the paradigm for the classical vaccine development. The last part of the book includes chapters on regulatory T cells and their therapeutic potential followed by the last chapter on the challenges and use of gene therapy to treat human disease.

Omaha, NE   Manzoor M. Khan

Chapter 1
Overview of the Immune Response

Introduction

The immune system is a defense system that protects from infectious organisms and cancer. It is made up of various cells, proteins, tissues and organs. The cells participating in immune response are designed to recognize and eliminate invading agents. If not eliminated, these invading microorganisms may cause disease. The recognition is a very specific process that enables the body to recognize nonself molecules so that the second phase of the process, an immune response, may initiate. Under normal circumstances, an immune response is not generated against self, which are the body's own proteins and tissues. The immune system for each individual is unique, and it employs small and efficient tools to recognize invading organisms that lack a central control. However, they are widely distributed in the body. The recognition of the nonself is not perfect and absolute detection of every pathogen is not required. The immune system is able to recognize molecules that it has never seen before and can produce an effective response against them. It has been suggested that this system is scalable, resilient to subversion, robust, very flexible and degrades.

After the recognition of the nonself, the effector phase is generated, which is characterized by the generation of a response against the invading microorganism in which various cells and molecules participate, resulting in the neutralization and/or the elimination of the pathogen. Some memory is retained of that pathogen and a second exposure to the same organism results in the development of a memory response. This response has a quick onset and is fiercer, resulting in a more efficient elimination of the pathogen.

Components

Innate Immunity

This is a general protection that is also termed natural immunity that is present at birth in all individuals. A species is armed with innate immunity that provides an individual the basic resistance to disease. This is also called nonspecific immunity

M.M. Khan, *Immunopharmacology*, DOI: 10.1007/978-0-387-77976-8_1,
© Springer Science+Business Media, LLC 2008

and is the initial defense against infections. It is characterized as broad-spectrum responses, with limited repertoire of recognition molecules and a lack of memory component. Since it is the first line of defense, it is present at birth, it is nonspecific and it does not allow an increase in resistance after repeated infections. It destroys vast amounts of microorganisms in a short time, with which an individual comes in contact every day, and protects from causing disease. There are three types of barriers for the innate immunity, physical barriers, chemical barriers and cellular barriers.

Physical Barriers

The physical barriers include the skin, mucus membranes, epidermis and dermis. The skin maintains a low pH because of lactic and fatty acids. The mucus membranes in the respiratory system, urogenital system and gastrointestinal (GI) system create a substantial surface barrier. Epidermis and dermis constitute additional physical barriers. The dermis also produces sebum, which maintains an acidic pH due to its lactic and fatty acid content.

Chemical Barriers

A number of endogenous chemicals provide effective barriers in innate immunity. They include hydrolytic enzymes of saliva, low pH of stomach and vagina and proteolytic enzymes in the small intestine. Additional examples include cryptidins, α-defensins, β-defensins, interferons (IFNs), and surfactant proteins A and D. Cryptidins and α-defensins are produced in the small intestine and β-defensins are produced by the skin and respiratory tract.

Cellular Barriers

The cellular barriers include macrophages, eosinophils, phagocytes and natural killer (NK) cells. Some of these cells internalize macromolecules that they encounter in circulation or in tissues. This internalization takes place either by pinocytosis, receptor-mediated endocytosis or phagocytosis. The pinocytosis involves nonspecific membrane invagination. In contrast, receptor-mediated endocytosis involves specific macromolecules that are internalized after they bind to respective cell surface receptors. Endocytosis is not cell-specific and is carried out probably by all cells.

As opposed to endocytosis, phagocytosis is more cell-specific and results in the ingestion of particulate as well as whole microorganisms. The cells involved in phagocytosis include monocytes and macrophages, neutrophils and dendritic cells. Furthermore, fibroblasts and epithelial cells can also be induced to assume phagocytic activity.

Innate Mechanisms

The common microbial patterns render recognition by organisms and their self/nonself recognition is based on innate mechanisms. The common microbial patterns include lipopolysaccharides (LPS), mannose, fucose, technoic acid and N-formyl peptides. These common microbial patterns are called pattern recognition molecules (PRMs) or pattern recognition receptors (PRRs).

Adaptive/Acquired Immune Response

This immune response occurs when the body encounters an antigen and/or a pathogen. With this response the body protects itself from future encounters with the same antigen/pathogen so that they will not cause disease. The response is more complex than the innate immune response. It requires the recognition and processing of the immune response. After the antigen is recognized, the adaptive immune response employs humoral and cellular responses specifically designed to eliminate the antigen. This response also includes a memory component that allows improved resistance against that specific antigen during subsequent infections. T lymphocytes, B lymphocytes and macrophages participate in the acquired immune response. The lymphocytes (T and B cells) are central to all acquired immune responses, because of their specificity in recognizing the pathogens. This recognition can take place either inside the tissue or in blood or tissue fluids. B cells recognize antigens by synthesizing and releasing antibodies that specifically recognize antigens. T lymphocytes do not secrete antibodies but have a wide range of regulatory and effector functions.

Immunogens and Antigens

Immunogens are any agents capable of inducing an immune response. Their characteristics include foreignness, high molecular weight and chemical complexity.

Antigens are any agents capable of binding specifically to the components of immune response. They include carbohydrates, lipids, nucleic acids and proteins. Most immunogenic molecules that induce an immune response require both T and B lymphocytes. Because T lymphocytes mature in thymus, these immunogens are called thymus-dependent antigens. Certain types of molecules can induce the production of antibodies without the apparent participation of T lymphocytes. Such molecules are referred to as thymus-independent antigens. The portion of an antigen that binds specifically with the binding site of an antibody or a receptor on lymphocytes is termed the epitope.

Antibodies

The antibodies are proteins produced during the immune response. They identify and eliminate foreign objects such as bacteria and viruses. They are synthesized in response to specific antigens and bind only to the antigen against which they are produced. Antibodies are glycoproteins and are also called immunoglobulins. They are synthesized and secreted by B lymphocytes that produce antibodies after their activation resulting from exposure to an antigen. The antibodies can circulate freely or in the bound form attached to the cells that possess Fc receptors.

The antibody molecule (monomer) is a Y-shaped structure (Fig. 1.1). It consists of two identical light and heavy chains, which are connected by disulfide bonds. In the native state, the chains are coiled into domains, each of which consists of 110 amino acids. Both light and heavy chains are made up of constant and variable regions. The two identical light (L) chains and two identical heavy (H) chains are held together by disulfide bonds. Both light and heavy chains have constant and variable regions. Two major classes of L chains are kappa and lambda and the ratio of κ and λ chains varies from species to species. Papain splits the immunoglobulin molecules into three fragments of about equal size. Two fragments are antigen-binding fragment (Fab) and the third fragment is fragment crystallizable (Fc). The variable region of both the heavy and light chains forms the antigen-binding site (Fab). The Fab region varies according to the specificity of the antibody. The antibodies are very diverse molecules and their differences reside predominantly in the variable region. This variability enables each antibody to recognize a particular antigen.

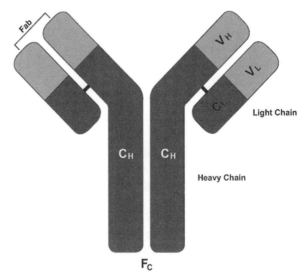

Fig. 1.1 The immunoglobulin molecule; each immunoglobulin molecule is composed of two identical heavy (CH + VH) and two identical light chains (CL + VL). The antigen binds to the Fab region, which varies according to the specificity of the antibody. The rest of the domains (blue) are constant. The classes of the antibody molecules differ based on the Fc region of the heavy chain (*see* Color Insert)

There are five different types of heavy chains, which correspond with five different classes of antibodies. The constant region of the heavy chain is identical in all antibodies of the same class. The classes of the immunoglobulin molecules differ based on the Fc region of the heavy chain, which are responsible for the different functions performed by each class. Thus the constant region confers on each class of antibody its effector function.

As depicted in Table 1.1, there are five different classes of H chains. Each chain differs in antigenic reactivity, carbohydrate content and biological function. The nature of the H chain confers to the molecule its unique biological properties. A distinctive set of glycoforms characterizes each immunoglobulin. The glycoforms render broad differences in the frequency, form and locality of oligosaccharides, which are responsible for the diversity of immunoglobulins. Since these glycoform populations can be identified on a regular basis, any alteration in their characteristics suggests a disease state and could be a potential therapeutic tool. The oligosaccharides possess critical recognition epitopes. This provides the immunoglobulins with additional functional repertoire. The effector function of immunoglobulins is thus regulated by these sugar molecules.

The antibody molecule has two distinct functions. One is to bind to the pathogen such as the virus or the bacteria against which the immunoglobulins were produced and the second is to recruit other cells and molecules, such as phagocytes or neutrophils, to destroy the pathogen to which the antibody is bound.

Table 1.1 Classes (Isotypes) of H Chains

Immunoglobulin Class (Isotype)	Heavy Chain
IgM	μ
IgG	γ
IgA	α
IgD	δ
IgE	ε

Classes of Immunoglobulins

There are five different classes of immunoglobulins; IgG, IgM, IgA, IgD and IgE. There structures are shown in Fig. 1.2.

IgG: Immunoglobulin G is present in lymph fluid, blood, cerebrospinal fluid and peritoneal fluid. It is composed of 2 γ chains of 50 kDa and 2 L chains (κ or λ) of 25 kDa with a total molecular weight of 150 kDa. The functions of IgG include agglutination and formation of precipitate, passage through placenta and thus conferring immunity to fetus, opsonization, antibody-dependent cell-mediated cytotoxicity (ADCC), activation of complement, neutralization of toxins, immobilization of bacteria and neutralization of virus.

IgM: The gene segment that encodes the μ constant region of the heavy chain occupies the front position among other constant region gene segments and consequently IgM is the first immunoglobulin produced by mature B cells. It has a

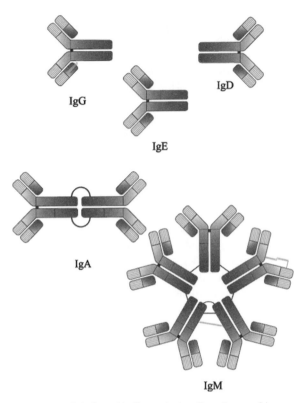

Fig. 1.2 Classes of immunoglobulins; this figure depicts five classes of immunoglobulins, IgG, IgE, IgD, IgA, IgM. IgA can be present as a monomer or a dimer molecule whereas IgM exists as a pentamer (*see* Color Insert)

molecular weight of 900 kDa and is a five-chain structure. All chains consist of 2 L and 2 H chains and have five antigen-binding sites. It is synthesized in appreciable amounts in children and adults after immunization or exposure to thymus-independent antigens. Elevated levels usually indicate a recent infection. IgM does not pass across the placenta but is synthesized by the placenta. Its elevated levels in the fetus are indicative of congenital infection. It is the best agglutinating and complement-activating antibody and possesses high avidity and is sometimes referred to as a natural antibody since it is often bound to specific antigens, even when there was no prior immunization.

IgA: Immunoglobulin A is a major immunoglobulin in external secretions (saliva, mucus, sweat, gastric fluid and tears). It is a major immunoglobulin of colostrum and milk, it has a molecular weight of 165 kDa and is present in both monomeric and dimeric forms. IgA is present as two isotypes, IgA1 (90%) and IgA2 (10%). The bone marrow B cells produce IgA1, which is present in serum. The B cells located in the mucosa synthesize IgA2 which is present in secretions. Chemically the heavy and light chains of IgA2 are bound by noncovalent bonds and are not connected by disulfide bridges. It plays an important role in mucosal infections, bactericidal activity and antiviral activity. Plasma cells produce polymeric IgA and mucosal epithelial cells express polymeric Ig receptor, resulting in high levels of IgA in mucosal areas. This is followed by its transportation across mucosal epithelial cells where it is separated from its receptors, resulting in its release into secretions. Its effects are achieved after interaction with specific receptors including FcR1, Fcα/μ R and CD71. However, certain pathogens block the protective properties of IgA.

IgD: Immunoglobulin D causes differentiation of B cells to a more mature form and is expressed on the surface of B lymphocytes. It is present in a monomeric form with a molecular weight of 180 kDa.

IgE: Immunoglobulin E is associated with type 1 hypersensitivity reaction and allergic disease. Its molecular weight is 200 kDa. It also plays a role in host defense against parasitic infections. IgE binds to specific Fc receptors on the cell surface of mast cells, basophils, eosinophils, macrophages, monocytes and platelets. Two main types of Fc receptors for IgE include FcεRI and FcεRII. The former is a high-affinity receptor whereas the later, also termed CD23, is a low-affinity receptor. FcεRI receptors are present on mast cells and basophils, whereas FcεRII are present on B cells, although their expression can be induced on other cell types including monocytes, macrophages, eosinophils and platelets by the TH_2 cytokine interleukin (IL)-4. IgE serves as a stimulus for the upregulation of both receptors. Binding of IgE to its receptors on mast cells results in the release of various endogenous mediators including several cytokines, and the symptoms can vary from a mild allergic response to potentially life-threatening anaphylactic shock. Normal physiological levels of IgE are low, but under atopic conditions, its levels rise as a result of an isotype switch from IgG to IgE in response to an antigen and under the influence of TH_2 cell-derived cytokines.

Cell Cooperation in Antibody Response

After exposure to an antigen, its recognition by the immune system is followed by either the production of an immune response or the development of tolerance, depending on the circumstances. The immune response could be humoral, cell-mediated or both. On second and subsequent encounters with the same antigen, the type of response is determined by the outcome of the first response. However, the quantity and the quality of both responses are very different.

Primary and Secondary Antibody Responses

After administration of an antigen for the first time, there is an initial lag phase where antibodies are not produced. This is followed by a period in which the antibody titer rises logarithmically to a maximum and subsequently declines. The decline is due to either the breakdown or the clearance of the antibodies.

The primary and secondary responses differ in the following four ways:

1. *Time course.* The secondary response has a shorter lag phase and an extended plateau and decline.
2. *Antibody levels.* The antibody levels are 10 times higher in the secondary response compared to the primary response.
3. *Antibody class.* The major proportion of the primary response is made up of IgM, whereas the secondary response consists almost entirely of IgG.
4. *Antibody affinity.* The affinity of the antibodies is much greater in the secondary response as opposed to the primary response, which is termed "affinity maturation."

Cells Involved in the Immune Response

The immune system is composed of various different cell types and organs that are involved in specifically recognizing nonself antigens to eliminate them. Phagocytes are an important defense that participates in both innate and acquired immune responses. The lymphoid cells render the high degree of specificity involved in the recognition of nonself antigens and are part of the acquired immune response. All cells participating in the immune response arise from pluripotent stem cells and are divided into the lymphoid lineage—consisting of lymphocytes—and the myeloid cells—consisting of phagocytes (monocytes and neutrophils) and other cells.

There are three different kinds of lymphocytes that have specific functions: T cells, B cells and NK cells. T cells develop in the thymus while B cells develop in the adult bone marrow. The thymus and the bone marrow are the primary lymphoid organs where lymphocytes acquire specific cell surface receptors that give them the ability to recognize antigens. NK cells are cytotoxic lymphocytes that develop in the bone marrow. The phagocytes are made up of either monocytes (macrophages) or polymorphonuclear granulocytes, which include neutrophils, eosinophils and basophils.

Lymphoid Cells

All lymphoid cells originate during hematopoiesis from a common lymphoid progenitor in the bone marrow. Their formation is known as lymphopoiesis. B cells mature in the bone marrow while T cells mature in the thymus. The bone marrow and the thymus are called primary lymphoid organs. This is followed by migration via circulation into the secondary lymphoid tissue (spleen, lymph nodes, tonsils and unencapsulated lymphoid tissue). The average human adult has about 10^{12} lymphoid cells, and lymphoid tissue as a whole represents about 2% of the total body weight. Lymphoid cells represent about 20% of the total white blood cells present in the adult circulation. After culmination of the immune response, many mature lymphoid cells live a very long life as memory cells.

Morphology: Lymphocytes possess a large nucleus with little to no basophilic cytoplasm. Differences are seen in the nuclear (N) to cytoplasmic ratio, the degree of cytoplasmic staining with histological dyes and the presence or absence of azurophilic granules.

Markers: Most of the lymphocytes express specific cell surface makers on their cell surface. Some are present for a short duration while others are responsible for their characterization. Such molecules can be used to distinguish various cell subsets. The selected antigenic markers on leukocytes are depicted in Table 1.2.

T Cells

T lymphocytes develop in the thymus and are consequently called T cells. However, their precursors are found in the bone marrow. The presence of the T-cell receptor (TCR) distinguishes T cells from other lymphoid cells. There are presently two defined types of TCR; TCR-2 is a heterodimer of two disulfide-linked polypeptides α and β with a molecular weight of 90 kDa and TCR-1 is structurally similar but consists of γ and δ polypeptide. Each chain has a constant and a variable region. These chains are characterized by an intrachain disulfide bridge and some sequence homology with the immunoglobulin domains. Both receptors are associated with a complex of polypeptides making up the CD3 complex. Thus, a T cell is defined by either TCR-1 or TCR-2, which is associated with CD3. Approximately 95% of blood T cells express TCR-2 and up to 5% have TCR-1. TCR-1 cells are also called $\gamma\delta$ T cells. They are made up of one γ and one δ chain and are mostly present in the gut mucosa with intraepithelial lymphocytes (IEL). These cells do not recognize antigens in the context of major histocompatibility complex (MHC) molecules. In the peripheral blood the $\gamma\delta$ T cell population is composed of Vγ9/Vδ2 T cells and is unique in the sense that this is specific for a nonpeptide microbial metabolite, HMB-PP, which is a precursor of isopentenyl pyrophosphate. The $\gamma\delta$ T cell population produces a rapid response after recognizing HMB-PP. The TCR-2 cells are composed of two glycoprotein chains α and β and are subdivided into two distinct populations, the TH cells, which are CD4$^+$, and the Tc cells, which are CD8$^+$. CD4$^+$ T cells recognize antigens in association with MHC class II molecules, while CD8$^+$ T cells recognize antigens in association with MHC class I molecules.

There are three types of TH cells, TH$_0$, TH$_1$ and TH$_2$, based on the production of different cytokines they secrete. Following activation, TH cells become TH$_0$ cells, which have features of both TH$_1$ and TH$_2$ cells, and are capable of secreting IL-2, IL-3, IL-4, IL-5, IL-6, IL-10, IL-13, IFN-γ, tumor necrosis factor (TNF)-α, TNF-β and granulocyte–monocyte colony-stimulating factor (GM-CSF). After further activation, TH$_0$ cells differentiate into either TH$_1$ or TH$_2$ cells. This classification is on the basis of cytokines secreted by these subsets. TH$_1$ cells secrete IL-2, IL-3, IFN-γ, TNF-α, TNF-β and GM-CSF. The classification of IL-2 as a TH$_1$ cytokine is misleading since all TH cells secrete IL-2 during early stages. TH$_1$ cells are involved in inflammation, cytotoxicity and delayed-type hypersensitivity reactions. TH$_2$ cells secrete IL-4, IL-5, IL-6, IL-10, IL-13, TNF-β and GM-CSF and support B-cell activation, isotype switching and IgE production.

Table 1.2 Selected Antigenic Markers on Leukocytes

Antigen	Molecular Weight (kDa)	Distribution	Function
CD1	43–49	Dendritic cells, B cells	T-cell response
CD2	45–58	T cells, NK cells	T-cell activation
CD3	20–28	T cells, NKT cells	TCR expression and signal transduction
CD4	55	T cells, NKT cells	MHC class II-restricted immune recognition
CD5	58	T cells, B cells	Modulation of TCR and BCR signaling
CD8	32–34 each monomer α and β	T cells, T$_c$ Class I-restricted T cells	Coreceptor
CD11a	180	Leukocytes	α chain of LFA-1 (adhesion molecule)
CD11b	160	Monocytes, granulocytes	α chain of complement receptor CR3 (MAC-1)
CD11c	150	Monocytes, granulocytes	α chain of p150, 95 (complement receptor/adhesion molecule)
CD16	50–80	NK cells, macrophages, neutrophils	Fc receptor subunit (low affinity). Phagocytosis, adenovirus and ADCC
CD21	130 (soluble), 145 (membrane-bound)	B cells, follicular dendritic cells	Receptors for various antigens. Involved in signal transduction
CD22	140	Mature B cells	Adhesion and signaling
CD23	45	B cells, follicular dendritic cells, monocytes	IgE synthesis regulation, induction of inflammatory cytokines
CD25	55	Mitogen-induced T cells, monocytes/macrophages. Anti-IgM-induced B cells	α subunit of IL-2 receptor
CD28	90	Most peripheral T cells. CD3$^+$ thymocytes	Costimulator for T-cell activation
CD29	130	Leukocytes	β subunit of VLA-1 integrin
CD32	40	B cells, monocytes, granulocytes	IgG molecule (Fc region)-antigen-binding receptor
CD40	48 monomer	B lineage cells, follicular dendritic cells, endothelial cells, macrophages	B-cell growth, differentiation, isotype switching. Induction of cytokine release and adhesion molecules
CD45	180–220	Hematopoietic cells	T- and B-cell activation
CD45R	220, 205	B cells, T-cell subsets, granulocytes, monocytes	Restricted leukocyte common antigen
CD80	60	Activated B and T cells, macrophages	Costimulator of T-cell activation
CD86	80	B cells, monocytes, dendritic cells	Costimulator for T-cell activation

Regulatory T Cells

The Treg cells were previously known as the suppressor T cells and are pivotal in maintaining immune tolerance. Their principal functions include a negative feedback after the generation of an immune response to limit unintended damage and to protect from autoimmunity. These cells are described in detail in Chapter 9.

Memory T Cells

Memory T cells persist even after the invading pathogen has been eliminated. There are two types of memory T cells, central memory T cells and effector memory T cells. They are either $CD4^+$ or $CD8^+$ and respond quickly after they see the antigen a second time.

Natural Killer T Cells

Activation of natural killer T (NKT) cells results in the performance of functions similar to $CD4^+$ and $CD8^+$ T cells. They release cytokines including IL-4, IFN-γ, GM-CSF and others, and are also cytotoxic. They are mostly not MHC restricted, instead they recognize antigens that are glycolipids and require CD1d. NKT cells bridge innate and acquired immune responses. They express both $\alpha\beta$ TCR and markers present on NK cells such as NK1.1. However, they include $NK1.1^+$, $NK1.1^-$, $CD4^+$, $CD8^+$, $CD4^-$ and $CD8^-$ cells. They also express CD16 and CD56 antigens that are present usually on NK cells. NKT cells are classified into Type 1 NKT, Type 2 NKT, and NKT-like cells. NKT-like cells are not CD1d restricted and may have MHC or other restrictions. Abnormal NKT cell function may be associated with cancer and autoimmune disease.

B Cells

These lymphocytes are unique due to their ability to secrete immunoglobulins. The letter "B" refers to "bursa of Fabricius," an organ where B cells mature in birds. In humans, B cells are produced in the bone marrow. The development of B cells occurs at various stages, which include progenitor B cells, early pro-B cells, late pro-B cells, large pre-B cells, small pre-B cells, immature B cells and mature B cells. Each stage is characterized by rearrangement of certain genes and expression of receptors. In particular, immature B cells start to express IgM receptors and mature B cells also express IgD receptors.

The cell surface of each receptor possesses unique receptors called B-cell receptors (BCR), which have a membrane-bound immunoglobulin and will bind to a particular antigen. Following encounters with an antigen and recognition of a second signal from T helper cells, B cells differentiate into plasma cells. Memory B cells are also formed from activated B cells, which are antigen-specific, and quickly respond when they encounter the same antigen a second time, producing secondary immune response. All antigens do not require a second signal from TH cells to activate B cells, which is termed T-independent activation.

A single B cell has approximately 1.5×10^5 antibody molecules on its cell surface, which are all specific for a particular antigen. Other molecules expressed on mature B cells include B220, MHC class II molecules, CD21, CD32, CD35, CD40, CD80 and CD86. B220 is a form of CD45—CD45R— and is used to identify B cells despite not being exclusive for B cells. CD40 interacts with its ligand on TH cells, and this interaction is crucial for the development of B cells to differentiate into either antibody-secreting cells or memory cells.

After recognizing antigens through membrane-bound antibodies, there is B cell proliferation and differentiation for about 4–5 days. This results in the production of plasma and memory cells. One of the five classes of antibodies are produced and secreted by plasma cells that do not possess membrane-bound antibodies. Plasma cells survive for about 1–2 weeks.

Natural Killer Cells

After B and T cells, NK cells are the third largest class of lymphocytes that were originally identified due to their spontaneous killing ability of tumor cells. They develop in bone marrow from a common lymphoid progenitor and require IL-15, C-KIT and FLT-3. NK cells share effector function and the ability to produce cytokines with T cells. They were previously referred to as null cells because they do not express either T cell or B cell receptors, do not secrete antibodies and do not possess antigen-recognizing receptors. Their morphology is also different from B and T cells since they are large granular lymphocytes. These cells make up about 5–10% of the lymphocytes in human peripheral blood.

NK cells participate in innate immunity and are the first responders against infection and possibly tumors. They are $CD3^-$ and also lack immunoglobulin receptors; however, they carry CD56 antigen, which is used to identify these lymphocytes. NK cells also express CD16, which is a low-affinity receptor for IgG and is not present on mature T cells. They are involved in antibody-mediated cellular cytotoxicity and apoptosis. The binding of CD16 of NK cells to the Fc portion of IgG facilitates the release of cytoplasmic granules, which cause the destruction of the target cell. Natural cytotoxicity receptors (NCRs) are exclusively expressed on NK cells and include NKp30, NKp44, NKp46 and NKp80. NKp44 and NKp46 are important in defense against viruses, as they bind to influenza hemagglutinin. Another NK cell surface receptor, 2B4, binds to CD48 and may play a role in the defense against Epstein–Barr virus. A defect in 2B4 function is associated with the X-linked lymphoproliferative syndrome. In addition, NK cells possess receptors for various cytokines including IL-2, IL-12, IL-15, IL-21, IFN-α and IFN-β.

The killing ability of NK cells is associated with the expression of MHC class I molecules. According to the "missing-self hypothesis," NK cells search for the presence of MHC class I molecules that are ubiquitously expressed. A decrease in the expression of MHC class I molecules on a cell allows NK cells to kill the target as it is released from the influence of MHC class I molecules. The ability of NK cells to kill tumors and virally infected cells resides in several inhibitory receptors called immunoglobulin-like receptors and CD94/NKG2 heterodimers. The immunoglobulin-like receptors include KIR, immunoglobulin-like transcripts (ILT) and leukocyte Ig-like receptors (LIR). These receptors bind to human leukocyte antigens (HLAs) and after a cascade of signal transduction inhibit NK cell stimulation. In addition to these inhibitory receptors, there are also stimulatory receptors that rely on perforin and IFN-γ for their function. NK cells produce a number of other cytokines and chemokines including TNF-α, IL-5, IL-13, GM-CSF, MIP-1 (α and β) and RANTES.

These cells are crucial in fighting viral infections as part of the early innate response. Severe systemic viral infections, specifically herpes virus, may result from a lack or malfunction of NK cells. Patients infected with human immunodeficiency virus (HIV) have low numbers of NK cells. These cells also play a role in killing tumors. Patients with Chediak–Higashi syndrome have an increased risk of lymphomas, and this disease is associated with impaired NK cells, macrophages and neutrophils. A number of cytokines including IL-2, IL-12, IL-15, IL-21, IFN-α and IFN-β induce NK cells, which results in their proliferation, margination, cytokine production and cytotoxicity.

Antigen-Presenting Cells

Antigen-presenting cells (APCs) are a heterogeneous population of cells with extraordinary immunostimulatory capacity. Some play an important role in the induction of the function of T helper cells and some communicate with other lymphocytes. Cytokines can render the ability to present antigens to various cell types. This results in the expression of MHC class II molecules, which are sometimes lacking on some cells such as endothelial cells. The different types of APCs include macrophages, dendritic cells, B cells and interdigitating cells. APCs are mostly derived from bone marrow and are distributed in lymphoid tissues and in the skin. These three types of major APCs are also called professional APCs.

Macrophages

Macrophages participate in innate as well as acquired immune response. As opposed to T and B cells, they are not characterized by any specific cell surface receptors and play an important role in normal tissue repair and aging. They are phagocytes that continuously remove self-proteins that are degraded and presented to T cells in the context of MHC class II molecules. However, this does not result in the activation of T cells since in the absence of infection, the expression of MHC class II molecules on macrophages is low and the presence of B7, a costimulatory molecule, is almost negligible. Following infection, there is an upregulation of MHC class II and B7 molecules.

Also termed the mononuclear phagocytic system, monocytes circulate in the blood and macrophages in the tissues. In the bone marrow during hematopoiesis, the progenitor cells for granulocytes–monocytes differentiate into promonocytes. The promonocytes then leave the bone marrow and enter the bloodstream. In the blood, promonocytes mature into monocytes. Monocytes/macrophages are derived from the bone marrow stem cells. After monocytes enter damaged tissue via endothelium by chemotaxis, they differentiate into macrophages. Monocytes undergo multiple changes during differentiation into macrophages; they enlarge severalfold, the number of intracellular organelles is increased, they are able to produce hydrolytic enzymes and their phagocytic ability is augmented. As shown in Table 1.3, macrophages are classified according to their tissue distribution where their functions are diverse and tissue-specific.

The most important role of macrophages is antigen presentation. However, macrophages also play several other important roles in immune response, which include inflammatory response, antitumor activity, microbicidal activity, lymphocyte activation and tissue reorganization. Most of their physiopathological effects are mediated via cytokines. In addition to the microbicidal activity, they release oxygen-dependent free radicals, cytotoxins, antimicrobials and oxygen-independent hydrolases. For tissue reorganization, they secrete collagenases, elastases and angiogenesis factors.

Oxygen-dependent killing mechanisms involve the production of a number of reactive oxygen and nitrogen intermediates, which possess potent antibacterial activity, by the activated macrophages. The reactive oxygen intermediates are produced as a result of a process called the respiratory burst and include superoxide anion (O_2^-), hydroxyl radicals (OH^-), hydrogen peroxide (H_2O_2) and hypochlorite anion (CLO). The reactive nitrogen intermediates include nitric oxide (NO), nitrogen dioxide (NO_2) and nitrous acid (HNO_2). Oxygen-independent killing involves TNF-α, defensins, lysozymes and hydrolytic enzymes.

Table 1.3 The Types of Macrophages and Their Distribution

Name	Tissue Type
Alveolar macrophages	Lungs
Kupffer cells	Liver
Histiocytes	Connective tissue
Microglial cells	Brain
Osteoclasts	Bone
Mesangial cells	Kidney

Dendritic Cells

Dendritic cells are very important APCs along with mononuclear phagocytes. They are distributed in small quantities in various tissues that come in contact with the external environment including the skin, inner lining of the nose, lungs, stomach and intestine. The blood contains immature dendritic cells. After activation, dendritic cells migrate to the lymphoid tissue to initiate an acquired response following their interaction with lymphoid cells, B and T cells. There are two most common types of dendritic cells, myeloid dendritic cells and lymphoid dendritic cells. Myeloid dendritic cells secrete IL-12, are very similar to monocytes and can be divided into at least two subsets, mDC1 and mDC2. mDC1 stimulates T cells, whereas mDC2 may have a role in fighting wound infection. The lymphoid dendritic cells are similar to plasma cells and produce high levels of IFN-α. The dendritic cells are of hematopoietic origin, having myeloid dendritic cells of myeloid origin and lymphoid dendritic cells of lymphoid origin. The types and distribution of dendritic cells is shown in Table 1.4.

The skin dendritic cells are called Langerhans cells, are of myeloid origin and are CD34$^+$ with high levels of HLA-DR. The hematopoietic progenitors for other myeloid dendritic cells are also CD34$^+$. Furthermore, they are derived from cells of either CD14$^+$CD1a$^-$ or CD14$^-$CD1a$^+$ lineage. The dendritic cells derived from CD14$^+$CD1a$^-$ are interstitial and/or peripheral blood dendritic cells with phagocytic properties. Their maturation is induced by IL-4, IL-13, GM-CSF and TNF. The dendritic cells derived from CD14$^-$CD1a$^+$ are epidermal dendritic cells and their maturation is induced by transforming growth factor (TGF)-β.

The progenitor cells for dendritic cells are present in bone marrow, which initially transform into immature dendritic cells. The presence of toll-like receptors (TLRs) on these cells, mDC (TLR2, TLR4), lymphoid dendritic cells (TLR7, TLR9), along with other receptors, collectively called PRRs, allows the immature dendritic cells to search for viruses and bacteria in their environment. A contact with an invading pathogen allows these cells to quickly develop into mature dendritic cells. Although immature dendritic cells can process antigens, only mature dendritic cells can present the antigenic fragment in the context of MHC molecule to T cells. During T-cell activation, a number of coreceptors including CD40, CD80 and CD86 are simultaneously upregulated by dendritic cells. A chemokine receptor, CCR7, that allows migration of dendritic cells from blood to secondary lymphoid organs is also upregulated by these cells. Dendritic cells communicate with other cells by cell–cell interaction. For example, CD40L present on lymphocytes binds to CD40 receptors on dendritic cells, resulting in cross talk. Dendritic cells promote the generation of

Table 1.4 Classification of Dendritic Cells

Type	Distribution
Circulating dendritic cells	Blood, lymph
Langerhans cells	Skin, mucous membrane
Interdigitating dendritic cells	Secondary lymphoid tissue (with T cells), thymus
Interstitial dendritic cells	Gastrointestinal tract, heart, liver, lungs, kidney

Table 1.5 Features of Immature Dendritic Cells

Presence of CD1a molecules on cell surface
Augmented expression of MHC class II molecules
A lack or very low expression of costimulatory molecules
IL-10 inhibits maturation
A lack or very low expression of CD25, CD83 and p55

Table 1.6 Features of Mature Dendritic Cells

Antigen is taken up by macrophage mannose receptor and DEC-205 receptors
Present antigen in the context of MHC class I and class II molecules
T-cell binding and costimulatory molecules including CD40, CD54, CD58, CD80 and CD86 have
    ample expression
Macrophage-restricted molecules are not present. Produce IL-12 in large quantities
Resistant to p55, CD83, S100b, IL-10 dendritic cells – restricted molecules

TH$_1$ cells via IL-12 production after they come in contact with a pathogen. The characteristics of mature and immature dendritic cells are further pointed out in Tables 1.5 and 1.6.

Lymphoid dendritic cells promote negative selection in the thymus. This may be attributed to their ability to induce fas-mediated apoptosis. Based on their ability to cause apoptosis and their ability to eliminate self-reactive T cells, lymphoid dendritic cells exhibit a regulatory function instead of a stimulatory immune effector function. Myeloid dendritic cells also have differential effects. For example, T cells can be primed to selectively activate TH$_1$ responses by CD14-derived myeloid dendritic cells. Naïve B cells can be activated in the presence of CD40L and IL-2 to secrete IgM by CD34$^+$, CD14-derived myeloid dendritic cells. This effect on naïve B cells is not observed with CD1a-derived dendritic cells.

Nonprofessional Antigen-Presenting Cells

In addition to the three major classes of APCs (Table 1.7) already described, a number of other cells including fibroblasts, thyroid epithelial cells, thymic epithelial cells, glial cells, endothelial cells (vascular) and pancreatic β cells can be induced to present antigens to T cells. These cells do not constitutively express MHC class

Table 1.7 Characteristics of Major Antigen-Presenting Cells (APCs)

	Macrophages	B Cells	Dendritic Cells
Distribution	Lymphoid and connective tissues. Body cavities	Blood, lymphoid tissue	Skin, lymphoid and connective tissue
Antigens expressed	MHC class I, low levels of MHC class II (inducible with stimulus), low levels of B7 (inducible with stimulus)	MHC class I, MHC class II (further induced by stimulus), low levels of B7 (inducible by stimulus)	MHC class I, MHC class II, high expression of B7
Types of antigens presented	Extracellular antigen in the context of MHC class II molecules	Extracellular antigens. Presentation involves immunoglobulin receptors and MHC class II molecules	Extracellular and intracellular antigens in the context of MHC class I and II molecules

II molecules, but after their exposure to certain cytokines such as IFN-γ, the MHC molecules are expressed, which enables them to present antigen to naïve T cells.

Polymorphonuclear Leukocytes

Neutrophils

Approximately 90% of circulating granulocytes are neutrophils. They possess a multilobed nucleus and a granulated cytoplasm and are produced in the bone marrow by hematopoiesis and then migrate to the bloodstream. Neutrophils are terminally differentiated and after entering into the tissues they live for only a few days. However, their life span can be increased in the presence of IL-2 as they express IL-2βR. At the site of injury/inflammation, neutrophils appear first and as a result of infection their production and release are augmented in the bone marrow. The migration of neutrophils from the blood to the tissues is regulated by adhesion molecules, which will be described later. Briefly, the neutrophils first bind to vascular endothelium, which allows their transport into the tissue by the creation of gaps in the blood vessels. The path of their migration is directed by various molecules including IL-8, IFN-γ and C5a, which bind to neutrophils via their specific receptors that are present on their cell surface membrane. This whole process is called chemotaxis. The gathering of neutrophils at the site of infection/injury/inflammation is further aided by the production of various chemotactic factors.

Neutrophils are phagocytes as is the case for macrophages. They employ a number of bactericidal substances and lytic enzymes in addition to both oxygen-dependent and oxygen-independent pathways to kill microbes. Three types of granules, primary granules, secondary granules and tertiary granules, contain various proteins that are released as a result of stimulus-induced degranulation. The primary granules, which are termed azurophilic granules, contain myeloperoxidase, defensins, cathepsin G and bactericidal/permeability-increasing protein. The secondary granules (specific granules) contain lactoferrin and cathelicidin, and tertiary granules contain gelatinase and cathepsin. Neutrophils can regulate the function of monocytes and lymphoid cells via various cytokines including IL-1β, IL-1ra, IL-8, TGF-β, TNF-α as well as preformed granules. The production of cytokines by neutrophils is variable depending on the stimulus.

Eosinophils

Eosinophils develop in the bone marrow where they also mature. They are then released into the blood followed by their migration into the tissue spaces. Eosinophils comprise 1–5% of blood leukocytes in nonatopic individuals. They are present in thymus, spleen, lymph nodes, uterus and lower GI tract. Under normal conditions, they are not present in the skin, lungs or esophagus. They can live up to 72 h in the tissue. Their migration to the site of inflammation or parasitic infection is directed by leukotriene B4 and chemokines eotaxin (CCL11) and RANTES (CCL5). They play a specialized role in immunity to helminth infection by releasing special granules. Eosinophils are also capable of phagocytosis and killing ingested microorganisms, but this is not their primary function.

Their immune function is mediated via production of cationic granule proteins, reactive oxygen species, leukotrienes, prostaglandins, elastase, a plethora of cytokines (IL-1, IL-2, IL-4, IL-5, IL-6, IL-8, IL-13, TNF-α) and growth factors [Platelet-derived growth factor (PDGF), TGF-β, vascular endothelial growth factor (VEGF)]. The cytotoxic granules released by eosinophils in response to a stimulus include major basic protein, eosinophil cation protein, eosinophil peroxidase and eosinophil-derived neurotoxin.

Eosinophils are attracted by proteins released by T cells, mast cells and basophils [eosinophil chemotactic factor of anaphylaxis (ECF-A)]. They bind schistosomulae coated with IgG or IgE, degranulate and release major basic protein, which is toxic. Eosinophils also release histaminase and aryl sulfatase, which inactivates histamine and Slow reacting substance of anaphylaxis (SRS-A). This results in anti-inflammatory effects and inhibits migration of granulocytes to the site of injury.

Basophils and Mast Cells

Basophils are found in very small numbers in the circulation (0.01–0.3% of leukocytes) and are nonphagocytic granulocytes. They are of bone marrow origin and contain large cytoplasmic granules. After activation, basophils secrete a number of mediators including histamine, leukotrienes, proteoglycans and proteolytic enzymes, as well as several cytokines. These substances play a pivotal role in allergic responses.

Mast cells are similar to basophils, but they are derived from different precursor cells in the bone marrow. The precursor cells for both basophils and mast cells express CD34. Mast cells are released from the bone marrow in an undifferentiated state into the blood, and they differentiate upon reaching the tissue. Mast cells are distributed in a wide variety of tissues including skin, lungs, digestive tract, genitourinary tract, nose and mouth. There are two types of mast cells, the connective tissue mast cells (CTMCs) and mucosal mast cells (MMCs). The activity of MMCs is dependent on T cells. Mast cells play a crucial role in allergic disease and inflammation. They express receptors for IgE (FcϵR1), and binding of the antibody to its receptor followed by a second exposure to the specific antigen results in massive degranulation of mast cells. The mediators released by mast cells include histamine, prostaglandin D2, leukotriene C4, heparin, serine proteases and a plethora of cytokines. In addition to their role in allergic disease and anaphylactic shock, mast cells are implicated in autoimmune diseases, including rheumatoid arthritis and multiple sclerosis (MS), and induction of peripheral tolerance. Mast cells also participate actively in the innate immune responses to many pathogens. They play a role as innate effector cells in augmenting the initial events in the development of adaptive immune responses. Mast cells may also play a role in combating viral infections through various direct or indirect pathways. These pathways involve the activation of mast cells by pathogens including TLRs and coreceptors.

Cellular Migration

Cell Adhesion Molecules

The migration of molecules generated in blood and leukocytes into the tissues is controlled by the vascular endothelium. The passage through the endothelial cells lining the walls of blood vessels is required for leukocytes so that they can enter an area of injury, inflammation or into the peripheral lymphoid organs. This process is called extravasation. Leukocyte-specific cell adhesion molecules are expressed on endothelial cells. Their expression could be constitutive while others are expressed in response to cytokines during an inflammatory response. Many different types of circulating leukocytes including lymphocytes, monocytes and neutrophils express cell surface receptors that recognize and bind to cell adhesion molecules on vascular endothelium. There are many types of adhesion receptors on the cell surface of which four major families include most of these receptor types. They are classified as integrins, selectins, immunoglobulin super-family and cadherins. In addition to cellular migration, they are involved in a number of important cellular functions including growth, differentiation, mutagenesis and cancer metastasis. They also transmit information into the cell from the cellular matrix.

Integrins

Integrins are noncovalently linked heterodimeric membrane proteins with two subunits, an α and a β chain. These transmembrane proteins are constitutively expressed, but require activation for binding to their ligands. After activation, a signal is transmitted from the cytoplasm resulting in a modification in the extracell-ular domains of integrins. This modification in the conformation of these domains results in an increased affinity of the integrins for their ligands. When bacterial peptides stimulate leukocytes, "inside out" signaling occurs, resulting in an increase in the affinity of the leukocyte integrins for members of the immunoglobulin family. The binding of an integrin to its ligand results in "outside in" signaling, which may regulate cell proliferation and apoptosis.

Integrins are expressed on leukocytes, and different subsets of leukocytes express different integrins. Fifteen different α and eight different β units have been identi-fied, which can combine in various ways to express different integrin receptors. Integrins have three activation states, basal avidity, low avidity and high avidity. They play a pivotal role in immunity, cancer, homeostasis and wound healing, which is mediated bidirectionally via transmitting signals across the plasma membrane. The affinity to the ligand is regulated by global conformational changes, which relate to the interdomain and intradomain shape shifting. The downward move-ments of the C-terminal helices play a critical role in integrin conformational signaling.

Selectins

Selectins are a family of membrane glycoproteins that are divalent cation-dependent and bind to specific carbohydrates containing sialylated moieties. They are composed of leukocyte (L)-selectins, endothelial (E)-selectins and platelet (P)-selectins. L-selectins are expressed on most leukocytes, whereas vascular endothelial cells express E-selectins and P-selectins. Initial binding of leukocytes with vascular endothelium involves selectins and consequently they play an important role in leukocyte trafficking.

Circulating leukocytes bind to the endothelium via selectins expressed on vascular endothelium or leukocytes. Selectins have a low binding affinity but cause the leukocytes to slow down as they roll on endothelial cells. During this slowdown period, chemoattractants activate leukocytes. This activation increases the affinity of their integrin (β2) receptors for ligands on activated endothelial cells. The leukocytes migrate between endothelial cells of venules to the site of injury or inflammation as a result of a chemotactic signal emanating outside the venules. The functions of selectins were identified by using knockout mice for each gene. L-selectin-deficient mice have an impaired homing of lymphocytes to lymph nodes. In P-selectin-deficient mice, leukocytes roll at sites of inflammation but do not roll along normal blood vessels. E-selectin-deficient mice are normal but both P- and E-selectin-deficient mouse leukocytes do not roll even at sites of inflammation.

Immunoglobulin Superfamily

These are calcium-independent transmembrane glycoproteins that contain a variable number of immunoglobulin-like domains. This group includes intercellular adhesion molecules (ICAM), vascular cell adhesion molecules (VCAM), neural cell adhesion molecules (NCAM) and platelet–endothelial cell adhesion molecules (PECAM). ICAM and VCAM are present on vascular endothelial cells and bind to various integrin molecules or other immunoglobulin superfamily cell adhesion molecules. NCAM-1 are found predominantly in the nervous system, are involved in neuronal patterns and mediate homophilic interactions. By associating laterally with fibroblast growth factor receptor, NCAM stimulates tyrosine kinase activity associated with that receptor, resulting in the induction of outgrowth of neurites. PECAM-1 play a role in inflammation and immune response.

Cadherins

Cadherins are calcium-dependent and establish molecular links between adjacent cells. They include neural cadherins (NC), epithelial cadherins (EC), placental cadherins (PC), protocadherins and desmosomal cadherins. Cadherins mediate homophilic interactions and form zipper-like structures at adherens junctions and are linked to the cytoskeleton through the catenins. The ability of catenins to interact with the intracellular domain confers their adhesive properties. Cadherins play a crucial role in embryonic development and tissue organization.

All adhesion molecules after binding to their ligands and/or due to clustering of their receptors as a result of interaction with the ligand transduce signals across the membrane. Interactions with downstream signaling and the cytoskeleton take place after conformational changes in the receptors, which are followed by the reorganization of the cytoskeleton. There is phosphorylation and dephosphorylation of a number of molecules as a result of a signaling cascade initiating from the receptor activation, which causes conformational changes in a number of cytoplasmic kinases. The signal causes the synthesis of new proteins such as cytokines, soluble adhesion molecules and metalloproteases as a result of induction in gene expression. Defective interactions between adhesion molecules cause disease.

Molecules That Recognize Antigen

The acquired immune response is designed to recognize foreign antigens, which is achieved by two distinct types of molecules that are involved in this process: the immunoglobulins and the T-cell antigen receptors. The immunoglobulins or antibodies are a group of glycoproteins present in a number of bodily fluids including the serum and are secreted in large amounts by plasma cells that have differentiated from precursor B cells. B lymphocytes possess membrane-bound immunoglobulins that have similar binding specificity as that of the terminally differentiated plasma cells. Recognition of an antigen by B lymphocytes triggers the production of antibodies. The TCRs are present only on the cell surface of T lymphocytes, and they do not produce any soluble molecules similar to the immunoglobulins.

The antigen receptors of T and B lymphocytes belong to the immunoglobulin supergene family and are derived from a common ancestor. Immunoglobulins are composed of two identical heavy (H) chains and two identical light (L) chains. The TCR has an antigen-binding portion consisting of either α and β chains or γ, δ and ϵ chains. Circulating antibodies are structurally identical to B-cell antigen receptors but lack the transmembrane and intracytoplasmic sections.

T-Cell Receptor

The TCR is composed of integral membrane protein that recognizes the antigen and as a consequence its activation produces an immune response to eliminate the antigen. This process results in the development of CD4$^+$ or CD8$^+$ cells from the precursor T cells. It involves other cell surface receptors and downstream signal transduction mechanisms.

The TCR is a member of the immunoglobulin superfamily and is composed of an N-terminal immunoglobulin variable domain, an immunoglobulin constant domain, a transmembrane region and a short cytoplasmic tail. Three hypervariable complementarity-determining regions are present in the variable domain of both TCR-α and TCR-β. Additional areas of hypervariability are present in the variable region of the β chain, which are not considered complementarity-determining regions due to their inability to contact the antigen. The processed antigen is recognized by complementarity-determining region-3, and additional interactions involve the N-terminal part of the antigen with complementarity-determining region-1 of the α chain and the C-terminal part of the antigen with complementarity-determining region-1 of the β chain. The MHC is recognized by complementarity-determining region-2. The TCR falls into two groups, TCR-1 (γδ) and TCR-2 (αβ). More than 95% of the TCR are TCR-2. Both types of T cells arise from hematopoietic precursor cells.

γδ T-Cell Receptor

The structure of γδ TCR (TCR-1) is similar to αβ heterodimer. The generation of the γ chain involves VJ recombination, and that of the δ chain involves V(D)J recombination. However, they differ from TCR-2 in terms of the types of antigens recognized, the mode of antigen presentation and recognition and signal transduction pathways. Antigen recognition by γδ TCR is similar to antigen recognition by antibodies. They recognize intact protein antigens as well as nonprotein antigens. These TCRs exhibit a limited diversity, which may imply that they have few natural ligands. Most of their known ligands are not of foreign origin; instead, they are of host origin. There is a correlation between Vγ and Vδ genes expressed by a γδ T cell and its function. This may be due to its function being determined by the interaction between particular TCR-V domains and function. Furthermore, complementarity-determining region-3 of the TCR-δ chain may be involved in determining the receptor specificity without any consideration for Vγ and/or Vδ.

αβ T-Cell Receptor

αβ T-cell receptor (TCR-2) is a heterodimer with a molecular weight of 90 kDa and is made up of two peptide chains, an α chain (45 kDa) and a β chain (40 kDa). Each chain is composed of distinct constant and variable regions. These regions contain disulfide intrachain, which has partial homology with the immunoglobulin domains. The α chain is generated by VJ recombination and the β chain is generated by V(D)J recombination. Each of the V regions contains three hypervariable regions that make up the binding site. TCR-2 exhibits tremendous diversity, which is due to a unique combination of segments at the intersection of the specific regions, plus palindromic and random nucleotide additions. They recognize only membrane-bound processed antigens in the context of MHC molecules.

T-Cell Activation, T-Cell Receptor Complex and Signal Transduction

Activation of CD4$^+$ T cells fits a "two-signal model" although multiple variations exist between different T-cell types. The interaction of the TCR and CD28 on the T cell with MHC complex and B7 family members on the APCs provides stimulus for the induction of CD4$^+$ T cells. CD28 molecule is essential for this activation since its absence results in anergy. This is followed by downstream signal transduction emanating from the TCR and CD28. Binding of the TCR to the antigenic fragment presented in context with MHC molecules on the APCs provides the first signal. This signal is necessary to activate the antigen-specific TCR on a T cell. As described, APCs include macrophages, B cells and dendritic cells. Usually, in case of a naïve response, dendritic cells are involved. MHC class I molecules present short peptides to CD8$^+$ cells, whereas MHC class II molecules present longer peptides to CD4$^+$ cells. The receptors on the antigen are stimulated by some stimuli, resulting in costimulation, which serves as the second signal. These stimuli include antigens of the pathogens, necrotic bodies and heat shock proteins. CD28 is a costimulatory molecule expressed constitutively on naïve T cells. CD80 and CD86 expressed on the APCs serve as costimulators for naïve T cells. In addition, the T cells acquire other receptors such as ICOS and OX40 after stimulation. The expression of ICOS and OX40 is CD28-dependent. T cells only respond to the antigen after the production of a second signal, and this mechanism has been put in place to avoid autoimmune responses.

The TCR is associated with CD3 (Fig. 1.3), resulting in the formation of a TCR–CD3 membrane complex. The signal transduction is carried out by accessory molecules after the processed antigen comes in contact with the T cell. CD3 is composed of five polypeptide chains, which together form three dimers. These include a heterodimer of $\gamma\varepsilon$, a heterodimer of $\delta\varepsilon$ and either a homodimer of two ζ chains or a heterodimer of $\zeta\eta$ chains. The γ, δ, and ε chains of CD3 belong to the immunoglobulin superfamily as these chains possess an immunoglobulin-like extra domain, a transmembrane region and a cytoplasmic domain. Copies of a sequence motif called immunoreceptor tyrosine-based activation motifs (ITAMs) are present in the cytoplasmic tails of the γ, δ, ε and ζ subunits. ITAMs serve as substrates for tyrosine kinase as well as binding sites for SH2 domains for other kinases. Members of both the Src family (Lck) and Syk family (ZAP-70) of kinases play an important role in the recruitment of protein kinases to activate TCR. An ion pair is formed by the transmembrane domains of TCR-α and CD3δ, which is essential for the assembly and expression of the TCR. The transmission of signals from antigen-induced TCR to downstream transduction pathways also involves lateral association between transmembrane domain helices. CD45-associated proteins are involved in binding kinases with their intracellular domain and CD45 through transmembrane domain–transmembrane domain interaction. These proteins promote induction of kinases by the phosphatase CD45. The phosphorylation of intracellular tyrosines on TCR-interacting molecule (pp30) is crucial for the recruitment of

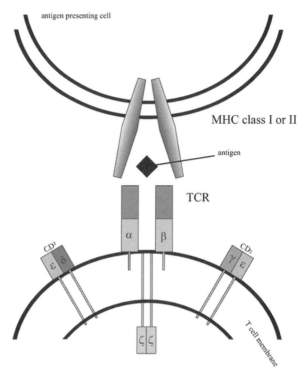

Fig. 1.3 The structure of T-cell receptor (TCR) and antigen presentation. The TCR (TCR-2) is a heterodimer composed of two transmembrane glycoprotein chains α and β, which are disulfide-linked polypeptides. TCR-1 is structurally similar but composed of γ and σ polypeptides. The TCR is also associated with the CD3 complex. The latter is composed of six polypeptide glycoprotein chains known as CD3γ, CD3σ, CD3ε and another protein known as ζ. When the TCR recognizes antigen, the CD3 complex is involved in signal transduction. The antigen is recognized by the TCR only in context with MHC molecules after it is taken up and processed by the antigen-presenting cells (APCs) [A part of this diagram is based on one published by Janeway and Travers, 1996] (*see* Color Insert)

phosphatidylinositol-3-kinase to the membrane. TCR–CD3 complex may be present not only in a monovalent, but also as several different multivalent forms on unin-duced resting T cells. These multivalent receptors play a critical role in recognizing low doses of antigens and are responsible for sensitivity. In contrast, a wide dynamic range is conferred by coexpressed monovalent TCR–CD3 complex.

After activation of the TCR, there is induction of Src family tyrosine kinase (p56^{lek}), which phosphorylates phospholipase Cγ1. This is followed by the hydrol-ysis of phosphatidylinositol 4,5-bisphosphate, resulting in the production of diacyl-glycerol (DAG) and inositol trisphosphate (IP$_3$). Protein kinase C is activated by DAG, which phosphorylates Ras. Ras is a GTPase and its phosphorylation induces Raf and initiation of MAP kinase signaling pathway. IP$_3$ is involved in calcium-dependent activation of IL-2 gene expression via nuclear factor of activated T cells (NFAT).

Another signal transduction mechanism involves phosphoinositide 3-kinases (PI3 Ks), which phosphorylate phosphatidylinositol 4,5-bisphosphate, resulting in the activation of phosphatidylinositol trisphosphate activating pathways. These pathways overlap with Phospholipase C (PLC) gamma and are PI3 K-specific. PI3 K-specific pathways involve Akt-mediated inactivation of FOX0 transcription factors. Furthermore, this pathway also includes glucose uptake and metabolism, which is transcription-independent. P110 delta also plays a role in this process, which has an isoform of PI3 K and primary source of its activity. In addition to TCR and CD3, other molecules including CD4, CD8, LFA-1, LFA-2, CD28 and CD45R, termed accessory molecules, play a role in antigen recognition and induction of T cells.

CD4 is present on helper T cells and CD8 is present on cytotoxic T cells. CD4 works as a monomer with four Ig domains in its extracellular protium, whereas CD8 functions as a dimer involving either an α and a β chain or two identical α chains. One extracellular IgV domain in their extracellular protium is present in CD8α and CD8β molecules. CD8$\alpha\alpha$ and CD8$\alpha\beta$ are structurally similar by having notably different functions. The CD8$\alpha\alpha$ does not support conventional positive selection and its expression does not require recognition by TCR in the context of MHC molecules. In contrast, CD$\alpha\beta$ has a role as a coreceptor for the TCR on MHC class I-restricted mature T cells and thymocytes. The specific modulation of TCR activation signals to promote their survival and differentiation is mediated by CD8$\alpha\alpha$. These CD$\alpha\alpha$ receptors are expressed on agonist-triggered immature thymocytes, antigen-induced CD$\alpha\beta$ T cells and mucosal T cells. Antigenic stimulation through TCR results in the expression of CD8$\alpha\alpha$. A number of other molecules play an accessory role as well. LFA-1, an adhesion molecule, strengthens the interaction of the T cell with the APC. CD28, present on helper T cells, binds to B7 on APCs, which serves as a costimulatory signal, resulting in the induction of T lymphocytes. In the absence of CD28–B7-mediated signal, T lymphocytes do not respond, resulting in anergy.

The specificity of antigen recognition and the resulting T-cell response is dependent on the interaction between the TCR, the MHC peptide complex and the APC. The selection of the T-cell repertoire in thymus and the induction of peripheral mature T cells seem to be dependent on the binding kinetics of the TCR and the MHC peptide complex. In particular, the half-life of the TCR–MHC interaction plays a significant role in T cell activation. The significance of this kinetic binding parameter is observed in the capacity of T cells to respond to invading pathogens, autoimmunity and tumor-produced antigens. This protects the body from reacting to self. The MHC complex, which binds to the TCR with a short half-life, does not result in T cell activation and in some cases may even have an inhibitory effect in response to a stimulus. In contrast, prolonged half-lives also disrupt T-cell induction, suggesting a set half-life is required for producing an optimal response. This restricted immune response serves as a deterrent against autoimmune disease. The kinetics of the TCR–MHC interaction also determines the nature of the response by the peripheral mature T cells.

The regulation of T-cell responses during T-cell development and in mature T cells is dependent on the modulation of TCR expression. The rate constants for synthesis, endocytosis, recycling and degradation are critical for TCR expression levels. There is a slow and constitutive cycling of TCR between the plasma membrane and the intracellular compartment in resting T cells. The di-leucine-based (diL) receptor-sorting motif in the TCR subunit CD3γ is required for constitutive TCR cycling. The quality control of the TCR may be regulated by this event. TCR downregulation is dependent on an enhancement in endocytic rate constant induced by the TCR. The endocytosis of triggered TCR results, either from ubiquitination of the TCR as protein tyrosine kinase is activated or Protein kinase C (PKC)-dependent induction of the diL motif. Furthermore, PKC/CD3 γ-dependent pathways are responsible for the endocytosis of nontriggered TCR. The signaling is inhibited by TCR downregulation. Alternatively, internal stores of TCR can be used by the immune system to recognize antigens.

Pre-T-Cell Receptor

CD4 and CD8 receptors are not present on early T-cell progenitors in the thymus. These cells express a nascent TCR-β chain and an invariant pre-TCR-α chain and CD3 receptors. This is followed by the generation of a functional TCR-β chain on these pre-TCR-expressing thymocytes. A process of β chain selection induces the expression of CD4 and CD8 antigens on these thymocytes and their differentiation into distinct subsets. The T-cell development utilizes the pre-TCR as a key molecular sensor. The rearrangement of the TCR-α and CD4$^+$ and CD8$^+$ cell survival and proliferation are induced by the pre-TCR. Other signaling molecules including CD3ε, Syk, Fyn, Lck and Zap-70 participate in pre-TCR-mediated differentiation and signaling. Pre-TCR-mediated β-selection is further supported by IL-7 and notch ligand and may require an exogenous ligand on thymic stroma. However, most of the data suggest that this may not be the case, and this signaling may be stimulus-independent. It is now accepted that the pre-TCR is involved in the immune response in an autonomous and ligand-independent manner. The autonomous signaling is a function of CD4$^-$CD8$^-$ thymocytes where the pre-TCR signaling takes place. At this time, the structural basis for the oligomerization of pre-TCRs is not known.

Antigen Recognition

Antigens are recognized by the immune system by utilizing antibodies generated by B cells and by TCR on the T cells. However, the mechanisms by which T cells and B cells recognize antigens are different. Both the antibodies and the T cells are capable of recognizing a wide range of antigens. The antibodies and the TCR have many similar features. They both possess variable (V) and constant (C) regions, and the process of gene recombination, which produces the variable domains from V, D and J gene segments, is also similar for each type of receptor. However, T cells and antibodies recognize antigens differently: antibody recognizes antigens in solution or on cell surfaces in their native forms, while TCRs do not recognize antigens in their native forms. The antigens are recognized by T cells only after they are processed and then only in the context of MHC molecules. Antigens recognized by T cells are first taken up by macrophages or other APCs and processed so that the determinant recognized by the T-cell antigen receptor is only a small fragment of the original antigen.

A second difference between the antibody and the TCR is that the antibody can be produced in two forms, secreted or membrane-bound, whereas the TCR is always membrane-bound. The secreted antibody has a dual function where the V domains bind antigens and the C domain binds to the Fc receptors.

Antigen–Antibody Binding

The binding of antigen to antibody involves the formation of multiple noncovalent bonds between the antigen and amino acids of the antibody. Antigen–antibody complex results from hydrogen bonding, electrostatic bonding, van der Waals bonds and hydrophobic bonds. A considerable binding energy is produced as a result of multiple types of bonds.

The Structure of Antigens

Antigens are three-dimensional structures and present many different configurations to B cells, resulting in a high number of different possible antibodies. Furthermore, different antibodies to an antigen often bind to overlapping epitopes. This allows the binding of different antibodies to a particular antigenic region of the molecule, without binding exactly to the same epitope.

Technically, antibodies may be produced against any part of the antigen, but this is usually not the case. Certain areas of the antigen are particularly antigenic and the majorities of the antibodies bind to these regions, which are called immunodominant regions. These regions are present at exposed areas on the outside of the antigen, where there are loops of polypeptide lacking a rigid tertiary structure and which could be very mobile.

T-Cell Antigen Recognition

As already stated, T cells do not see unprocessed antigens. Another requirement for antigen recognition by T cells is that they should recognize antigen on the surface of other cells, called APCs. These cells could also be virally infected cells, which can then be killed by cytotoxic T cells. The antigen is taken up by the APC and is degraded, so only a small fragment of the antigen can be presented to the TCR in the context of MHC molecules. Interference in the internal processing of the antigen in the APCs will result in their inability to present the antigen.

The interaction between the T cells and the targets is said to be genetically restricted. For example, cytotoxic T cells specific for a particular virus will recognize it only in the infected cells of their own MHC haplotype. This is also the case for the recognition of helper T cells. Consequently, the T cell recognizes both the antigen and the MHC molecules on other cells.

The processed antigen must be physically associated with the MHC molecules. The processed antigens interact with MHC molecules and with the TCR via amino acid residues.

Major Histocompatibility Complex

The rejection of grafted tissues provided clues about the existence of a very diverse MHC. The immune system must discriminate "nonself" from "self," which results in rejection of the grafted tissues and the response to invading pathogens. This function is achieved via the molecules of the MHC. The MHC is based in multiple loci, which determine the acceptability of the grafted tissue. It also plays a crucial role in the development of immune response since T cells recognize antigens only in the context of MHC molecules. The MHC complex is present on chromosome 17 in mice and on chromosome 6 in humans where it is called HLA. Highly polymorphic loci constitute the MHC complex and are closely linked. In humans, MHC region on chromosome 6 is about 3.6 Mb, which contains 140 genes between flanking genetic markers MOG and COL11A2. Half of them have known immunological functions. Every antigen, both nonself and self, is recognized by T cells in the context of MHC molecules. $CD4^+$ T cells recognize antigens in the context of class II MHC molecules, whereas CD8 T cells recognize antigens in the context of class I MHC molecules. During embryogenesis, T cells recognizing self antigens in the context of MHC molecules are eliminated, whereas T cells recognizing foreign antigens in the context of MHC molecules are retained. Breakdown in this process of self-recognition and elimination can cause autoimmune diseases. In contrast, failure to recognize foreign antigens may result in an immunodeficient state, resulting in infections and development of tumors.

Major Histocompatibility Complex Molecules

The MHC region is divided into three subgroups, called MHC class I molecules, MHC class II molecules and MHC class III molecules. MHC class I and class II molecules are membrane-bound glycoproteins, which are similar in structure and function. However, the class I and class II molecules are distinguishable on the basis of their structure, tissue distribution and function. Class I molecules include the HLA-A, HLA-B and HLA-C. They are present on all nucleated cells and encode peptide-binding proteins and antigen-processing molecules including Tapasin and TAP. Class II molecules include HLA-DR, HLA-DQ and HLA-DP. They are present on APCs and encode heterodimeric peptide-binding proteins and proteins that regulate adding of peptides to MHC class II molecule proteins in the

Table 1.8 Comparison of Class I and II HLA Molecules

Properties	Class I	Class II
Antigen	HLA-A, -B, -C	HLA-D, -DR, -DQ, -DP
Tissue distribution	All nucleated cells	Antigen-presenting cells
Function	Present processed antigenic fragments to CD8 T cells. Restrict cell-mediated cytolysis of virus-infected cells	Present processed antigenic fragment to CD4 cells. Necessary for effective interaction among immunocompetent cells

lysosomal compartment. Class III molecules include the complement system. These are soluble molecules and do not act as transplantation antigens, nor do they present antigen to T cells. A comparison of the characteristics of MHC class I and MHC class II molecules is shown in Table 1.8.

Structure of Class I Human Leukocyte Antigen Molecules

MHC class I molecules are heterodimers made up of an α chain and β_2 microglobulin. The α chain is composed of three polymorphic domains, α_1, α_2 and α_3 with a peptide-binding groove between α_1 and α_2 chains. The peptides derived from cytosolic proteins bind to this groove. Eight β-pleated sheets and two α helices form this groove, and both chains are noncovalently linked. The α chain contains 338 amino acid residues. Class I HLA molecules are present on all nucleated cells. For an antigen to be recognized by a CD8 T lymphocyte, the antigen must be recognized in combination with a class I molecule.

Class II Molecules

MHC class II molecules are made up of two membrane-spanning proteins; each chain has a size of 30 kDa and is made up of two globular domains. These domains are called α-1, α-2, β-1 and β-2. Each chain possesses an immunoglobulin-like region next to the cell membrane. MHC class II molecules are highly polymorphic.

Class II HLA molecules have a limited cellular distribution as they are found on APCs, which include B lymphocytes, macrophages and dendritic cells. In addition, some cells that do not normally express class II molecules (such as resting T cells, endothelial cells, thyroid cells as well as others) can be induced to express them. The function of class II molecules is to present processed antigenic peptide fragments to CD4 T lymphocytes during the initiation of immune responses. Just as CD8 T lymphocytes recognize peptide fragments only in the context of class I molecules, CD4 T lymphocytes recognize peptide fragments only in the context of class II molecules. The HLA-A, -B, and -C genetic loci determine the class I molecules that bear the class I antigens, and the HLA-DR, -DQ and -DP genetic subregions, each of which contains several additional loci, determine the class II molecules that bear the class II antigens.

Bibliography

Abbas AK, Lichtman AH, Pober JS. 1994. Cellular and Molecular Immunology, WB Saunders Company, Philadelphia.

Alacorn B, Swamy M, Van-Santen HM, Schamel WW. 2006. T cell antigen receptor stoichiometry: Pre-clustering for sensitivity. EMBO-Rep. 7:490–495.

Allison TJ, Garboczi DN. 2002. Structure of $\gamma\delta$ T cell receptors and their recognition of non-peptide antigens. Mol Immunol. 38:1051–1061.

Arnold JN, Wormald MR, Sim RB, Rudd PM, Dwek RA. 2007. The impact of glycosylation on the biological function and structure of human immunoglobulins. Ann Rev Immunol. 25:21–50.

Banchereau J, Steinman RM. 1998. Dendritic cells and the control of immunity. Nature. 392: 245–252.

Barclay A. 2003. Membrane proteins with immunoglobulin-domains – a master superfamily of interaction molecules. Sem Immunol. 15:215.

Baron P, Constantin G, D'Andrea A, Ponzin D, et al. 1993. production of TNF and other proinflammatory cytokines by human mononuclear phagocytes stimulated with myelin P2 protein. Proc Natl Acad Sci USA. 90:4414–4418.

Bazzoni F, Cassatella MA, Rossi F, Ceska M, et al. 1991. Phagocytosing neutrophils produce and release high amounts of the NaP-1/interleukin 8. J Exp Med. 173:771–774.

Benichou G. 1998. The presentation of self and allogeneic MHC peptides to T lymphocytes. Hum Immunol. 59:540–561.

Butcher E, Picker LJ. 1996. Lymphocyte homing and homeostasis. Science. 272:60–62.

Cantrell D. 1996. T-cell antigen receptor signal transduction pathways. Ann Rev Immunol. 14:259.

Carayannopoulos L, Capra JD. 1993. Immunoglobulins: Structure and function. In: Paul WE, Ed. Fundamental Immunology. 3rd Ed., Raven Press, New York.

Carreno LJ, Gonzales PA, Kalergis AM. 2006. Modulation of T cell function by TCR/pMHC binding kinetics. Immunobiology. 211:47–64.

Cassatella MA. 1995. The production of cytokines by neutrophils. Immunol Today. 16:21–26.

Cassatella MA. 1999. Neutrophil-derived proteins: selling cytokines by the pound. Adv Immunol. 73:369–509.

Castellino F. 1997. Antigen presentation by MHC Class II molecules: Invariant chain function, protein trafficking and the molecular basis of diverse determinant capture. Hum Immunol. 54:159–169.

Cella M, Sallusto F, Lanzavecchia A. 1997. Origin, maturation and antigen presenting function of dendritic cells. Curr Opin Immunol. 9:10–16.

Chien YH, Bonneville M. 2006. Gamma delta T cell receptors. Cell Mol Life Sci. 63:2089–2094.

Chtanova T, MacKay CR. 2001. T cell effector subsets: extending the Th_1/TH_2 paradigm. Adv Immunol. 78:233–266.

Colucci F, Caligiuri MA, DiSanto JP. 2003. What does it take to make a natural killer? Nat Rev Immunol. 3:413–425.

Cooper MA, Fehniger TA, Calilgiuri MA. 2001. The biology of human natural killer-cell subsets. Trends Immunol. 22:633–640.

Davis MM, Chien Y. 1995. Issues concerning the nature of antigen recognition by $\alpha\beta$ and $\gamma\delta$ T cell receptors. Immunol Today. 16:316–318.

Dawicki W, Marshall JS. 2007. New and emerging roles for mast cells in host defence. Curr Opin Immunol. 19:31–38.

Dianzani U, Malavasi F. 1995. Lymphocyte adhesion to endothelium. Crit Rev Immunol. 15: 167–178.

Fernette PS, Wagner DD. 1996. Adhesion molecules. NEJM. 334:1526–1529.

Fitzsimmons CM, McBeath R, Joseph S, Jones FM, Walter K, Hoffman KF, et al. 2007. Factors affecting human IgE and IgG responses to allergin-like schistosoma mansoni antigens: Molecular structure and patterns of in vivo exposure. Int Arch Allergy Immunol. 142: 40–50.

Gangadhavan D, Cheroutre H. 2004. The CD8 isoform CD8 alpha alpha is not a functional homologue of the TCR co-receptor CD8 alpha beta. Curr Opin Immunol. 16:264–270.

Geisler C. 2004. TCR trafficking in resting and stimulator T cells. Crit Rev Immunol. 24:67–86.

Gleich G, Adolphson C. 1986. The eosinophilic leukocyte: Structure and function. Adv Immunol. 39:177–253.

Godfrey DI. 2004. Natural killer T cells. Nat Rev Immunol. 4:231.

Goldsby RA, Kindt TJ, Osborne BA. 2000. Kuby Immunology, WH Freeman and Company, New York.

Greenwald RJ, Freeman GJ, Shapre AH. 2005. The B7 family revisited. Ann Rev Immunol, 23:515–548.

Grimbaldeston MA, Metz M, Yu M, Galli SJ. 2006. Effector and potential immunoregulatory role of mast cells in IgE-mediated acquired immune responses. Curr Opin Immunol. 18:751–760.

Gurish MF, Boyce JA. 2006. Mast cells: Ontogeny, homing and recruitment of a unique innate effector cell. J All Clin Immunol. 117:1285–1291.

Hart DN. 1997. Dendritic cells: Unique leukocyte populations which control the primary immune response. Blood. 90:3245–3287.

Hofmeyr SA. 1997. An overview of the immune system. http://www.cs.unm.edu/~immsec/html-imm/introduction.html. Retrieved 8/7/07.

Hynes RO. 2004. The emergence of integrins: A personal and historical perspective. Matrix Biol. 23:333–340.

Janeway CA. 1995. Ligands for the T cell receptors: Hard times for avidity models. Immunol Today. 16:223–225.

Janeway CA, Travers P. 1996. Immunobiology, the immune system in health and disease. Current biology limited. London, San Francisco and Philadelphia, Garland publishing incorporated, New York and London.

Jerud ES. 2006. Natural killer T cells: Roles in tumor immunosurveillance and tolerance. Trans Med Hemother. 33:18–36.

Kohler G, Milstein C. 1975. Continuous cultures of fused cells secreting antibody of predefined specificity. Nature. 256:459.

Kuhns MS, Davis MM, Garcia KC. 2006. Deconstructing the form and function of the TCR/CD3 complex. Immunity. 24:133–139.

Laky K, Fleischacker C, Fowlkes BJ. 2006. TCR and notch signaling in CD4 and CD8 T-cell development. Immunol Rev. 209:274–283.

Lanier LL. 2005. NK cell recognition. Ann Rev Immunol. 23:225–274.

Lanzavecchia A, Lezzi G, Viola A. 1999. From TCR engagement to T cell activation: A kinetic view of T cell behavior. Cell. 96:1.

Lehner PJ, Trowsdale J. 1998. Antigen presentation: coming out gracefully. Curr Biol. 8: R605–R621.

Li H, Llera A, Mariuzza RA. 1998. Structure function studies of T cell receptor-superantigen interactions. Immunol Rev. 163:177–186.

Liszewski MK, Yokoyama WM, Atkinson JP. 2000. Innate immunity: Innate immune cells. Immunology/Allergy, II Innate immunity, ACP Medicine, New York.

Luo B, Springer TA. 2006. Integrin structures and conformational signaling. Curr Opin Cell Biol. 18:1–8.

Luo B, Carmen CV, Springer TA. 2007. Structural basis of integrin regulation and signaling. Ann Rev Immunol. 25:619–647.

Maruotti N, Crivellato E. Cantatore FP, Vacca A, Ribatti D. 2007. Mast cells in rheumatoid arthritis. Clin Rheumatol. 26:1–4.

McDonald PP, Bald A, Cassatella MA. 1997. Activation of the NF-kB pathway by inflammatory stimuli in human PMN. Blood. 89:3421–3433.

Mellman I, Turley SJ, Steinman RM. 1998. Antigen processing for amateurs and professionals. Trends Cell Biol. 8:231–242.

MHC Sequencing Consortium. 1999. complete sequence and gene map of a human major histocompatibility complex. Nature. 401:921–923.

Morgan R, Costello R, Durcan N, Kingham P, Gleich G, McLean W, Walsh M. 2005. Diverse effects of eosinophil cationic granule proteins in IMR-32 nerve cell signaling and survival. Am J Respir Cell Mol Biol. 33:169–177.

O'Brien RL, Roark CL, Aydintug MK, French JD, et al. 2007. Gamma delta T cell receptors functional correlations. Immunol Rev. 215:77–88.

Okkenhaug K, Ali K, Vanhaesebroek B. 2007. The antigen receptor signaling, a distinctive role for the p10 delta isoform of PI3 K. Trends Immunol. 28:80–87.

Orange JS, Ballas ZK. 2006. Natural killer cells in human health and disease. Clin Immunol. 118:1–10.

Pamer E, Cresswell P. 1998. Mechanisms of MHC Class I-restricted antigen processing. Ann Rev Immunol. 16:323–358.

Perussia B, Chen Y, Loza MJ. 2005. Peripheral NK cell phenotypes: Multiple changing of faces of an adapting, developing cell. Mol Immunol. 42:385–295.

Pribila JT, Quale AC, Mueller KL, Shimizu Y. 2004. The integrins and T cell mediated immunity. Ann Rev Immunol. 22:157–180.

Prussin C, Metcalfe DD. 2003. IgE, mast cells, basophils and eosinophils. J Allergy Clin Immunol. 111:S486–S494.

Reid CD. 1997. The dendritic cell lineage in haemopoiesis. Br J Hematol. 96:217–223.

Rogge L. 2002. A genomic view of helper T cell subsets. Ann N Y Acad Sci. 975:57–67.

Rothenberg M, Hogan S. 2006. The eosinophil. Ann Rev Immunol. 24:147–174.

Sallusto F, Lanzavecchia A. 2002. The instructive role of dendritic cells on T cell responses. Arthritis Res. 4:S127–S132.

Schwarz BA, Bhandoola A. 2006. Trafficking from the bone marrow to the thymus: a prerequisite for thrombopoiesis. Immunol Rev. 209:47–57.

Springer TA. 1990. Adhesion receptors of the immune system. Nature. 346:425–434.

Springer TA. 1993. Adhesion receptors in inflammation: A précis. In: Thomas ED, Carter SK, Eds. Application of Basic Scineces to Hemoatopoiesis and treatment of disease, Raven Press, New York, pp. 231–239.

Springer TA. 1994. Traffic signals for lymphocyte recirculation and leukocyte emigration: The multistep paradigm. Cell. 76:301–314.

Steinman RM, Cohn ZA. 1973. Identification of a novel cell type in peripheral lymphoid organs of mice. I Morphology, quantitation, tissue distribution. J Exp Med. 137:1142–1162.

Takhar P, Smurthwaite L, Coker HA, Fear DJ, et al. 2005. Allergen derives class switching to IgE in the nasal mucosa in allergic rhinitis. J Immunol. 174:5024–5032.

Theoharides TC, Kalogeromitros D. 2006. The critical role of mast cells in allergy and inflammation. Ann NY Acad Sci. 1088:78–99.

Trulson A, Bystrom J, Engstrom A, Larsson R, Venge P. 2007. The functional heterogeneity of eosinophil cationic protein is determined by a gene polymorphism and post-translational modifications. Clin Exp Allergy. 37:208–218.

Venge P, Bystrom J, Carlson M, Hakansson L, Karawacjzyk M, Peterson C, Seveus L, Trulson A. 1999. Eosinophil cationic protein (ECP): Molecular and biological properties and the use of ECP as a marker of eosinophil activation in disease. Clin Exp Allergy. 29:1172–1186.

Vernon P, Allo V, Riviere C, Bernard J, et al. 2007. Major subsets of human dendritic cells are efficiently transduced by self-complementary Adeno-associated virus vectors 1 and 2. J Virol. 81:5385–5394.

Wiersma EJ, Collins C, Fazel S, Shulman MJ. 1998. Structural and functional analysis of J chain deficient IgM. J Immunol. 160:5979–5989.

Woof JM, Mestecky J. 2005. Mucosal immunoglobulins. Immunol Rev. 206:64–82.

Yamasaki S, Takashi S. 2007. Molecular basis for a pre-TCR mediated autonomous signaling. Trends Immunol. 28:39–43.

Young J, Peterson C, Venge P, Cohn Z. 1986. Mechanism of membrane damage mediated by human eosinophil cationic protein. Nature. 321:613–616.

Chapter 2
Role of Cytokines

Introduction

Cytokines are small glycoproteins produced by a number of cell types, predominantly leukocytes, that regulate immunity, inflammation and hematopoiesis. They regulate a number of physiological and pathological functions including innate immunity, acquired immunity and a plethora of inflammatory responses. The discovery of cytokines was initiated in the 1950s, but the precise identification of their structure and function took many years. The original discoveries were those of IL-I, IFN and nerve growth factors (NGFs); however, these cytokines were purified and given their names years later. Elucidation of the precise physiological, pathological and pharmacological effects of some of the cytokines is still in progress. The modern techniques of molecular biology were principally responsible for their complete identification and as a consequence, several hundred cytokine proteins and genes have been identified, and the process still continues.

Cytokines are produced from various sources during the effector phases of natural and acquired immune responses and regulate immune and inflammatory responses. They are also secreted during nonimmune events and play a role unrelated to the immune response in many tissues. Generally, their secretion is a brief, self-limited event. They not only are produced by multiple diverse cell types, but also act upon many different cell types and tissues. Cytokines often have multiple effects on the same target cell and may induce or inhibit the synthesis and effects of other cytokines. After binding to specific receptors on the cell surface of the target cells, cytokines produce their specific effects. Multiple signals regulate the expression of cytokine receptors. The target cells respond to cytokines by new mRNA and protein synthesis, which results in a specific biological response.

Interleukin-1

Interleukin-1 was originally discovered as a factor that induced fever, caused damage to joints and regulated bone marrow cells and lymphocytes, it was given several different names by various investigators. Later, the presence of two distinct proteins, IL-1α and IL-1β, was confirmed, which belong to a family of cytokines, the

M.M. Khan, *Immunopharmacology*, DOI: 10.1007/978-0-387-77976-8_2,
© Springer Science+Business Media, LLC 2008

IL-1 superfamily. Ten ligands of IL-1 have been identified, termed IL-1F1 to IL-1F10. With the exception of IL-1F4, all of their genes map to the region of chromosome 2. IL-1 plays an important role in both innate and adaptive immunity and is a crucial mediator of the host inflammatory response in natural immunity. The major cell source of IL-1 is the activated mononuclear phagocyte. Other sources include dendritic cells, epithelial cells, endothelial cells, B cells, astrocytes, fibroblasts and Large granular lymphocytes (LGL). Endotoxins, macrophage-derived cytokines such as TNF or IL-1 itself, and contact with $CD4^+$ cells trigger IL-1 production. IL-1 can be found in circulation following Gram-negative bacterial sepsis. It produces the acute-phase response in response to infection. IL-1 induces fever as a result of bacterial and viral infections. It suppresses the appetite and induces muscle proteolysis, which may cause severe muscle "wasting" in patients with chronic infection. IL-1β causes the destruction of β cells leading to type 1 diabetes mellitus. It inhibits the function and promotes the apoptosis of pancreatic β cells. Activation of T-helper cells, resulting in IL-2 secretion, and B-cell activation are mediated by IL-1. It is a stimulator of fibroblast proliferation, which causes wound healing. Autoimmune diseases exhibit increased IL-1 concentrations. It suppresses further IL-1 production via an increase in the synthesis of PGE_2.

IL-1s exert their effects via specific cell surface receptors that include a family of about nine members characterized as IL-1R1 to IL-1R9. All family members with the exception of IL-1R2 have an intracellular TLR domain. Each type of receptor in the family has some common and some unique features. The ligands (Table 2.1) for all of these receptors have not yet been identified.

Kineret (Anakinra)

Kineret is a human IL-1 receptor antagonist and is produced by recombinant DNA technology. It is nonglycosylated and is made up of 153 amino acids. With the exception of an additional methionine residue, it is similar to native human IL-1Ra. Human IL-1Ra is a naturally occurring IL-1 receptor antagonist, a 17-kDa protein, which competes with IL-1 for receptor binding and blocks the activity of IL-1.

Table 2.1 IL-1 Ligands and Their Receptors

Ligand Name	Receptor Name
IL-1F1	IL-1R1, IL-1R2
IL-1F2	IL-1R1, IL-1R2
IL-1F3	IL-1R1, IL-1R2
IL-1F4	IL-1R5
IL-1F5	IL-1R6
IL-1F6	Unknown
IL-1F7	IL-1R5
IL-1F8	Unknown
IL-1F9	IL-1R6
IL-1F10	IL-1R1

Kineret is recommended for the treatment of severely active rheumatoid arthritis for patients 18 years of age or older. It is recommended for patients who have not responded well previously to the disease-modifying antirheumatic drugs. It reduces inflammation, decreases bone and cartilage damage and attacks active rheumatoid arthritis. The drug can be used alone or in combination with other antirheumatic drugs. However, it is not administered in combination with TNF-α antagonists. Kineret also improves glycemia and β-cell secretory function in type 2 diabetes mellitus. It is administered daily at a dose of 100 mg/day by subcutaneous injection.

The most serious side effects of Kineret are infections and neutropenia. Injection site reactions are also common. Other side effects may include headache, nausea, diarrhea, flu-like symptoms and abdominal pain. The increased risk of malignancies has also been observed.

Interleukin-2

IL-2, a single polypeptide chain of 133 amino acid residues, is produced by immune regulatory cells that are principally T cells. When a helper T cell binds to an APC using CD28 and B7, $CD4^+$ cells produce IL-2. IL-2 supports the proliferation and differentiation of any cell that has high-affinity IL-2 receptors. It is necessary for the activation of T cells. Resting T lymphocytes (unstimulated) belonging to either the $CD4^+$ or the $CD8^+$ subsets possess few high-affinity IL-2 receptors, but following stimulation with specific antigen, there is a substantial increase in their numbers. The binding of IL-2 with its receptors on T cells induces their proliferation and differentiation.

IL-2 is the major growth factor for T lymphocytes, and the binding of IL-2 to its specific receptors on TH cells stimulates the proliferation of these cells and the release of a number of cytokines from these cells. IL-2 is required for the generation of $CD8^+$ cytolytic T cells, which are important in antiviral responses. It increases the effector function of NK cells. When peripheral blood lymphocytes are treated with IL-2 for 48–72 h, lymphokine-activated killer (LAK) cells are generated, which can kill a much wider range of targets including the tumor cells. IL-2 enhances the ability of the immune system to kill tumor cells and may also interfere with the blood flow to the tumors. It not only induces lymphoid growth but also maintains peripheral tolerance by generation of regulatory T cells. IL-2 knockout mice produce a wide range of autoantibodies and many die of autoimmune hemolytic anemia, which suggests that it plays a role in immune tolerance.

Interleukin-2 Receptors

The IL-2 receptor occurs in three forms with different affinities for IL-2; the three distinct subunits are the α, β and γ chains. The monomeric IL-2Rα possesses low affinity, the dimeric IL-2Rβγ has intermediate affinity and the trimeric IL-2Rαβγ has high affinity (Table 2.2). The α chain is not expressed on resting T cells but

Table 2.2 IL-2 Receptors

	Low Affinity	Intermediate Affinity	High Affinity
Affinity constant (M)	10^{-8}	10^{-7}	10^{-11}
Dissociation constant (M)	10^{-8}	10^{-9}	10^{-11}
Subunits	IL-2Rα	IL-2Rβγ	IL-2Rαβγ

only on activated T cells and is also called TAC (T cell activation) receptor. Both β and γ chains are required for the signal transduction mediated via IL-2 receptors. The low-affinity and high-affinity IL-2 receptors are expressed by activated CD4⁺ and CD8⁺ T cells and in low numbers on activated B cells. The intermediate-affinity IL-2 receptors are expressed on NK cells and in low numbers on resting T cells.

When IL-2 binds to high-affinity receptors, it becomes internalized following receptor-mediated endocytosis. After high-affinity binding, there is an increase in the stimulation of phosphoinositol turnover, redistribution of protein kinase C from the cytoplasm to the cell membrane, and an increased expression of IL-2 receptors, with low-affinity receptors being preferentially increased.

Clinical Uses of Interleukin-2

Immunotherapy for Cancer

Proleukin (Aldesleukin)

Proleukin is a recombinant human IL-2 that received approval for the treatment of renal cell carcinoma in 1992 and for the treatment of metastatic melanoma in 1998. It is also being evaluated for the treatment of non-Hodgkin's lymphoma (NHL). The therapy is restricted to patients with normal cardiac and pulmonary functions.

The treatment generally consists of two treatment cycles, each lasting for 5 days and separated by a rest period. Every 8 h a dose of 600,000 IU/kg (0.037 mg/kg) is administered. The IV infusion period is 15 min and a maximum of 14 doses are administered. After a rest period of 9 days, another 14 doses are administered. Additional treatment can be given following an evaluation after 4 weeks.

The most frequent adverse reactions associated with the administration of proleukin include fever, chills, fatigue, malaise, nausea and vomiting. It has also been associated with capillary leak syndrome (CLS). CLS is defined as a loss of vascular tone and effusion of plasma proteins and fluids into the extravascular space. This leads to hypotension and decreased organ perfusion, which may cause sudden death. Other side effects include anaphylaxis, injection site necrosis and possible autoimmune and inflammatory disorders.

Lymphokine-Activated Killer Cell Therapy

IL-2 has been tested for antitumor effects in cancer patients as part of LAK therapy. LAK cell therapy involves infusion into cancer patients of their own (autologous) lymphocytes after they have been treated in vitro with IL-2 for a minimum of 48 h to generate LAK cells. IL-2 needs to be administered with LAK cells in doses ranging from 10^3 to 10^6 U/m^2 body area or from 10^4 to 10^5 U/kg body weight.

Interleukin-2 and AIDS

HIV is a retrovirus that infects CD4$^+$ cells. After HIV becomes integrated into the genome of the CD4$^+$ cells, activation of these cells results in the replication of virus, which causes lysis of the host cells. Patients infected with HIV, and with AIDS, generally have reduced numbers of helper T cells and the CD4:CD8 ratio may be as low as 0.5:1 instead of the normal 2:1. As a consequence, very little IL-2 is available to support the growth and proliferation of CD4$^+$ cells despite the presence of effector cells, B cells and cytolytic T cells.

Proleukin has not been approved for the treatment of HIV; however, studies show that proleukin in combination with antiretroviral therapy significantly increases the number of CD4$^+$ cells. Low-frequency doses of subcutaneous proleukin at maintained intervals increased CD4$^+$ cell levels. The CD4 count increased from 520 cells/μl to 1005 cells/μl, and the mean of CD4$^+$ cells present from 27 to 38%. The overall effects of proleukin administration in combination with other anti-HIV drugs are being studied to determine the regulation of immune response as well as a delay in the progression of HIV disease.

Interleukin-4

IL-4 is a pleiotropic cytokine produced by TH$_2$ cells, mast cells and NK cells. Other specialized subsets of T cells, basophils and eosinophils also produce IL-4. It regulates the differentiation of antigen-activated naïve T cells. These cells then develop to produce IL-4 and a number of other TH$_2$-type cytokines including IL-5, IL-10 and IL-13. IL-4 suppresses the production of TH$_1$ cells. It is required for the production of IgE and is the principal cytokine that causes isotype switching of B cells from IgG expression to IgE and IgG4. As a consequence, it regulates allergic disease. IL-4 leads to a protective immunity against helminths and other extracellular parasites. The expression of MHC class II molecules on B cells and the expression of IL-4 receptors are upregulated by IL-4. In combination with TNF, IL-4 increases the expression of VCAM-1 and decreases the expression of E-selectin, which results in eosinophil recruitment in lung inflammation.

IL-4 mediates its effects via specific IL-4 receptors that are expressed on a number of tissues including hematopoietic cells, endothelium, hepatocytes, epithelial cells, fibroblasts, neurons and muscles. The receptor is composed of an α chain, which is the high-affinity receptor, but its signaling requires a second chain, a γ chain (γC), which is also a component of IL-2 receptors. However, the presence of a γ chain does not significantly increase the affinity of the receptor complex for IL-4. IL-4 causes the heterodimerization of the α chain with the γ chain, resulting in IL-4 receptor-dependent signaling pathway. As is the case with other cytokines, the signaling pathways activated after the binding of IL-4 to its receptors are insulin receptor substrate (IRS-1/2) and Janus family tyrosine kinases–signal transducers and activators of transcription (JAK–STAT) pathways. However, for IL-4, the specificity results from the activation of STAT-6.

The antibodies to IL-4 inhibit allergen-induced airway hyperresponsiveness (AHR), globlet cell metaplasia and pulmonary eosinophilia in animal models. Inhibition of IL-4 by soluble IL-4 receptor (SIL-4R, Nuvance) has proven to be very promising in treating asthma. Clinical trials with recombinant SIL-4R administered by a single weekly dose of 3 mg via nebulization have been effective in controlling the symptoms of moderate persistent asthma.

Interleukin-5

IL-5 is secreted predominantly by TH$_2$ lymphocytes. However, it can also be found in mast cells and eosinophils. It regulates the growth, differentiation, activation and survival of eosinophils. IL-5 contributes to eosinophil migration, tissue localization and function, and blocks their apoptosis. Eosinophils play a seminal role in the pathogenesis of allergic disease and asthma and in the defense against helminths and arthropods. The proliferation and differentiation of antigen-induced B lymphocytes and the production of IgA are also stimulated by IL-5. TH$_2$ cytokines IL-4 and IL-5 play a central role in the induction of airway eosinophilia and AHR. It is a main player in inducing and sustaining the eosinophilic airway inflammation.

IL-5 mediates its biological effects after binding to IL-5R, which is a membrane-bound receptor. The receptor is composed of two chains, a ligand-specific α receptor (IL-5Rα) and a shared β receptor (IL-5Rβ). The β chain is also shared by IL-3 and GM-CSF, resulting in overlapping biological activity for these cytokines. The signaling through IL-5R requires receptor-associated kinases. Two different signaling cascades associated with IL-5R include JAK/STAT and Ras/mitogen-activated protein kinase (MAPK) pathways.

IL-5 is usually not present in high levels in humans. However, in a number of disease states where the number of eosinophils is elevated, high levels of IL-5 and its mRNA can be found in the circulation, tissue and bone marrow. These conditions include the diseases of the respiratory tract, hematopoietic system, gut and skin. Some other examples include food and drug allergies, atopic dermatitis, aspirin sensitivity and allergic or nonallergic respiratory diseases.

Another way of interfering with IL-5 or IL-5R synthesis is by the use of antisense oligonucleotides. Antisense oligonucleotides are short synthetic DNA sequences that can hybridize specifically to the mRNA of the cytokine or its receptors. This will result in the inhibition of the transcription and processing of mRNA. The administration of IL-5-specific antisense oligonucleotides results in reduced lung eosinophilia in animal models. However, there is no complete inhibition of antigen-specific late-phase AHR, suggesting that in addition to IL-5, other pathways may also be involved in airway hyperreactivity.

Interleukin-6

IL-6 is a proinflammatory cytokine, which is a member of the family of cytokines termed "the IL-6 type cytokines." The cytokine affects various processes including the immune response, reproduction, bone metabolism and aging. IL-6 is synthesized by mononuclear phagocytes, vascular endothelial cells, fibroblasts and other cells in response to trauma, burns, tissue damage, inflammation, IL-1 and, to a lesser extent, TNF-α. Pathogen-associated molecular patterns (PAMPs) binding to the TLRs present on macrophage result in the release of IL-6. This cytokine is synthesized by some activated T cells as well. It is also secreted by osteoblasts to stimulate osteoclast formation. Acute-phase response and fever are caused by IL-6, which is also the case for IL-1 and TNF-α. It affects differentiation of B cells and causes neutrophil mobilization. IL-6 is elevated in patients with retroviral infection, autoimmune diseases and certain types of benign or malignant tumors. It stimulates energy mobilization in the muscle and fatty tissue, resulting in an increase in body temperature. IL-6 acts as a myokine — a cytokine produced by muscles — and muscle contraction occurs as a result of elevated IL-6 concentrations. The expression of IL-6 is regulated by various factors, including steroidal hormones, which could be at both transcriptional and posttranscriptional levels. IL-6 mediates its effects via binding to cell surface receptors, IL-6R, which are active in both membrane-bound and soluble forms.

Interleukin-9

Originally described as a mast cell growth factor due to its ability to promote the survival of primary mast cells and as an inducer of IL-6 production, IL-9, which is secreted by TH$_2$ cells, stimulates the release of a number of mediators of mast cells and promotes the expression of the high-affinity IgE receptors (FcϵR1α). IL-9 augments TH$_2$-induced inflammation and enhances mucus hypersecretion and the expression of its receptors is increased in asthmatic airways. It also promotes eosinophil maturation in synergy with IL-5. IL-9 activates airway epithelial cells by stimulating the production of several chemokines, proteases, mucin genes and ion channels. It is important to point out that, as opposed to the IL-4-induced isotype switching and production of IgE or the IL-5-mediated stimulation of eosinophil maturation, IL-9 induces actions of other cytokines. It is an essential cytokine for asthmatic disease as biopsies from asthmatic patients show an increase in the expression of IL-9 compared to healthy individuals, and therefore it is an important therapeutic target for clinical intervention.

Interleukin-10

First identified as an inhibitor of IFN-γ synthesis in TH$_1$ cells, IL-10 is an important immunoregulatory cytokine. It is an anti-inflammatory cytokine that was first called human cytokine synthesis inhibitory factor. IL-10 is secreted by macrophages, TH_2 cells and mast cells. Cytotoxic T cells also release IL-10 to inhibit viral infection-stimulated NK cell activity. IL-10 is a 36-kDa dimer composed of two 160-amino-acid-residue-long chains. Its gene is located on chromosome 1 in humans and consists of five exons. IL-10 inhibits the synthesis of a number of cytokines involved in the inflammatory process including IL-2, IL-3, GM-CSF, TNF-α and IFN-γ. Based on its cytokine-suppressing profile, it also functions as an inhibitor of TH_1 cells and by virtue of inhibiting macrophages, it functions as an inhibitor of antigen presentation. Interestingly, IL-10 can promote the activity of mast cells, B cells and certain T cells.

There are several viral IL-10 homologs: Epstein–Barr virus (BCRF.1), cytomegalovirus, herpesvirus type 2, orf virus and Yaba-like disease virus. Now the IL-10 family of cytokines includes not only IL-10 but also its viral gene homologs and several other cytokines including IL-19, IL-20, IL-22, IL-24, IL-26, IFN-λ1, IFN-λ2 and IFN-λ3. IL-10 mediates its effects after binding to two receptor chains, IL-10R1 (α) and IL-10R2 (β). These receptors are members of the class II or IFN receptor family. The interaction of IL-10 with its receptors is highly complex and the IL-10R2 (β) chain is essential for the production of its effects. Several hundred genes are activated after interaction of IL-10 with its receptors. The tyrosine kinases JAK1 and Tyk2 are activated by the interaction of IL-10 with its receptors, which results in the induction of transcription factors STAT1, STAT3 and STAT5, and eventual gene activation.

The major immunobiological effect of IL-10 is the regulation of the TH_1/TH_2 balance. TH_1 cells are involved in cytotoxic T-cell responses whereas TH_2 cells regulate B-cell activity and function. IL-10 is a promoter of TH_2 response by inhibiting IFN-γ production from TH_1 cells. This effect is mediated via the suppression of IL-12 synthesis in accessory cells. IL-10 is involved in assisting against intestinal parasitic infection, local mucosal infection by costimulating the proliferation and differentiation of B cells. Its indirect effects also include the neutralization of bacterial toxins.

IL-10 is a potent inhibitor of IL-1, IL-6, IL-10 itself, IL-12, IL-18, CSF and TNF. It not only inhibits the production of proinflammatory mediators but also augments the production of anti-inflammatory factors including soluble TNF-α receptors and IL-1RA. IL-10 downregulates the expression of MHC class II molecules (both constitutive and IFN-γ-induced), as well as that of costimulatory molecule, CD86, and adhesion molecule, CD58. It is an inhibitor of IL-12 production from monocytes, which is required for the production of specific cellular defense response. IL-10 enhances the expression of CD16, CD32 and CD64 and augments the phagocytic activity of macrophages. The scavenger receptors, CD14 and CD163, are also upregulated on macrophages by IL-10. It is a stimulator of NK cells, enhances their cytotoxic activity, and also augments the ability of IL-18 to stimulate NK cells. Based on its immunoregulatory function, IL-10 and ligands for its receptors are tempting candidates for therapeutic intervention in a wide variety of disease states, including autoimmune disorders, acute and chronic inflammatory diseases, cancer, infectious disease, psoriasis and allergic disease.

Modest but significant improvement has been observed in patients with chronic hepatitis C, Crohn's disease, psoriasis and rheumatoid arthritis after subcutaneous administration of IL-10 in human clinical trials. The systemic administration of IL-10 produces general immune suppression, inhibition of macrophage and T-cell infiltration, less secretion of IL-12 and TNF-α by monocytes and suppression of nuclear factor (NF)-κB induction. In patients with acute myelogenous leukemia, IL-10 increases the serum levels of TNF-α and IL-1β. The use of IL-10 for human cancer therapy is under investigation and despite its immunosuppressive effects it may serve a role as a facilitator in preconditioning tumors to be recognized by immune effector cells.

Interleukin-11

IL-11, a member of the IL-6 superfamily, is produced by bone marrow stroma and activates B cells, plasmacytomas, hepatocytes and megakaryocytes. The gene for IL-11 is located on chromosome 19. IL-11 induces acute-phase proteins, plays a role in bone cell proliferation and differentiation, increases platelet levels after chemotherapy and modulates antigen–antibody response. It promotes differentiation of progenitor B cells and megakaryocytes. The recovery of neutrophils is accelerated by IL-11 after myelosuppressive therapy. IL-11 also possesses potent anti-inflammatory effects due to its ability to inhibit nuclear translocation of NF-κB. Additional biological effects of this cytokine include epithelial cell growth, osteoclastogenesis and inhibition of adipogenesis. The effects of IL-11 are mainly mediated via the IL-11 receptor α chain. IL-11 forms a high-affinity complex in association with its receptor and associated proteins and induces gp130-dependent signaling.

Oprelvekin (Neumega)

Recombinant human IL-11 (oprelvekin) is a polypeptide of 177 amino acids. It differs from natural IL-11 due to lack of glycosylation and the amino-terminal proline residue. Oprelvekin is administered by subcutaneous injection, usually 6–24 h after chemotherapy, at a dose of 25–50 μg/kg per day. The drug has a half-life of about 7 h. It is used to stimulate bone marrow to induce platelet production in nonmyeloid malignancies in patients undergoing chemotherapy. The common side effects of oprelvekin include fluid retention, tachycardia, edema, nausea, vomiting, diarrhea, shortness of breath and mouth sores. Other side effects include rash at the injection site, blurred vision, paresthesias, headache, fever, cough and bone pain. Rarely, CLS may occur.

Interleukin-13

IL-13 belongs to the same α-helix superfamily as IL-4, and their genes are located 12 kb apart on chromosome 5q31. It was originally identified for its effects on B cells and monocytes, which included isotype switching from IgG to IgE, inhibition of inflammatory cytokines and enhancement of MHC class II expression. Initially, IL-13 appeared similar to IL-4 until its unique effector functions were recognized. Nevertheless, IL-13 and IL-4 have a number of overlapping effects. IL-13 also plays an essential role in resistance to most GI nematodes.

It regulates mucus production, inflammation, fibrosis and tissue remodeling. IL-13 is a therapeutic target for a number of disease states including asthma, idiopathic pulmonary fibrosis, ulcerative colitis, cancer and others. Its signaling is mediated via IL-4 type 2 receptor. The receptor consists of IL-4Rα and IL-13Rα1 and IL-13Rα2 chains.

IL-13 induces physiological changes in organs infected with parasites that are essential for eliminating the invading pathogen. In the gut, it induces a number of changes that make the surrounding environment of the parasite less hospitable, such as increasing contractions and hypersecretion of glycoproteins from gut epithelial cells. This results in the detachment of the parasites from the wall of the gut and their subsequent removal. IL-13 response in some instances may not resolve infection and may even be deleterious. For example, IL-13 may induce the formation of granulomas after organs such as the gut wall, lungs, liver and central nervous system are infected with the eggs of *Schistosoma mansoni*, which may lead to organ damage and could even be life threatening.

IL-13 is believed to inhibit TH$_1$ responses, which will inhibit the ability of the host to eliminate the invading pathogens. The role of IL-13 in the etiology/pathogenesis of allergic disease/asthma has drawn broad attention. It induces AHR and goblet cell metaplasia, which result in airway obstruction and cause allergic lung disease. IL-13/chemokine interactions play a key role in the development of AHR and mucus production. IL-13 induces the expression of eotaxins. These chemokines recruit eosinophils into the site of inflammation in synergy with IL-5. Eosinophils release IL-13 and induce the production of IL-13 from TH$_2$ cells, which is mediated via IL-18. IL-13 then, through its effects on epithelial and smooth muscle cells, aids in the development of AHR and mucus production. In addition to its potent activation of chemokines, IL-13 is also an inducer of adhesion molecules involved in asthma.

Interleukin-18

IL-18 is a member of the IL-1 family that promotes the production of various proinflammatory mediators and plays a role in cancer and various infectious diseases. It was originally identified as IFN-γ-inducing factor and is produced by cells of both hematopoietic and nonhematopoietic lineages, including macrophages, dendritic cells, intestinal epithelial cells, synovial fibroblasts, keratinocytes, Kupffer cells, microglial cells and osteoblasts. The production of IL-18 is structurally homologous to that of IL-1β; it is produced as an inactive precursor of 24 kDa, which lacks a signal peptide. Endoprotease IL-1β-converting enzyme activates it after cleaving pro-IL-18, resulting in a biologically active cytokine. Caspase-1 plays an important role in the processing of IL-18, but is not exclusive since proteinase 3 can also perform the same function.

CC Chemokines

Also termed β-chemokines, CC Chemokines are composed of two adjacent cysteines near their amino-terminus. There are at least 27 different CC chemokines of which CCL9 and CCL10 are the same. Most of the members of this group possess four cysteines (C4-CC) but a small number have six cysteines (C6-CC). CC chemokines regulate the migration of monocytes, dendritic cells and NK cells. An important chemokine in this group is monocyte chemoattractant protein-1 (MCP-1, also called CCL2), which promotes the migration of monocytes from the bloodstream to the tissue where they differentiate to become macrophages. Other CC chemokines include MIPs, MIP-1α (CCL3) and MIP-1β (CCL4) and RANTES (CCL5). The effects of CC chemokines are mediated via specific cell surface receptors: 10 different types of these receptors (CCR1–CCR10) have been identified.

C Chemokines

Also termed γ-chemokines, C Chemokines are composed of only two cysteines: one on the N-terminus and the other a downstream cysteine. There are two chemokines in this group, lymphotactin-α (XCL1) and lymphotactin-β (XCL2). Their function is the attraction of T-cell precursors to the thymus.

CX3C Chemokines

Also termed δ-chemokines, CX3C Chemokines are composed of three amino acids between the two cysteines. This subgroup has only one member, fractalkine (CX3CL1). CX3C chemokines are secreted as well as present on the cell surface and serve both as a chemoattractant and as an adhesion molecule.

Biological Role of Chemokines

The primary function of chemokines is to induce the migration of leukocytes. A signal directs these cells toward the chemokines. During immunological surveillance, chemokines direct lymphocytes to the lymph nodes, which allows them to interact with the APCs and detect any invading pathogens. Such chemokines are called homeostatic chemokines and do not require a stimulus for their secretion. Some chemokines are proinflammatory in nature and require specific stimulus for their release. These stimuli include viral infection, bacterial products as well as other chemical agents. Proinflammatory cytokines including IL-1 and TNF-α promote their release. These chemokines are chemoattractants for neutrophils, leukocytes, monocytes and some effector cells, and they direct the migration of these leukocytes to the site of injury/infection. Some proinflammatory chemokines are also involved in wound healing similar to the proinflammatory cytokines. Chemokines are also capable of activating leukocytes to initiate an immune response and are involved in both innate and acquired immunity. Other chemokines play a role in development and are involved in angiogenesis and cell maturation.

Chemokine Receptors

Chemokine receptors are a family of G protein-coupled receptors that contain seven transmembrane domains. Chemokine receptors are present on the cell surface membrane of leukocytes. As was the case for chemokines, these receptors are also divided into four subgroups: CCR is specific for CC chemokines, CXCR for CXC chemokines, XCR1 for C chemokines and CX3CR1 for CX3C chemokines. The CC chemokine receptor family has eleven members, the CXC chemokine receptor family has seven members, and both the C chemokine receptor family and the CX3C chemokine receptor family have one member each. The signal transduction is mediated via the standard G protein-dependent pathway.

Chemokines and Disease States

Human Immunodeficiency Virus Infection

Human immunodeficiency virus requires CD4 and either CXCR4 or CCR5 to enter target cells. This allows the entry of HIV into $CD4^+$ T cells or macrophages, which eventually leads to the destruction of $CD4^+$ T cells and almost total inhibition of antiviral activity. Individuals who possess a nonfunctional variant of CCR5 and are homozygous for this gene remain uninfected despite multiple exposures to HIV. Clinical trials are under way to develop antagonists of these chemokine receptors as potential therapeutic agents for HIV infection and AIDS.

Diabetes with Insulin Resistance

Cytokines and chemokines have been implicated in insulin resistance. The cytokines which may play a role include IL-6 and TNF-α. CCR2 are present on adipocytes, and activation of inflammatory genes by the interaction of CCR2 with the ligand CCL2 results in impaired uptake of insulin-dependent glucose. Adipocytes also synthesize CCL2, resulting in the recruitment of macrophages. CCL3 may also be involved in insulin resistance.

Atherosclerosis

CCL2 is present in lipid-laden macrophages and atherosclerotic plaques that are rich in these macrophages. The production of CCL2 in endothelial and smooth muscle cells is stimulated by minimally oxidized low-density lipoproteins (LDLs). As a consequence, CCL2 is involved in the recruitment of foam cells to the vessel

Yokota T, Otsuka K, Mosmann T, Banchereau J, et al. 1986. Isolation and characterization of a human interleukin cDNA clone, homologous to mouse B-cell stimulatory factor 1, that expresses B-cell and T-cell stimulating activities. Proc Nat Acad Sci USA. 83:5894–5898.

Zdanov A. 2004. Structural features of the interleukin-10 family of cytokines. Curr Pharm Design. 10-:3873–3884..

Zheng LM, Ojcius DM, Garaud F, Roth C, et al. 1996. Interleukin-10 inhibits tumor metastasis through an NK cell dependent mechanism. J Exp Med. 184:579–584.

Zhou Y, McLane M, Levitt RC. 2001. TH_2 cytokines and asthma: Interleukin 9 as therapeutic target for asthma. Resp Res. 2:80–84.

Zimmermann N, Hershey GK, Foster PS, Rothenberg ME. 2003. Chemokines in asthma: Cooperative interaction between chemokine and IL-13. J All Clin Immunol. 111:227–242.

Zohlnhofer D, Ott I, Mehilli J, Schoming K, et al. 2006. Stem cell mobilization by granulocyte-colony-stimulating factor in patients with acute myocardial infarction. JAMA. 295: 1003–1010.

Zurawski G, de Vries JE. 1994. Interleukin 13, an interleukin-4 like cytokine that acts on monocytes and B cells, but not on T cells. Immunol Today. 15:19–26.

Chapter 3
Cytokine Receptors and Signaling

Introduction

The biological effects of cytokines result after they interact with their highly specific cell surface receptors distributed on various tissues. The ligand–receptor interaction is quite limited to produce the biological responses since the number of cytokine receptors varies from as low as 10^2/cell to as high as 10^5/cell and the range for their affinity is about 10^{10} M^{-1}. The cytokine receptors are generally composed of multiple polypeptide chains, all of which may interact for the successful transmission of the cytokine-induced signal. There are five families of cytokine receptors, which include immunoglobulin superfamily receptors, class I cytokine receptor family (hematopoietin receptor family), class II cytokine receptor family (IFN receptor family), TNF receptor family and chemokine receptor family (Fig. 3.1).

Immunoglobulin Superfamily Receptors

The immunoglobulin superfamily includes various cell surface receptors as well as soluble proteins. Their function includes recognition, binding and adhesion among cells. These receptors share features with immunoglobulins as they all possess an immunoglobulin domain or fold. This group contains receptors not only for cytokines but also for antigens and costimulatory molecules involved in the immune response. Although they share a similar immunoglobulin fold, there are differences in their distribution, composition and biological functions.

One prominent member of this group, the IL-1R/TLR superfamily of receptors, plays an essential role in innate immunity and inflammation. The autoimmune and inflammatory diseases are associated with these signaling pathways. Both IL-1 and IL-18 act through this family of receptors and as a result promote a wide range of proinflammatory mediators. The IL-1R pathway is an ancient pathway of host defense. Another receptor homologous to IL-1R is the TLR family, which is very important in innate immunity and includes receptors for lipoteichoic acid (TLR2), LPS (TLR4), bacterial flagellin (TLR5) and CpG motifs in DNA (TLR9).

The downstream signaling molecules for this receptor family are shared by IL-1R, IL-18R and TLR family, which include MyD88 (an adaptor molecule),

M.M. Khan, *Immunopharmacology*, DOI: 10.1007/978-0-387-77976-8_3,
© Springer Science+Business Media, LLC 2008

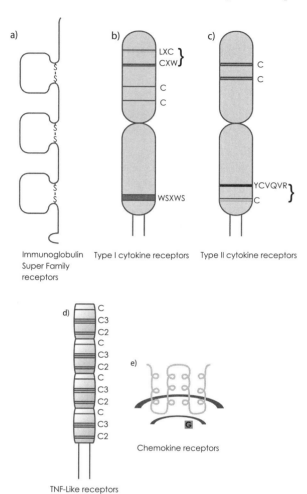

Fig. 3.1 Families of cytokine receptors; the cytokine receptors are classified into five major families: immunoglobulin superfamily receptors, type I cytokine receptors, type II cytokine receptors, TNF-like receptors and chemokine receptors. The drawings illustrate their general biochemical structure (*see* Color Insert)

IRAKs (IL-1R-associated protein kinases), TAK1 (TGF-β-activated kinase), TAB1 and 2 (TAK1 and TAK2 binding protein) and TRAF6 (TNF receptor-associated factor 6). MyD88 is recruited after the binding of a ligand to IL-1R or TLR, resulting in the binding of IRAK4 and IRAK1, which are the IL-1R-associated kinases. In this process, the activation of IRAK4 hyperphosphorylates IRAK1, which allows the interaction of TRAF with the complex. This complex then binds to another complex, which includes TAK1, TAB1 and TAB2, resulting in the phosphorylation of TAB2 and TAK1 and migration of TAK1, TAB1, TAB2 and TRAF6 to the cytosol. The activation of TAK1 in the cytosol results in the phosphorylation of IκB kinase (IKK), which degrades IκB and frees NF-κB. This signaling also results in the induction of MAPKs and Janus kinases (JNKs) via activation of TAK1.

Class I Cytokine Receptor Family

The class I cytokine receptor family is the largest and most divergent family of cytokines, which include approximately 50 different interleukins, lymphokines, hematopoietins, growth hormones and neuropoietins, and the majority of cytokine receptors belong to this family. They are composed of 200 amino acids and contain two motifs; one is made up of four cysteine residues (CCCC) and the other is composed of a conserved sequence of the amino acids tryptophan–serine–(any amino acid)–tryptophan–serine (WSXWS). The presence of two fibronectin type III modules gives it a "barrel-like" shape and the cytokine binding pocket is the trough formed in between two barrel-like modules. These receptors are heterodimers: one binds to the cytokine and the other is involved in downstream signal transduction. Class I cytokine receptors are composed of three subfamilies and each subfamily member has identical signal transduction mechanisms through similar subunits. However, their cytokine-specific subunits are different, but redundancy and antagonistic properties of some cytokines result from similar signal transduction units.

The presence of at least two chains is the common characteristic of many of these receptors. The cytokines having two chain structures exhibit dual affinity although there are some exceptions. The examples of receptors with two chain structures include sharing of γ–γ subunit of IL-2 with IL-4, IL-7, IL-9, IL-15 and IL-21, common β chains of IL-3, IL-5 and GM-CSF and common α chain for IL-4 and IL-13. gp130, which is a second chain required for IL-6-α activity, is also a trigger for several other cytokines. The binding of the cytokine to the double chain renders a dual affinity as is the case for IL-2, IL-3, IL-5, IL-7 and GM-CSF.

Class II Cytokine Receptor Family

The class II cytokine receptor family is also called the IFN receptor family. These receptors have either one (γ-IFN) or two 200-amino-acid domains; the domains contain cysteine in both carboxy-terminal domains, and are also related to fibronectin type III domains, which are composed of 100 amino acids. Each fibronectin type III domain contains conserved Cys, Pro and Trp residues, which are responsible for the formation of a folding pattern of seven β strands. These β strands are similar to the constant domain of the immunoglobulins. The class II cytokine receptor family lacks the WSXWS motif present in class I cytokine receptors. Type I IFN-α/β and IFN-γ are made up of two receptor chains, one for ligand binding and the other for signaling. This family includes receptors for IFN-α1, IFN-α2, IFN-γ1, IFN-γ2, IFN-κ (type I IFN), IFN-λ (IL-28/-29) and molecules related to IL-10 (IL-19, IL-20, IL-22, IL-24 and IL-26). The type I IFN receptors (IFNAR) have different components classified as IFNAR1 and IFNAR2. They have a number of cognate ligands including 13 IFN-α subtypes, β, ω, ε, κ and others. The type I IFN receptors are from the type II IFN-γ (IFNGR1 and IFNGR2) and the type III IFNs (IFNLR and IL-10Rβ).

Two ligand-binding IFN-γ receptor chains 1 in conjunction with two IFN-γ receptor chains 2 and associated signal-transducing proteins constitute functional IFN-γ receptor (IFN-γR). This is a class of receptors that binds the agonist in the small angle of a V produced by the two Ig-like folds, which form the extracellular domain. The response to IFN-γ depends on the IFNγR2 chain since there is an ample amount of IFNγR1 chain. The levels of the IFNγR2 chain are strictly regulated depending on the activation and differentiation state of the cell despite their constitutive expression. The intrinsic kinase/phosphatase activity is not associated with either IFNγR chain and consequently they associate with signaling machinery for signal transduction. The binding motifs for the Janus tyrosine kinase 1 and STAT1 are present in the intracellular domain of IFNγR1. The JAK1 and STAT1 motifs are necessary for the phosphorylation and signal transduction of the receptors, as well as for the induction of the biological response. A noncontiguous binding motif for JAK2 kinase recruitment is present in the intracellular region of IFNγR2 and participates in signal transduction.

IFN-λ is similar to type I IFNs in possessing antiviral and proliferation-inducing activities but has distinct receptors, where CRF2 proteins are involved in the formation of heterodimers and mediating cytokine-induced signaling activity. IFN-λ is present in three forms, IFN-λ1 (IL-29), IFN-λ2 (IL-28A) and IFN-λ3 (IL-28B). The ligand-binding chains for IFN-λ, IL-22 and IL-26 are different from IL-10, but all of them use IL-10 receptor 2 as a common second chain to form active receptor complexes. After a class II receptor ligand binds to its receptor, the receptor complex is fully assembled and the JAKs that are associated with the intracellular domains of class II receptors are induced. This results in the phosphorylation of tyrosine residues in the receptor chain, which serves as docking sites for the transcription factors, STATs. Several different STATs are activated by these cytokines, which can then combine with other cytosolic proteins before migration to the nucleus.

Tumor Necrosis Factor Receptor Family

The TNF receptor family members include TNF-α, TNF-β, gp39 (CD40-L), CD27-L, CD30-L and NGF. TNF-α has two types of receptors TNF-R1 (CD120a, p55/p60) and TNF-R2 (CD120b, p75/80), which have differential distribution. TNF-R1 is ubiquitously distributed while TNF-R2 is restricted to immune cells. TNF-R1 is activated by both the membrane-bound and soluble trimeric form of TNF, whereas TNF-R2 responds to the membrane-bound form. These receptors possess four extracellular repeats of a cysteine-rich domain, which usually contain six conserved cysteines. Following binding to the ligand, trimers may also be formed by TNF receptors as their tips intrude between TNF monomers resulting in the dissociation of silencer of death domains (SODD) from the intracellular death domain. SODD is an inhibiting protein, and its dissociation allows binding of TRADD (TNF receptor death domain-associated protein), an adaptor protein, to the death domain. This serves as a platform for further association of transcription factors, which is followed by initiation of one of the three possible mechanisms, activation of NF-κB, activation of MAPK pathways or induction of death signaling. NF-κB is activated as a result of recruitment of TRAF2 (TNF receptor-associated factor-2) and receptor-interacting protein (RIP) by TRADD. Subsequently, protein kinase IKK is recruited by TRAF2, which is activated by RIP, a serine–threonine kinase. IκB kinase also phosphorylates and degrades IKBα, which is an inhibitory protein for NF-κB. This allows NF-κB to migrate to the nucleus and direct gene expression of a number of inflammatory proteins and inhibit apoptosis.

TNF also activates MAPKs including stress-related JNK and p38-MAPK, but extracellular signal-regulated kinases (ERKs) are not significantly activated. JNK-inducing upstream kinases are induced by TRAF2 resulting in the activation and translocation of JNK into the nucleus where it activates other transcription factors resulting in cell proliferation and differentiation.

Lastly, TNF-R1 is involved in apoptosis. However, this is not a major function as opposed to its vast role in inflammatory processes. This is due to the interference of its antiapoptotic effects by NF-κB and Fas, another family member that exhibits more potent apoptotic effects. The apoptotic effects result from the binding of TRADD to Fas-associated death domain (FADD) protein, which is followed by the recruitment of cysteine protease caspase-8. High concentrations of caspase-8 induce cell apoptosis after its autoproteolytic activation and cleaving of effector caspase. The Fas antigen member of the TNF receptor gene family has been called the "death gene," and Fas ligands related to TNF and CD40-L trigger apoptosis. CD40 is a natural ligand protein present on activated T-cells and is critical for T cell-dependent B-cell activation. The regulation of apoptosis is also the function of the other members of this family, including NGF receptors, CD27 and CD30.

Chemokine Receptor Family

The chemokine receptor family is also termed the G protein-coupled receptor superfamily of proinflammatory cytokines. It includes IL-8, MIP-α, f-Met-Leu-Phe and C5a as ligands. These receptors have seven predicted transmembrane domains and use G proteins as signal transducers. G proteins couple this family of receptors, which activate adenylate cyclase and phospholipase C. These are structurally related receptors and are divided into four families: CXCR binds CXC chemokines, CCR binds CC chemokines, CX3CR1 binds only CX3C, and XCR1 binds two chemokines, XCL1 and XCL2. The ligand-binding specificity of a chemokine receptor is dependent on the N-terminal portion of chemokine receptors. All chemokine receptors contain approximately 350 amino acids with a short acidic N-terminal, seven helical transmembrane domains, each of which contains three intracellular and extracellular loops and a serine- and threonine-containing intracellular C-terminus. Signal transduction of chemokine receptors is achieved via

G proteins, which after activation induce phospholipase C, resulting in the production of inositol triphosphate and DAG. A number of signaling pathways are activated as a result of production of these two second messengers.

Cytokine Receptor-Associated Transcription Factors

Signal Transducers and Activators of Transcription

A number of transcription factors are involved in downstream signaling pathways in response to cytokines, hormones and growth factors. One such prominent family of transcription factors is STAT. There are seven members of the STAT family, STAT1, STAT2, STAT3, STAT4, STAT5a, STAT5b and STAT6, which are made up of 750–850 amino acids and have molecular weights ranging from 90 to 115 kDa. They are constitutively present in an inactive form in the cytoplasm and are activated by their respective agonists including cytokines after they bind to their receptors. After their activation, STATs participate in gene activation and expression, where they play a dual role. In the cytoplasm they bind to the receptor-associated kinases and consequently relay the message from the cell surface to the nucleus and after translocating to the nucleus and binding to DNA, they activate transcription. The association of STAT proteins with human disease is shown in Table 3.1.

STAT1

STAT1 is activated by various ligands including cytokines, IFN-α, IFN-γ, IL-4, IL-5, IL-13, growth factors and hormones. It has four conserved domains: N-terminal, C-terminal, SH2 domain and DNA-binding domain, which is typical of all STAT family members. The SH2 domain is distinct for each STAT and a stable SH2–phosphotyrosine bond is formed after the SH2 domain binds to its specific phosphotyrosine domain. This SH2–pTyr bond is responsible for STAT activation

Table 3.1 STAT Proteins and Human Disease

STAT	Ligand	STAT Deficiency-Associated Human Disease
STAT1	IFN	Alzheimer's disease, cancer, celiac disease, ischemic heart disease, Inflammatory Bowel Disease (IBD), rheumatoid arthritis
STAT2	IFN	Cancer, IBD
STAT3	IL-2, IL-6, IL-10	Cancer, IBD, MS
STAT4	IL-12	Chronic obstructive pulmonary disease (COPD), rheumatoid arthritis
STAT5a	PRL	Cancer, diabetes, ischemic heart disease
STAT5b	GH	Growth retardation
STAT6	IL-4, IL-13	Allergic disease/asthma, cancer, ischemic heart disease

by recruiting STAT to its specific cytokine receptor, binding to JAK–STAT and dimerization of STAT–STAT. The N-terminal domain has a tyrosine residue, which is phosphorylated by activated JAKs, resulting in the dimerization and activation of STAT. The DNA-binding domain is connected to the SH2 domain and to the DNA sequence resulting in gene transcription. The C-terminal is important for transcriptional activity. STAT1 is activated by JAK as it phosphorylates the transcription factor assisting in its recruitment to the receptor, dimerization and nuclear translocation. This results in the formation of a physical link between the STAT and the associated receptor.

Most of the biological actions of IFN-γ are STAT1-dependent, although IFN-γ receptors activate additional signaling pathways and can regulate gene expression without the involvement of STAT1. Binding of IFN-γ to its receptors results in the phosphorylation and oligomerization of the receptors. As opposed to many growth factors that possess intrinsic kinase activity, cytokine receptors require JAKs for phosphorylation. The transphosphorylation of JAK at the tyrosine residues results in the recruitment of STAT1 to the cytokine receptor where the binding of the SH2 domain of the receptor to the pTyr residues of the JAKs takes place. The formation of either homo (STAT1:STAT1) or hetero (STAT1:STAT2) dimers occurs after the activated JAKs phosphorylate STAT1. These dimers migrate to the nucleus to bind to the target sequences on the DNA, inducing the transcription of STAT1-dependent genes. These genes include STAT1, ICAM1, RANTES and interferon regulatory factor (IRF-1). Interferon-stimulated gene factor (ISF3γ) is required for the binding of STAT1:STAT2 heterodimer to the IFN-stimulated response element (ISRE) on the DNA. The binding of other cytokines, which utilize STAT1 as a transcription factor, to their receptors results in the same pathway.

In addition to JAKs, STAT1 is also directly phosphorylated by protein kinase C. This process is mediated by inositol triphosphate, Ca^{2+} release, formation of Ca^{2+}–calmodulin complex and release of calcineurin. Calcineurin dephosphorylates NF-AT resulting in its translocation to the nucleus and subsequent activation of STAT1, in addition to other genes.

STAT1 plays an important role in biological function, since in its absence tissues are no longer able to respond to IFN-γ, becoming susceptible to viral infections. STAT1 is also a tumor suppressor due to its ability to promote growth arrest and apoptosis. It mediates the expression of inhibitors of cyclin-dependent kinases. Furthermore, it suppresses genes necessary for entry into the cell cycle. STAT1 mediates the effects of biological therapies of leukemia as well as the induced differentiation of leukemia cells. It is also a key negative regulator of angiogenesis. In lung and prostate cancers, there is a loss in the ability of IFN-γ to activate STAT1. Increased survival of the newly diagnosed breast cancer patients and a decrease in relapse are associated with enhanced phosphorylation and DNA binding by STAT1. The involvement of STAT1 in asthma is evident from the expression of ICAM1 and RANTES after its activation since they are involved in the recruitment of inflammatory cells at the site of inflammation. STAT1 is also associated with the inflammatory process in Alzheimer's disease.

STAT2

STAT2, a critical transcription factor for type I IFNs, is present primarily in cytoplasm and is tyrosine phosphorylated following the binding of IFN-α to its cell surface receptors. After tyrosine phosphorylation, STAT2 migrates to the nucleus and binds to DNA along with STAT1 and IRF9. IFN-α/β mediate their antiviral effects via STAT1 and STAT2 but not via STAT3. STAT2 is critical for IFN responsiveness in target cells. The antiviral and growth inhibitory effects of IFN are not observed in the absence of STAT2. For the IFN-stimulated gene factor 3 (ISGF3), a potent transactivational domain is provided by STAT2. STAT2:1 and STAT2:3 are ISGF3-independent and bind a γ-activated sequence-like (GAS) element.

STAT3

STAT3 was originally identified as a DNA-binding activity from IL-6-stimulated hepatocytes that selectively interacted with an enhancer element in the promoter of acute-phase gene, known as the acute-phase response element. Further characterization revealed that STAT3 was closely related to STAT1 and was activated by the IL-6-type cytokine family, which signals through gp130 and/or related receptors. Many unrelated agonists such as oncogenes, growth factors and IFNs may also activate STAT3.

STAT3 exists in two isoforms, a long form (STAT3a) and a short form (STAT3β). STAT3a possesses a STAT family DNA-binding domain, a major serine phosphorylation site at S727, a major tyrosine phosphorylation site at 705 and an SH2 domain. It is dimerized as a result of pTyr–SH2 interactions after it is phosphorylated and it could also heterodimerize with STAT1, and after phosphorylation STAT3 migrates to the nucleus. IL-6 and a number of other cytokines and growth factors can cause the induction of STAT3 through their respective receptors.

IL-10R produces a strong anti-inflammatory response via STAT3, which serves as an antagonist to proinflammatory signals that are activated during the innate immune response. However, anti-inflammatory response is not unique to the IL-10R-dependent signal transduction, but can be produced by a number of cytokines that activate STAT3. STAT3β (short form) lacks the C-terminal domain, which is induced by phosphorylation of serine 727 and is the domain for transcriptional activation. STAT3 is also known as an acute-phase response factor and has ubiquitous distribution. Many forms of cancers including breast cancer, head and neck cancer, prostate cancer and glioblastoma exhibit an enhanced STAT3 activity. Induction of STAT3 is also associated with inflammatory and autoimmune diseases such as acute lung injury, pulmonary fibrosis and Crohn's disease.

STAT4

STAT4 is very important in mediating proinflammatory immune response, and on the basis of its restrictive distribution of mRNA expression in myeloid and lymphoid tissues, it is evident that STAT4 is a distinct transcription factor. Cytoplasmic STAT4 exists in a latent form and its homodimers are formed after induction by cytokines via their specific receptors. Similar to other STATs, this is followed by migration of the dimer to the nucleus, binding to its niche in DNA and subsequent gene expression.

STAT4 is responsible for IL-12-mediated functions and for the development of TH_1 cells that secrete IFN-γ from naïve CD4$^+$ T cells. After IL-12 induces its receptors, STAT4 binds to IL-18γ1 promoter, which results in an increase in acetylated histones H3 and H4 transiently. The transient hyperacetylation induced by STAT4 inhibits DNA methyltransferase recruitment and the consequent suppression of the IL-18γ1 locus.

STAT4 is induced not only by IL-12 but also by IFN-α and IL-23. IL-12 and IL-23 are produced in response to various pathogenic organisms and regulate innate and acquired immune responses. IL-12 binds to IL-12β1 and IL-12β2 receptors. A subunit called P40 is shared by IL-12 and IL-23. IL-12 and IL-23 activate the JAKs, JAK2 and Tyk2, STAT4 and other STATs.

The transduction activity of STAT4 involves several cell types and at different stages in the immune response. Its role in innate immunity is the production of IFN-γ and chemokines by macrophages, dendritic cells and NK cells. These early acting cells may also be involved in STAT-dependent tissue destruction. A lack of STAT4 shifts the TH_1/TH_2 balance to TH_2 cells and results in a decrease in the associated inflammatory response. A deficiency of STAT4 results in augmented inflammatory responses and susceptibility to various infections.

STAT4 is not only required for most of the biological responses produced by IL-12, which includes IFN-γ production, but also essential for the normal differentiation of TH_1 cells and for the expression of TH_1-specific genes. These genes include IL-12Rβ2, IL-18Rα, LT-α and selectins. STAT4 deficiency protects from T cell-mediated autoimmunity but not from predominantly antibody-mediated autoimmune diseases. Cytokines activating STAT4 produce biological responses that render protection against microbes such as *Mycobacterium tuberculosis, Toxoplasma gondii, Listeria monocytogenes, Trypanosoma cruzi, Schistosoma mansoni, Leishmania major* and others. However, unregulated stimulation of STAT4 produces inflammatory diseases such as arthritis, myocarditis, colitis, experimental autoimmune encephalomyelitis, diabetes and others.

STAT4 suppresses TH_2 cytokine production and supports TH_1 cytokine production. STAT4 activation in CD4$^+$ T lymphocytes results in the differentiation of TH_1 cells, which inhibit the development of IL-4-secreting TH_2 cells. A role of histamine type 1 receptors in histamine-mediated upregulation of STAT4 phosphorylation has been proposed. Histamine shifts TH_1/TH_2 cytokine balance from TH_1 to TH_2 cytokines, which may contribute to the etiology and pathogenesis of allergic disease/asthma.

STAT5

Among the seven members of the STAT family, STAT5, which exists in two isoforms STAT5A and STAT5B, is the most ubiquitously activated transcription factor as it is activated by a number of cytokines and growth factors including IL-2, IL-3, IL-4, IL-5, IL-7, IL-9, IL-15, GM-CSF, thrombopoietin, erythropoietin, prolactin and growth hormone. Its predominant role is the regulation of mast cells and IL-3; stem cell factor (SCF) activates STAT5 to regulate the development of mast cells, and both STAT5A and B are involved in this function. Cytokine receptor-induced JAKs cause dimerization, migration to nucleus, and DNA binding after tyrosine phosphorylation of STAT5. Its regulation of expression of cytokine target genes takes place after STAT5 binds to IFN-GAS motifs. This is followed by STAT5 contacting coactivators and parts of the transcription apparatus, resulting in transcriptional activation. Serine/threonine kinase Pim-1, a target gene for STAT5, inhibits STAT5 activity by cooperating with SOCS1 and SOCS3.

STAT5 is required for the proliferation of T cells and B cells as well as the self-renewal of hematopoietic stem cells. Tumor development may result from aberrant regulation of STAT5 activity. STAT5 is constitutively activated via cytokines or growth factor receptors and/or their associated tyrosine kinases and assists in the survival of malignant cells in various cancers, including leukemia, head, neck, prostate and breast cancers. Signaling activators such as PI3 kinase and Ras help STAT5, and apoptosis can be induced in these tumors by inhibiting STAT5, which may have therapeutic implications for treating certain forms of STAT5-sensitive tumors.

STAT6

STAT6, an important transcription factor, mediates IL-4 and IL-13 receptor-induced signals. It is composed of 850 amino acids and its gene is present on the 12q12–q24 region of the human chromosome. Like other STATs, STAT6 has four functional domains: N-terminal domain, C-terminal domain, a DNA-binding domain and a conserved Src homology 2 (SH2) domain. STAT6 plays an important role in TH_2 differentiation, and TH_2 responses are not present in STAT6-deficient mice. The binding of IL-4 or IL-13 to their respective receptors activates JAK–STAT6. IL-4 and IL-13 share the IL-4Rα chain of the IL-4 receptor. After IL-13 binds to IL-13Rα1, it recruits IL-4Rα, resulting in a high-affinity complex for ligand binding and signaling. This results in the activation of JAK1 and JAK3 tyrosine kinases followed by tyrosine phosphorylation. These steps create the docking site for STAT6, which is then phosphorylated and homodimerized, and migrates to the nucleus for gene expression. In addition to playing a critical role in TH_2 differentiation, which is a critical role in the pathogenesis of allergic disease/asthma, STAT6 also promotes eotaxin secretion, a crucial factor for the eosinophilic infiltration in the airways.

Janus Kinases

Four mammalian Janus kinase (JAK family members have been identified: JAK1, JAK2, JAK3 and Tyk2 with molecular weights ranging from 120 to 140 kDa.

JAK1, JAK2 and Tyk2 are expressed ubiquitously, whereas JAK3 is present only on hematopoietic cells. JAKs are activated by cytokines that share a common α-helical structure and use additional receptors for signal transduction. For ligand binding, the cytokine receptors are made up of polypeptides with a single trans-membrane domain as well as common extracellular motifs. There is no intrinsic catalytic activity present in the cytokine receptors; as a consequence, JAKs are required, which are present in the cytoplasmic region of the receptor and with their association the extracellular signal is transmitted intracellularly. The JAKs' association with the cytokine receptors require motifs called box 1 and box 2. Binding of the agonist to its receptors within minutes activates JAKs resulting in the dimerization or oligomerization of the subunits of the receptors, and preexisting dimers undergo conformational changes after agonist binding to the cytokine receptors. If the ligand binds to homodimeric receptors, only JAK2 is activated; however, binding to heterodimeric receptors activates a combination of JAKs. Following activation, the receptor subunits as well as other substrates including STATs are phosphorylated (Fig. 3.2).

JAKs are composed of an amino-terminal region (N), catalytically inactive kinase-like (KL) domain and a tyrosine kinase (TK) domain. The family is distin-

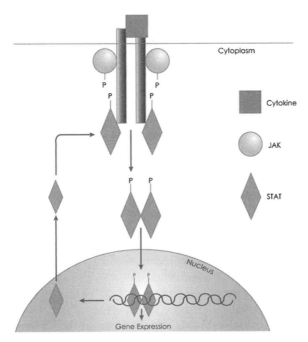

Fig. 3.2 The JAK/STAT signaling pathway; after cytokine binds to its receptors, the associated Janus kinase (JAK) is induced. This results in the phosphorylation of the receptor's cytoplasmic domain. STAT is recruited after the phosphorylation of the receptor's cytoplasmic domain, which after phosphorylation dimerizes and migrates into the nucleus. In the nucleus, STAT binds to its niche in the DNA and induces gene expression (*see* Color Insert)

guishable on the basis of the presence of an additional kinase-related domain. There are seven regions of homology in JAKs called JH1, JH2, JH3, JH4, JH5, JH6 and JH7, which are present from the carboxy- to the amino-terminus. JH1 corresponds to TK domains, JH2 to KL and JH3–JH7 to N regions. The N region of the JAKs is composed of 550 amino acids, and this region is responsible for binding to the cytokine receptors. The cytokine receptor-binding specificity is homed in JH7 and part of JH6, which is composed of the first 200 residues of the JAKs. The KL domain does not possess motifs required for the catalytic activity but shares similarities with TK. However, the KL domain can play a role that is not dependent on the TK domain and is considered to be a negative regulatory domain. The TK domain is responsible for the phosphorylation of the cytokine receptors, which is accomplished by the phosphorylation of tyrosine residues. These tyrosine residues are present in a loop in the TK domain, and the tyrosines that are phosphorylated are members of the JAK family.

The Janus Family Tyrosine Kinases–Signal Transducers and Activators of Transcription Signaling Pathway

The Janus family tyrosine kinases–signal transducers and activators of transcription signaling pathway is the best-understood cytokine receptor signaling cascade. Four Janus kinases (JAK1, JAK2, JAK3 and Tyk2) and seven STAT factors mediate the signal transduction of almost 40 cytokine receptors. The binding of cytokine to its receptors induces the associated JAK, resulting in the phosphorylation of the receptor's cytoplasmic domain. This phosphorylation permits the recruitment of STAT, which after phosphorylation dimerizes and migrates to the nucleus, where it binds to its niche in the DNA and induces gene expression. The JAK-binding sites are present close to the cell membrane where association of the JAKs with the cytoplasmic domains of cytoplasmic receptors takes place. After cytokine binds to its receptor, close proximity of JAK permits its induction. The binding sites for the SH2 domains of the STATs are created as a result of phosphorylation of cytokine receptor by the JAK. The migration to the nucleus follows tyrosine and occasionally serine phosphorylation of STAT by the JAKs as well as other kinases. Multiple cytokine receptors may be present on a single cell that may coordinate all the signals coming from multiple receptors. One JAK or a combination of JAKs are used by a class of receptors; for example, some receptors use only JAK1, some cytokines involved in hematopoiesis and proliferation use JAK2 and additional cytokines that have common γ-chain receptors use JAK1 and JAK3. The only exception is Tyk2, which is involved in signaling of many different classes of receptors. A summary of some transcription factors which interact with JAK and their stimulatory cytokine signals is shown in Tables 3.2 and 3.3.

Activated cytokine receptors in different cell types cause both cell type-specific as well as core transcription. The examples include the expression of a similar cohort of genes in any cell type induced by IFN-γ via STAT1, which overlaps with the gene expression caused by IFN-αβ. IFN-αβ signaling also utilizes STAT1, in

Table 3.2 Some Transcription Factors That Interact with JAK

JAK	Transcription Factor
JAK2	STAT3
JAK1, JAK2	STAT5
JAK3, Tyk2	
JAK2, JAK2, Tyk2	Raf-1
JAK1, JAK2	Grb-2
JAK1, JAK2, JAK2, Tyk2	SOCS-1
JAK1	PI3 K
JAK1	CPLA2
JAK1, Tyk2	SHP1
JAK1, JAK2	SHP2
Tyk2	Crk-L

Table 3.3 JAK/STAT Signal Transduction

Cytokines	JAK Kinase	STATs
IL-2, IL-7, IL-9	JAK1, JAK3	STAT3, STAT5
IL-3, IL-5, GM-CSF	JAK2	STAT5
IL-4	JAK1, JAK3 or JAK2	STAT6
IL-6, IL-11, G-CSF	JAK1, JAK2, Tyk2	STAT1, STAT3, STAT5
IL-10	JAK1, Tyk2	STAT1, STAT3
IL-12	JAK2, Tyk2	STAT4
IL-13	JAK1, JAK2, Tyk2	STAT6
IL-15	JAK1, JAK3	STAT3, STAT5
IFN-α, IFN-β	JAK1, Tyk2	STAT1, STAT2, STAT3
IFN-γ	JAK1, JAK2	STAT1

association with STAT2 and IRF9. These genes are called the "IFN signature" and are reflective of the activity of STAT1. An example of cell type-specific pathway is the IL-4 or IL-13-induced STAT6 pathway. The mechanism of activation of genes in T cells by IL-4 is more distinct than in macrophages or other cell types, which also applies to IL-13-mediated signaling. This suggests that specific gene expression is regulated by STATs based on their accessibility and with the involvement of other cofactors.

The importance of STATs in cytokine receptor-mediated signaling has been further elucidated by using STAT knockout mice. These studies have demonstrated that the genes regulated by IFN-γ that provides immunity against pathogens are dependent on STAT1. Similarly, the genes that are required for hematopoietic survival require STAT5a and STAT5b.

The signaling from receptors using the same JAK–STAT pathway in the same cell is different. For example, JAK1–STAT3 pathway is activated in macrophages when they bind to either IL-6 or IL-10, but the downstream signaling pathways in macrophages for IL-6 and IL-10 are different irrespective of the fact that both utilize JAK1–STAT3 pathway. IL-10, a negative regulator of inflammation, utilizes STAT3 and indirectly targets a few STAT3-regulated genes. However, IL-6, despite using JAK1–STAT3 pathway, does not induce the anti-inflammatory response. This has been attributed to the ability of the negative regulator of JAK–STAT activity SOC3 that controls IL-6R-mediated responses, since other cytokines that are SOC3-independent and utilize JAK1–STAT3 pathway can produce anti-inflammatory response.

The JAK–Cytokine Receptor Interaction

Seven JAK homology (JH) domains have been identified according to sequence similarities in JAK family members, which partially match the domain structure of JAK. The classical kinase domain is the JH1 domain at the C-terminus, preceded by the JH2 domain on the N-terminus, which is also called the pseudo-kinase domain, and does not have crucial residues for catalytic activity and for binding of nucleotides, although it has a kinase domain and modulates the kinase activity. The cytokine receptor binding is accomplished by JH3–JH7 regions, which are the N-terminal half of the JAKs. There is remarkable similarity of sequence between a segment of the N-terminal region of the JAKs and Ferm (four point-1, ezrin, radixin and moesin) domains. The Ferm domains are composed of three subdomains: F_1, F_2 and F_3. The F_1 subdomain has a ubiquitin-like β-grasp fold, the F_2 subdomain has an acyl-CoA-binding-protein-like fold and the F_3 subdomain shares phosphotyrosine-binding or pleckstrin homology domains. Ferm domains play an important role in the association of JAKs with cytokine receptors.

The proximal region of the cytokine receptors' membrane is the binding site for JAKs. The receptors have diverse sequences except sequence homology in the box 1 and box 2 regions, which are short stretches where the box 1 region is made up of eight amino acids that are proline-rich and the box 2 region is an aggregate of hydrophobic amino acid residues. Areas C-terminal of box 2 in some cytokine receptors participate in binding to JAK and its activation. The signaling involves the interface interaction between large segments of the receptor box 1/box 2 region and the N-terminal region of the JAKs resulting in correct alignment of JAKs; thus, agonist binding to the receptor causes dimerization and reconfiguration of the receptor and activation of both the JAKs and the receptors.

JAK association requires structural integrity of an α-helical structure. Two to three proline residues are present in the box 1 motif's C-terminal part. There is restructuring of certain receptor residues resulting from cytokine receptor–JAK interaction, which is termed "induced fit-like" interaction, causing defined interaction interfaces. The reorganization of the JAK–cytokine receptor binding interface is also possible as a result of the activation of JAKs. A high-affinity association of JAK with cytokine receptors results in its recruitment to membrane where there is no exchange of JAK molecules between different receptors.

In addition to serving as second messengers for cytokines, JAKs may also regulate cell surface expression of certain cytokine receptors. For example, the coexpression of JAK1, JAK2 or Tyk2 augments oncostatin M receptor (OSMR) expression. The expression of erythropoietin receptor (EPOR) requires JAK2 and also assists in its folding process in the endoplasmic reticulum. JAK2 and Tyk2 enhance the cell surface expression of the thrombopoietin receptors, and the expression of Tyk2 is required for stable cell surface expression of IFN-αR1 and IL-10R2.

Cytokine receptors compete for a limited amount of JAK as observed in the IL-12/ IFN-αR1 system. A receptor without a kinase is unable to send signals and may dilute the effects of a cytokine by serving similar to an artificial receptor used clinically as decoys as is the case in IL-4 antagonist in the treatment of asthma.

Mitogen-Activated Protein Kinases

Mitogen-activated protein kinases are signal transducers that play a pivotal role in various cellular functions. Such kinases include a large family of protein kinases involved in multiple cellular functions including cell proliferation, differentiation, survival and death and gene regulation. In addition to numerous biological functions, they also play an important role in immune response and inflammation, and their aberration leads to disease states. Mitogen-activated protein kinases are part of signal transduction cascades that are involved in delivering extracellular message via cytoplasm to the nucleus. Following activation, these proteins migrate from the cytoplasm to the nucleus where they regulate gene expression by activating a number of proteins and transcription factors. For activation, they require phosphorylation of tyrosine and threonine in the conserved threonine–X–tyrosine sequence in kinase subdomain VIII. The tyrosine and serine/threonine-specific phosphatases are involved in activation.

A MAP kinase module is composed of three kinases where MAP kinase kinase kinase (MAPKKK) will phosphorylate and induce a MAP kinase kinase (MAPKK), which will then phosphorylate and activate a MAP kinase. MAP kinase phosphorylates either transcription factors that are nonkinase proteins or other kinases that are called MAP kinase-activating protein kinases (MK). There are four distinct classes of MAP kinases, which include ERKs, C-Jun-N-terminal kinases, p38 isoforms and ERK5.

1. *Extracellular signal-regulated kinases*: The ERK family is activated by mitogens and is critical in transducing signals, resulting in cell proliferation. Extracellular signal-regulated kinase pathway consists of the MAPKKKs, A-Raf, B-Raf, C-Raf-1, the MAPKKs, MEK1 and MEK2, the MAPKs, ERK1 and ERK2, and the MAPKAPKs, MNK1, MNK2, MSK1, MSK2, RSK1, RSK2, RSK3, P70S5 K and P70S6 K. The sequence of the tripeptide motif for ERK is Thr-Glu-Tyr. Cytokines and growth factors activate A-Raf, B-Raf and C-Raf-1, which then phosphorylate MEK1 and MEK2 resulting in their activation. ERK1 and ERK2 are then phosphorylated by MEK1 and MEK2. The targets of ERK1 and ERK2 include STATs, ELK-1 and Ets (Fig. 3.3).

2. *C-Jun-N-terminal kinases*: The JNK/Stress-activated protein kinases (SAPKs) do not respond well to mitogens but are strongly activated by agents that induce cellular stress. These kinases phosphorylate C-Jun transcription factor. The sequence of the tripeptide motif for JNK is Thr-Pro-Tyr. The activators of cytokine and tyrosine kinase receptors transduce signal to the upstream MAPKKKs. These

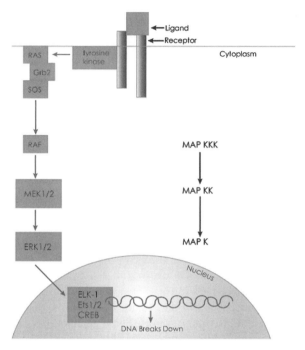

Fig. 3.3 The ERK/MAP kinase signaling pathway; cytokines and growth factors activate tyrosine kinase to which the adaptor protein Grb2 binds. This localizes SOS to plasma membrane. RAS is then activated by SOS. Activated RAS then binds to RAF, which forms a transient membrane-anchoring signal. Active RAF kinase phosphorylates MEK. The activated MEK phosphorylates ERK1/ERK2, which also migrates to the nucleus to phosphorylate ELK-1, Ets1/2 and CREB, resulting in the activation and expression of respective genes (*see* Color Insert)

MAPKKKs include ALK, MLK, TLP and TAK, which phosphorylate MAPKKs, MKK4 and MKK7. The phosphorylated MKK4 and MKK7 then activate JNK1, JNK2 and JNK3 by phosphorylation (Fig. 3.4). These kinases play a significant role in apoptosis and immunological diseases.

3. *p38 Isoforms*: As was the case for JNK/SAPKs, P38MAPKs do not respond well to mitogens but are strongly activated by agents that induce cellular stress. The p38 MAPKs include p38α, p38β, p38γ and p38δ. All of the family members contain a sequence TGY in their activation loop. They share the upstream MAPKKK activators with C-Jun-N-terminal kinases and are activated by cytokines, hormones and environmental stress. The sequence of the tripeptide motif for p38 is Thr-Gly-Tyr. They phosphorylate MKK3 and MKK6. p38α and p38β are activated by MKK3; however, all p38 MAP kinases are activated by MKK6 (Fig. 3.5). Among other biological functions, they play an important role in the immune response.

4. *Extracellular signal-regulated kinase 5*: ERK5 is similar to ERK1/ERK2 and has the Thr-Glu-Tyr (TEY) activation motif. It is activated by growth factors, mitogens and oxidative stress. MEK5, upstream activator of ERK5, is activated by MEKK2 and MEKK3. MEK5 phosphorylates and induces ERK5. The pathway

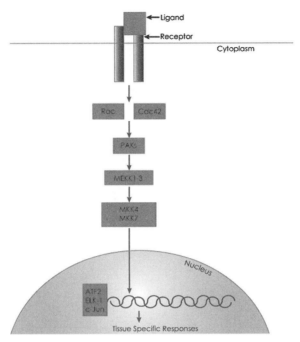

Fig. 3.4 The JNK/SAPK signaling pathway; various signals including cytokines activate the MAP kinases. The JNK/SAPK cascade is activated in response to inflammatory cytokines, heat shock or ultraviolet radiation. Two small G proteins, Rac and cdc42, mediate the activation of the MAP kinases. After activation, cdc42 binds to and activates PAK65 protein kinase. This results in the activation of MEKK, which eventually phosphorylates JNK/SAPK that migrates to the nucleus and activates the expression of several genes specifically the phosphorylation of c-Jun (*see* Color Insert)

involving MAPK ERK5 mediates growth factor and stress-induced signal transduction. It is a contributor to cell survival mechanisms. The inhibitors of ERK1/2 (classical MAP kinases) also inhibit ERK5. It mediates the effects of several oncogenes and its abnormal levels are associated with some forms of cancer. Selective activation of ERK5 induces a reporter gene driven by the IL-2 promoter without affecting CD69 expression.

Mitogen-Activated Protein Kinases in Immune Response

Toll-like receptors through intermediate signals including MyD88 adaptor protein, IRAK, TRAF6 and others cause the activation of MAP kinases and NF-κB resulting in the production of inflammatory cytokines, which include IL-1, IL-12 and TNF-α. The p38 MAPK pathway plays an important role in the production of IL-12 in dendritic cells and macrophages. In dendritic cells, a receptor–ligand pair upstream of p38 and JNK, CD40–CD40L interaction, is responsible for inducing IL-12 production. P38 MAPK also regulates TNF-α biosynthesis via MAP-kinase-activated protein kinases (MAPKAP), MK2 and MK3, at the posttranscriptional level.

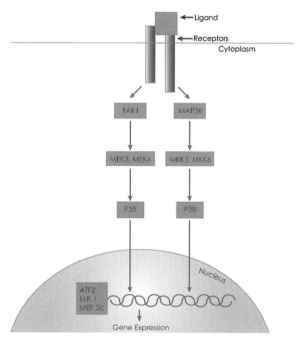

Fig. 3.5 The p38 kinase signaling pathway; inflammatory cytokines, osmotic stress and endotoxins activate this signaling pathway. The maximum activation of p38 requires two MAP2Ks, MKK3 and MKK6. Following activation, p38 translocates to the nucleus and phosphorylates ATF2, ELK1 and MEF-2C. p38 can also be activated independent of MAP2Ks by TAK1-binding protein via TAB1 (*see* Color Insert)

AU-rich elements (ARE) in TNF-α transcripts (3′-untranslated) are involved in the regulation of the translation of TNF-α by p38.

TNF-α and IL-1 activate both JNK and p38 MAP kinases; p38α plays a significant role in IL-1-mediated inflammatory responses. However, TNF, but not IL-1, requires MKK3-p38 for its cellular effects. The production of IL-6 and type-I IFNs requires JNK, which is also required for IL-1-induced effects in rheumatoid arthritis including joint inflammation and collagenase-3 expression. In IL-1 signaling pathway, JNK is activated by TRAF6 and TRAF2-induced JNK after activation of TNF receptors. While both JNK kinases, MKK4 and MKK7, are involved in stress-induced JNK activation, only MKK4 is involved in JNK activation following MEF treatment with IL-1 or TNF-α. The phosphorylation site in the sequence of the tripeptide motif is different for MKK4 and MKK7, as MKK7 phosphorylates threonine, whereas MKK4 preferentially phosphorylates tyrosine.

Effects of Mitogen-Activated Protein Kinases on T Cells

Induction of ERK is important for T-cell activation. The interaction of antigen with TCR in the context of MHC molecules results in the recruitment of various molecules including Grab2 and SLP76, both adaptor proteins to the cell surface.

This results in induction of the sos-ras-MEK-ERK pathway. The ERK pathway has also been implicated in thymocyte selection and maturation, TH_2 differentiation, and IL-4-induced STAT6 and IL-4 receptor phosphorylation.

p38 MAP kinases play a role in TH_1 differentiation as IL-12 and IL-18 are involved in p38 activation in T lymphocytes. IFN-γ produced by TH_1 cells is inhibited by imidazole antagonists of p38 kinases, which do not affect IL-4 secretion by TH_2 cells, but IFN expression is dependent on p38 regulation.

Naïve T cells have very limited JNK activity; however, its activity and expression is significantly augmented when T cells are activated, reaching a peak after 30–60 min of activation. Activation of JNK is also dependent on signals from activated TCR. Although not essential for the induction of JNK activity, CD28 costimulation significantly enhances its activity. JNK also may have an important role in TH_1 function where it is highly activated as opposed to TH_2 cells. JNK2 is essential for cytokine production and TH cell differentiation. In the absence of JNK2, JNK activity in TH_1 cells is reduced. Both JNK and p38 play an important role in the differentiation of TH_1 cells. This causes a feedback amplification of signal strength for JNK and p38 pathways resulting in the induction of Rac2 and GADD45. Rac2 is an upstream effector for JNK and p38 pathways, whereas GADD45 is involved in JNK signaling, which is specific to TH_1 cells. The induction of Rac2 and GADD45 significantly augments the signaling of JNK and p38, resulting in effector cytokine production by TH_1 cells.

Cell Surface Signals Activate Extracellular Signal-Regulated Kinase 1/2 and Other Mitogen-Activated Protein Kinases

The tyrosine kinase pathway is one of the most well-defined cell signaling systems that transduce messages from cell surface receptors to ERK1/2 transcription factors. Binding of the agonist to the cell surface receptors causes the autophosphorylation of the tyrosine residues in the receptor. Phosphorylation of the receptors initiates downstream signaling pathways. This results in the activation of a G protein Ras, which involves the recruitment of adaptor proteins shc and Gγb2 to the receptor and formation of a complex following linkage among SH2 domains and phosphotyrosine residues. This complex then incorporates the guanine nucleotide exchange factor (GEF), Son of Sevenless (SoS), resulting in the induction of Ras. The ultimate result is the exchange of GDP for GTP by Ras as GTP-liganded Ras can interact with Raf isoforms. This results in Raf-1-mediated signaling. G protein (Gαs)-mediated signaling has multiple effects on ERK activity, some of which are cAMP-dependent. PKA, which is cAMP-dependent, inhibits Raf-1 activity.

Nuclear Factor-κB

NF-κB is a family of pleiotropic transcription factors that regulate the expression of a wide variety of genes. These genes are involved in various biological functions including the immune and inflammatory response. The classical activated form of NF-κB is P50/P65 heterodimer. Generally, NF-κB is present in an inactive form in the cytoplasm where it is noncovalently associated with its inhibitor, IκB. This inhibitor prevents NF-κB-mediated signal transduction, which involves its migration to the nucleus and induction of specific gene expression after binding to the DNA. Some cytokines including IL-1 and TNF-α after binding to their specific cell surface receptors cause the phosphorylation of IκB, resulting in its degradation and removal of the inactivity barrier from NF-κB. This permits the migration of NF-κB into the nucleus where it binds to specific genes in the DNA resulting in their induction and expression (Fig. 3.6). One gene regulated by NF-κB is IκBα, which serves as a feedback inhibitor to maintain homeostasis. IL-1 and TNF-α induce NF-κB in a biphasic manner, which includes a transient phase and a persistent phase. The transient phase is mediated via IκBα and the persistent phase is mediated via IκBβ. The activation of NF-κB is complex and involves the activation of IKK, which phosphorylates IκB that is bound to NF-κB. Other transcription factors involved in this signaling cascade are cell-dependent.

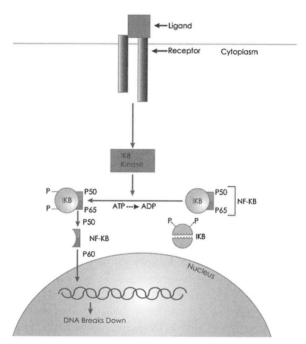

Fig. 3.6 The NF-κB signaling pathway. NF-κB is a P50/P65 heterodimer that is present in the cytoplasm in an inactive form as it is noncovalently associated with IκB. IκB is an inhibitor of NF-κB and prevents its migration and gene expression signal in the nucleus. Selected cytokines after binding to their specific receptors phosphorylate IκB resulting in its degradation, which removes the inactivity barrier from NF-κB. This is followed by the migration of NF-κB into the nucleus where it binds to specific genes and results in their induction and gene expression (*see* Color Insert)

Inhibitors of Cytokine Cell Signaling

Cytokine signaling is strictly regulated at several sites to avoid damages resulting from uncontrolled signal transduction initiated from activated cytokine receptors. A number of cytokine signaling inhibitors have been described.

SH2-Containing Phosphatase Proteins

SH2-containing phosphatase proteins (SHP) are expressed constitutively and are inhibitors of cytokine-mediated signal transduction. They act by dephosphorylating multiple transcription factors including JAKs and their receptors. Mammalian SHP have two family members SHP1 and SHP2, which are composed of two repetitive N-terminal SH2 domains and a C-terminal protein–tyrosine phosphatase domain. In general, activation of intermediate transcription factors involves the phosphorylation of serine, tyrosine or threonine residues. The phosphotyrosine residues of various cytokine receptors are targets of SHP1 and SHP2 via their SH2 domain. The signaling components that are inhibited by SHP1 after dephosphorylation include transcription factors for IL-4 receptors, JAK-2, EpoRs and SCF receptor (C-Kit). The role of SHP2 is that of an inducer of transduction pathway. However, it has been suggested that SHP2 can downregulate cytokine receptor-mediated signal transduction mechanisms through gp130 receptors.

Protein Inhibitors of Activated Signal Transducers and Activators of Transcription

Protein inhibitors of activated STATs (PIAS) are constitutively expressed, and their mechanism of action is inhibition of STAT pathways by sumoylation. The PIAS family members, PIAS1 and PIASx, inhibit STAT1, and PIAS3 and PIASy inhibit STAT3 and STAT4, respectively. There is a high degree of sequence conservation among these four family members. Other family members also exist, hZIMP7 and hZIMP10, but have limited similarity with other members. Protein inhibitors of activated STATs are transcriptional coregulators that mediate their effects by at least two mechanisms; they act as SUMOE3 ligases and increase sumoylation of the substrate resulting in the addition of SUMO moieties and consequently modifying their properties, while the other mechanism of action is independent of sumoylation. They act either by suppression or by induction, which depends on the target transcriptional factor. One of the resultant effects is the relocalization of transcriptional regulators to different subnuclear compartments.

Suppressors of Cytokine Signaling

Suppressors of cytokine signaling (SOCS) include eight members (SOCS1, SOCS2, SOCS3, SOCS4, SOCS5, SOCS6, SOCS7 and CIS), which are composed of a central SH2 (Src homology 2) domain, an N-terminal domain and a C-terminal SOC box domain. The N-terminal domain has variable length whereas the C-terminal domain is composed of a 40-amino-acid sequence. The SH2 domain, in phosphoty-rosine residues of cytokine receptors, is the target of SOCS proteins. These family members that bind to domains of cytokine receptors include SOCS2, SOCS3 and CIS. Alternatively, SOCS1 binds to its target sites of JAKs. The cytokine signaling is suppressed by multiple mechanisms, which include competition with STATs for the sites of receptor phosphorylation, inhibition of JAK activity and/or proteasomal degradation after binding to signaling proteins. Suppressors of cytokine signaling proteins are depicted as key negative regulators of cytokine signaling, most impor-tant of which is inhibition of the JAK–STAT pathway. Their synthesis is induced by various signals including IL-6, TNF, TGF-β and LPS.

SOCS2 and SOCS3 are also the negative regulators of cytokine receptor-mediated signal transduction. All three SOCS, SOCS3 in particular, inhibit the signals of the hematopoietin class of cytokine receptors. SOCS3 inhibits various cytokine signals including IL-2Rβ, gp130 receptors, GHR and EPOR and also regulates intestinal inflammation and energy homeostasis by leptin. It is also involved in the regulation of EPO signaling. Cytokine receptors aggregate after binding to their respective agonist, resulting in the activation of Janus kinases. Induction of Janus kinases causes receptor tyrosine phosphorylation and also of other proteins involved in signaling. This sequence also results in the activation of other signaling molecules such as the STATs. In contrast to SOCS1, which directly binds to the JAK and inhibits their catalytic activity, CIS binds to the cytokine receptor, resulting in the inhibition of STAT activity. SOCS3 acts by both mechanisms described for SOCS1 and CIS. SOCS3 does not possess a high affinity for JAKs and requires cytokine receptors to inhibit JAK kinase activity. The interaction of SOCS3 with the acti-vated cytokine receptor is required before recruitment to the signaling complex takes place, which results in the inhibition of cytokine-mediated activity after it binds to JAK kinases. Other molecules are also degraded by SOCS proteins including the interaction of SOCS box domain with elongins B and C, and these proteins via proteasomes target proteins for destruction, suggesting that SOCS proteins regulate signaling proteins and inhibit their catalytic activity as well as recruitment.

The transcriptional upregulation of SOCS1, SOCS3 and CIS is caused by STATs as they regulate the expression of SOCS1, and the binding sites for STAT1, STAT3 and STAT6 are present in the SOCS1 promoter. IL-6 or Leukemia inhibitory factor (LIF)-induced expression of the SOCS1 in RNA is inhibited by a dominant negative version of STAT3. Fibroblasts lacking STAT1 are unable to induce expression of the SOCS1 mRNA in response to IFN-γ. The effect of STAT1 is indirect on the SOCS1 gene as it is mediated via IRF-1. Signal transducers and activators of transcription also regulate the expression of the SOCS3 gene. STAT1/STAT3-binding element is present in the SOCS3 promoter that is responsible for LIF-induced activation of the SOCS3 promoter. Furthermore, SOCS3 expression is regulated by STAT5b. Four STAT5-binding sites are present in the promoter of the CIS gene, which is completely responsible for its activation by EPO.

Bibliography

Akira S. 2003. Toll-like receptor signaling. J Biol Chem. 278:38105–38108.

Alexander WS, Hilton DJ. 2004. The. role of suppressors of cytokine signaling (SOCS) proteins in regulation of the immune response. Ann Rev Immunol. 22:503–529.

Bazan JF. 1990. Structural. design and molecular evolution of a cytokine receptor superfamily. Proc Nat Acad Sci. USA. 87:6934–6938.

Bonni A, Brunet A, West AE, Datta SR, et al. 1999. Cell survival promoted by the Ras-MAPK signaling pathway by transcription-dependent and – independent mechanisms. Science. 286:1358–1362.

Bradley JR, Prober JS. 2001. Tumor necrosis factor receptor associated factors (TRAFs). Oncogene. 20:6482–6491.

Chadee DN, Yuasa T, Kyriakis JM. 2002. Direct activation of mitogen-activated protein kinase kinase kinase MEKK1 by the Ste20p homologue GCK and the adapter protein TRAF2. Mol Cell Biol. 22:737–749.

Chang F, Steelman LS, Lee JT, Shelton JG, et al. 2003. Signal transduction mediated by the Ras/Raf/MEK/ERK pathway from cytokines to transcription factors: potential targeting for therapeutic intervention. Leukemia. 17:1263–1293.

Chang L, Karin M. 2001. Mammalian MAP kinase signaling cascades. Nature. 410:37–40.

Chen G, Goeddel DV. 2002. TNF-R1 signaling: A beautiful pathway. Science. 296:1634–1635.

Chen W, Hershey GKK. 2007. Signal transducer and activator of transcription signals in allergic disease. J. Allergy Clin Immunol. 119:529–541.

Chen YR, Tan TH. 1999. Mammalian c-jun n-terminal kinase pathway and ste20-related kinases. Gene Ther Mol Biol. 4:83–98.

Cosman D, Lyman SD, Idzerda RL, Beckmann MP. 1990. A new cytokine receptor superfamily. Trends Biochem Sci. 15:265–270.

Dai X, Chen Y, Di L, Podd A, et al. 2007. Stat5 is essential for early B cell development but not for B cell maturation and function. J Immunol. 179:1068–1079.

D'Andrea AD. 1994. Cytokine receptors in congenital hematopoietic disease. NEJM. 330: 839–846.

deWeerd NA, Samarajiwa SA, Hertzog PJ. 2007. Type 1 interferon receptors: Biochemistry and biological functions. J Biol Chem. 282:20053–20057.

Egan PJ, Lawlor KE, Alexander PS, Wicks IP. 2003. Suppressor of cytokine signaling-1 regulates acute inflammatory arthritis and T cell activation. J Clin Inv. 111:915–924.

Elkasmi KC, Holst J, Coffre M, Mielke C, et al. 2006. General nature of the STAT3-activated anti-inflammatory response. J Immunol. 177:7880–7888.

Gadina M, Hilton D, Johnston JA, Morinobu A, et al. 2001. Signaling by type 1 and type II receptors: Ten years after. Curr Opin Immunol. 13:363–373.

Geijsen N, Koenderman L, Coffer PJ. 2001. Specificity in cytokine signal transduction: Lessons learned from the IL-3/IL-5/GM-CSF receptor family. Cytokine Growth Factor Rev. 12:19–25.

Haan C, Kreis S, Margue C, Behrmann I. 2006. JAKs and cytokine receptors – An intimate relationship. Biochem Pharmacol. 72:1538–1546.

Hazzalin CA, Mahadevan LC. 2002. MAPK-regulated transcription: A continuously variable gene switch? Nat Rev Mol Cell Biol. 3:30–40.

Heinrich PC, Behrmann I, Muller-Newen G, Schaper F, Graeve L. 1998. Interleukin-6 type cytokine signaling through the gp130/JAK/STAT pathway. Biochem J. 334:297–314.

Hershey GK. 2003. IL-13 receptors and signaling pathways: An evolving web. J All Clin Immunol. 111:677–690.

Ihle JN. 1995. Cytokine receptor signaling. Nature. 377:591–594.

Ihle JN, Witthum BA, Quelle FW, Yamamoto K, et al. 1995. Signaling through the hematopoietic cytokine receptors. Ann Rev Immunol. 13:369–398.

Ivashkiv LB, Tassiulas I. 2003. Can SOCS make arthritis better? J Clin Inv. 111:795–797.

Kato Y, Kravchenko VV, Tapping RI, Han J, et al. 1997. BMK1/ERK5 regulates serum-induced early gene expression through transcription factor MEF2C. EMBO J. 16:7054–7066.

Kiefer F, Tibbles LA, Anafi M, Janssen A, et al. 1996. HPK1, a hematopoietic protein kinase activating the SAPK/JNK pathway. EMBO J. 16:7013–7025.

Kondo M, Takeshita T, Ishii N, Nakamura M, et al. 1993. Sharing of the interleukin-2 (IL-2) receptor gamma chain between receptors for IL-2 and IL-4. Science. 262:1874–1877.

Koziczak-Holbro M, Joyce C, Gluk A, Kinzel B, et al. 2007. IRAK-4 kinase activity is required for interleukin-1 (IL-1) receptor- and toll-like receptor-7 mediated signaling and gene expression. J Biol Chem. 282:13552–13560.

Krebs DL, Hilton DJ. 2001. SOCS proteins: Negative regulators of cytokine signaling. Stem Cells. 19:378–387.

Kubo M, Hanada T, Yoshimura A. 2003. Suppressors of cytokine signaling and immunity. Nat Immunol. 4:1169–1176.

Levy DE, Lee C. 2003. What does STAT3 do? J Clin Inv. 109:1143–1148.

Li Y, shen BF, Karanes C, Sensenbrenner L, Chen B. 1995. Association between Lyn protein tyrosine kinase (p53/56 lyn) and the beta subunit of the granulocyte-macrophage colony-stimulating factor (GM-CSF) recepto9rs in a GM-CSF-dependent human megakaryocytic leukemia cell line (M-07e). J Immunol. 155:2165–2174.

Locksley RM, Killeen N, Lenardo MJ. 2001. The TNF and TNF receptor superfamilies: Integrating mammalian biology. Cell. 104:487–501.

Lynch RA, Etchin J, Battle TE, Frank DA. 2007. A small-molecule enhancer of signal transducer and activator of transcription 1 transcriptional activity accentuates the antiproliferative effects of IFN-γ in human cancer cells. Cancer Res. 67:1254–1261.

Martinez-Moczygemba M, Huston DP. 2003. Biology of common beta receptor-signaling cytokines: IL-3, IL-5, and GM-CSF. J All Clin Immunol. 112:653–665.

Miyajima A, Kitamura T, Harda N, Yokota T, et al. 1992. Cytokine receptors and signal transduction. Ann Rev Immunol. 10:295–331.

Moore KW, de-Waal R, Coffman RL, O'Garra A. 2001. Interleukin-10 and the interleukin-10 receptor. Ann Rev Immunol. 19:683–765.

Murray PJ. 2007. The JAK-STAT signaling pathway: Input and output integration. J Immunol. 178:2623–2629.

Ogata N, Kouro T, Yamada A, Koike M, et al. 1998. JAK2 and JAK1 constitutively associate with an interleukin-5 (IL-5) receptor alpha and betac subunit, respectively, and are activated upon IL-5 stimulation. Blood. 91:2264–2271.

Olosz F, Malek TR. 2002. Structural basis for binding multiple ligands by the common cytokine receptor gamma-chain. J Biol Chem. 277:12047–12055.

O'Neill LA. 2002. Signal transduction pathways activated by the IL-1 receptor/toll-like receptor superfamily. Curr Top Microbiol Immunol. 270:47–61.

O'Sullivan LA, Liongue C, Lewis RS, Stephenson SE, et al. 2007. Cytokine receptor signaling through the JAK-stat-socs pathway in disease. Mol Immunol. 44:2497–2506.

Pearson G, English JM, White MA, Cobb MH. 2001a. ERK5 and ERK2 cooperate to regulate NF-kappaB and cell transformation. J Biol Chem. 276:7927–7931.

Pearson G, Robinson F, Gibson TB, Xu BE, et al. 2001b. Mitogen-activated protein (MAP) kinase pathways: Regulation and physiological functions. Endocrine Rev. 22:153–183.

Pestka S, Krause CD, Sarkar D, Walter MR. 2004. Interleukin-10 and related cytokines and receptors. Ann Rev Immunol. 22:929–979.

Quelle FW, Sato N, Witthuhn BA, Inhorn RC, et al. 1994. JAK2 associates with the beta c chain of the receptor for granulocyte-macrophage colony-stimulating factor, and its activation requires the membrane-proximal region. Mol Cell biol. 14:4335–4341.

Radtke S, Haan S, Jorissen A, Hermanns HM, et al. 2005. The JAK1 SH2 domain does not fulfill a classical SH_2 function in JAK/STAT signaling but plays a role for receptor interaction and up-regulation of receptor surface expression. J Biol Chem. 280:25760–25768.

Ramana CV, Gil MP, Schreiber RD, Stark GR. 2002. STAT-1-dependent and – independent pathways in IFN-γ-dependent signaling. Trends Immunol. 23:96–101.

Ransohoff RM. 1998. Cellular responses to interferons and other cytokines: The JAK-STAT paradigm. NEJM. 338:616–168.

Rawlings JS, Rosler KM, Harrison DA. 2004. The JAK/STAT signaling pathway. J Cell Sci. 117:1281–1283.

Regis G, Conti L, Boselli D, Novelli F. 2006. IFN gamma R2 trafficking tunes IFN gamma-STAT1 signaling in T lymphocytes. Trends Immunol. 27:96–101.

Renauld JC. 2003. Class II cytokine receptors and their ligands, key antiviral and inflammatory modulators. Nat Rev Immunol. 3:667–676.

Rincon M, Pedraza-Alva G. 2003. JNK and P38 MAP kinases in CD4+ and CD8+ T cells. Immunol Rev. 192:131–142.

Roberts AW, Robb L, Rakar S, Hartley L, et al. 2001. Placental defects and embryonic lethality in mice lacking suppressor of cytokine signaling 3. PNAS. 98:9324–9329.

Sato N, Sakamaki K, Terada N, Arai K, Miyajima A. 1993. Signal transduction by the high-affinity GM-CSF receptor: Two distinct cytoplasmic regions of the common beta subunit responsible for different signaling. EMBO J. 12:4181–4189.

Schmidt C, Peng G, Li Z, Sclabas GM, et al. 2003. Mechanisms of proinflammatory cytokine-induced biphasic NF-KB activation. Mol Cell. 12:1287–1300.

Schroder K, Hertzog PJ, Ravasi T, Hume DA. 2003. Interferon-γ, an overview of signals, mechanisms and functions. J Leukocyte Biol. 75:163–189.

Sharrocks AD. 2006. PIAS proteins and transcriptional regulation – More than just SUMOE3 ligases. Genes Dev. 20:754–758.

Shelburne CP, McCoy ME, Piekorz R, Sexl VV, et al. 2002. STAT5: An essential regulatory of mast cell biology. Mol Immunol. 38:1187–1191.

Shuai K, Liu B. 2003. Regulation of JAK-STAT signaling in the immune system. Nat Rev Immunol. 3:900–911.

Shuai K. 2006. Regulation of cytokine signaling pathways by PIAs proteins. Cell Res. 16: 196–202.

Sims JE. 2002. IL-1 and Il-18 receptors, and their extended family. Curr Opin Immunol. 14: 117–122.

Stephanou A, Latchman DS. 2005. Opposing actions of STAT-1 and STAT-3. Growth. Factors. 23:177–182.

Symons A, Beinke S, Ley SC. 2006. MAP kinase kinase kinases and innate immunity. Trends Immunol. 27:40–48.

Takaki S, Kanazawa H, Shiiba M, Takatsu K. 1994. A critical cytoplasmic domain of the interleukin-5 (IL-5) receptor alpha chain and its function in IL-5-mediated growth signal transduction. Mol Cell Biol. 14:7404–7413.

Takaoka A, Hayakawa S, Yanai H, Stoiber D, et al. 2003. Integration of interferon-alpha/beta signaling to p53 responses in tumour suppression and antiviral defence. Nature. 424:516–523.

Taniguchi T, Ogasawara K, Takaoka A, Tanaka N. 2001. IRS family of transcription factors as regulators of host defense. Ann Rev Immunol. 19:623–655.

Taniguchi T, Takaoka A. 2002. The interferon-alpha/beta system in antiviral responses: A multimodal machinery of gene regulation by the IRF family of transcription factors. Curr Opin Immunol. 14:111–116.

Tavernier J, Devos R, Cornelis S, Tuypens T, et al. 1991. A human high affinity interleukin-5 receptor (IL5R) is composed of an IL5-specific alpha chain and a beta chain shared with the receptor for GM-CSF. Cell. 66:1175–1184.

Torii S, Nakayama K, Yamamoto T, Nishida E. 2004. Regulatory mechanisms and function of ERK MAP kinases. J Biochem. 136:557–561.

Wajant H, Henkler F, Scheurich P. 2001. The TNF receptor-associated factor family: Scaffold molecules for cytokine receptors, kinases and their regulators. Cell Signal. 13:389–400.

Wajant H, Pfizenmaier K, Scheurich P. 2003. Tumor necrosis factor signaling. Cell Death Differ. 10:45–65.

Waldmann T, Tagaya Y, Bamford R. 1998. Interleukin 2, interleukin 15 and their receptors. Int Rev Immunol. 16:205–226.

Watford WT, Hissong BD, Kannu JH, Muul L, O'Shea JJ. 2004. Signaling by IL-12 and IL-23 and the immunoregulatory role of STAT4. Immunol. Rev. 202:139–156.

Weston CR, Lambright DG, Davis RJ. 2002. Signal transduction. MAP kinase signaling specificity. Science. 296:2345–2347.

Wong PKK, Egan PJ, Croker BA, Donnell KO, et al. 2004. SOCS3 negatively regulates innate and adaptive immune mechanism in acute IL-1-dependent inflammatory arthritis. J Clin Inv. 116:1571–1581.

Yeh TC, Pellegrini S. 1999. The Janus Kinase family of protein tyrosine kinase and their ole in signaling. Cell Mol Life Sci. 55:1523–1534.

Yu Q, Thieu VT, Kaplan MH. 2007. STAT4 limits DNA methyltransferase recruitment and DNA methylation of the IL-18Rα gene during TH$_1$ differentiation. The EMBO J. 26:2052–2060.

Chapter 4
Immunosuppressive Agents

Introduction

Successful organ transplantation requires effective immunosuppression. Most of the modern advances in tissue transplantation have resulted from our precise understanding of the immune mechanisms involved in tissue rejection as well as from advances in surgical techniques to some degree. The last decade has witnessed the introduction of several potent immunosuppressive agents, which have led to considerable success in transplant medicine. The development of effective immunosuppressive therapy has been essential in tissue transplantation and for improvements in its positive outcome. Great advances have been made over the past 50 years in organ transplantation despite the complexities involved in immune suppression after tissue transplantation. The development of novel immunosuppressive agents has increased the graft survival rates; however, the side effects of these drugs resulting from their long-term use continue to present significant challenges. In addition to organ transplantation, immunosuppressive agents have found use in the prevention of Rh hemolytic disease of the newborn and for the treatment of some autoimmune diseases.

The initial organ transplantation was performed in 1933 when a kidney was transplanted from a cadaver. Total lymphoid irradiation was used for the immune suppression but the tissue was rejected and the patient eventually died. This was followed by the use of corticosteroids as immunosuppressive agents, but unfortunately steroids by themselves also did not produce positive results. In the early 1960s, cytotoxic agents were introduced for immune suppression; these were followed by the use of a combination of cytotoxic agents and corticosteroids until the mid-1980s when cyclosporine was discovered by Borel.

Despite all the advances in surgical techniques and the development of newer immunosuppressive drugs and antibodies, long-term survival of patients with organ transplants continues to provide a challenge for clinicians. The complexity of managing immunosuppression arises from many diverse factors related to the development of immune responses; for example, there are differential effects of immunosuppressive agents on primary versus secondary immune responses (they are more effective against the primary immune response). The primary immune response is observed during antigen processing, antigen-induced initial T-cell proliferation and

M.M. Khan, *Immunopharmacology*, DOI: 10.1007/978-0-387-77976-8_4,
© Springer Science+Business Media, LLC 2008

cytokine production. The secondary immune response is based on immunologic memory, which results in robust tissue rejection, and less impressive effects of immunosuppressive agents are observed. Another challenge that the immunosuppressive agents face is that their effects are antigen-dependent, that is, they could vary from antigen to antigen and require constant adjustments of the doses of the same drug to manage immune suppression.

The major classes of immunosuppressive drugs employed in clinical practice to avoid tissue rejection include calcineurin inhibitors, target of rapamycin (TOR) inhibitors, sphingosine-1-phosphate receptor (S1P-R) modulators, cytotoxic agents, glucocorticoids and monoclonal antibodies. These drugs need to be used on a life-long basis and have major undesirable side effects.

Calcineurin Inhibitors

Cyclosporine

Specific immunosuppressive therapy has its origin in the 1970s when cyclosporine was discovered by J.F. Borel in 1976. The drug was originally considered as an antifungal compound but its immunoregulatory activities were learned rapidly. The discovery was a result of a program of a Swiss pharmaceutical company to discover new antibiotics from fungal metabolites. Cyclosporine was isolated from the fungus *Tolypocladium inflatum Gams*. Its initial isolation was from a Norwegian soil sample and its chemical analysis revealed that cyclosporine was a cyclic undecapeptide. The structure and confirmational analysis was done with chemical degradation, X-ray crystallography and nuclear magnetic resonance imaging. Cyclosporine is a hydrophobic compound insoluble in water but soluble in other organic solvents.

Mechanism of Action

Cyclosporine enters through the cell membrane and binds to cyclophilins in the cytoplasm. Cyclophilins are a family of small proteins that selectively bind cyclosporine as well as its active analogs. Their distribution is abundant in lymphoid cells but they are present in most human tissues. The cyclosporine binding to cyclophilin results in the formation of a cyclosporine–cyclophilin complex. During an immune response, activation of T-cell receptor results in an increase in intracellular Ca^{2+}, which activates calcineurin, an enzyme called serine/threonine phosphatase, which is calcium-dependent. Under physiological conditions, calcineurin after its activation by Ca^{2+} dephosphorylates a cytosolic component of the nuclear factor of activated T cells (NFATc). After its dephosphorylation, NFATc migrates from the cytoplasm to the nucleus where it associates with the nuclear component of the nuclear factor of activated T cells (NFATn). This association of NFATc and NFATn results in the activation of the transcription of a number of genes including cytokine genes for IL-2, IL-3, IL-4, TNF-α, GM-CSF and others.

After the administration of cyclosporine, a cyclosporine–cyclophilin complex is formed, which binds to calcineurin, resulting in its inability to dephosphorylate

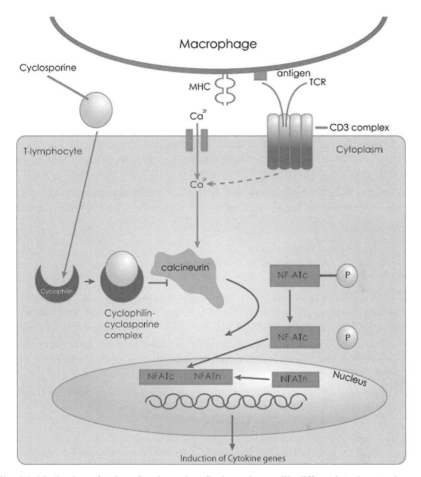

Fig. 4.1 Mechanism of action of cyclosporine. Cyclosporine readily diffuses into the cytoplasm of the target cells where it binds to cyclophilins. The cyclosporine–cyclophilin complex stably associates with calcineurin and inhibits calcineurin activity. Calcineurin is a Ca^{2+}-dependent enzyme— serine/threonine phosphatase— which after activation by Ca^{2+}, dephosphorylates a cytosolic component of NFAT (NFATc, cytosolic factor of activated T cells). After dephosphorylation, NFATc migrates from the cytoplasm to the nucleus where it associates with NFATn and induces transcription of several cytokine genes including IL-2. Cyclosporine inhibits calcineurin activity after associating with cyclophilins, resulting in the inhibition of IL-2 production and other cytokines (*see* Color Insert)

NFATc, and as a result, the transport of NFATc to the nucleus is prevented and consequently its association with NFATn does not proceed (Fig. 4.1). The association of NFATc with NFATn is essential for the initiation of IL-2 production, which is achieved through binding of NFATc–NFATn to the promoter of the IL-2 gene. As a result, IL-2 production is inhibited, which is necessary for the optimal function of the immune response. Cyclosporine does not inhibit cytokine-induced transduction mechanisms and also has no effect on antigen recognition by T cells in the context of MHC molecules.

Absorption, Distribution and Excretion

Cyclosporine is given orally or intravenously. Its oral bioavailability varies from 20 to 50%, and the peak concentrations in plasma are achieved within 3–4 h after its administration. Erythrocytes bind about 70% of the drug. It is note worthy that about 15–20% of cyclosporine is contained in leukocytes despite their small content in total blood. Cyclosporine is extensively distributed in compartments other than the vasculature, which may result in some of its toxic side effects. The half-life of cyclosporine is approximately 6 h. It is metabolized predominantly in the liver by CYP3A and is excreted in the bile. Only negligible amounts of the drug and its metabolites appear in the urine. More than 20 metabolites of cyclosporine have been identified but the metabolites have far less pharmacological activity and toxicity than the parent drug. Inhaled cyclosporine has been used after lung transplantation, which has helped avoid its undesirable side effects and may at least delay the onset of obliterative bronchiolitis.

Drug Interactions

The blood concentrations of cyclosporine are impacted by any drug that acts on microsomal enzymes, particularly the CYP3A system. Drugs that inhibit this enzyme would reduce the metabolism of cyclosporine and consequently will increase its blood concentrations. These drugs include antifungal agents, antibiotics, glucocorticoids, calcium channel blockers, protease inhibitors and others. In contrast, drugs that augment CYP3A activity will increase the metabolism of cyclosporine, resulting in reduced blood concentrations. These drugs include phenytoin, phenobarbital, trimethoprim–sulfamethoxazole and rifampin.

Toxicity

Renal and nephrotoxicity, which are seen in about 25–75% of patients treated with the drug, are the major side effects of cyclosporine. This could reduce glomerular filtration rate and renal plasma flow. There is also damage to proximal tubules and endothelial cells of small blood vessels. This causes hyperuricemia, which could result in hypercholesterolemia, increase in P-glycoprotein activity and worsening of gout. Hypertension is seen in most of the cardiac transplant patients and in about 50% of the renal transplants. About half of the patients receiving cyclosporine have elevated hepatic transaminase activity or concentration of bilirubin in plasma. Hirsutism and gingival hypoplasia are observed in 10–30% of the patients taking cyclosporine. Other side effects include peptic ulcers, pancreatitis, gum hyperplasia, convulsions, breathing difficulties, fever, vomiting and confusion. There is an increased vulnerability to fungal and viral infections, but there is a low incidence of malignancies if cyclosporine is administered alone and not with other immunosuppressive agents.

Clinical Uses

Cyclosporine is used to prevent organ rejection after tissue transplantation. It is also used for rheumatoid arthritis, psoriasis and dry eyes (keratoconjunctivitis sicca). For tissue transplantation, it is mostly used in combination with other immunosuppressive agents, and its doses vary based on the clinical circumstances. Its nephrotoxicity limits its use before the tissue is grafted, which poses a challenge in renal transplants where rejection must be differentiated from renal toxicity.

Tacrolimus

Tacrolimus is a 23-membered lactone chain isolated in 1984 from *Streptomyces tsukubaensis*, although it was originally found in a soil fungus. It is a macrolide antibiotic, and its name is derived from "Tsukuba macrolide immunosuppressant."

Mechanism of Action

Tacrolimus suppresses peptidyl-prolyl isomerase activity by binding to the immunophilin FK506-binding protein-12 (FKBP-12), and the tacrolimus–FKBP-12 complex binds to calcineurin and inhibits calcineurin phosphatase activity. As a result, calcineurin is unable to dephosphorylate NFATc and thus its migration to nucleus is blocked where its association with NFATn is necessary for the activation of key cytokine genes. Therefore, its mechanism of action is similar to cyclosporine although tacrolimus binds to a separate set of immunophilins in the cytoplasm. Tacrolimus, like cyclosporine, inhibits the secretion of key cytokines and inhibits T-cell activation (Fig. 4.2).

Absorption, Distribution and Excretion

Tacrolimus is given orally (twice-daily dose regimen) or as an injection. A modified release (MR) oral dosage form of tacrolimus has been developed for administration once a day to overcome noncompliance, which is the major problem in acute graft rejection in solid transplant recipients. Tacrolimus is not completely absorbed by the GI tract and its rate of absorption could vary. It binds to plasma protein at a rate of 75–99% with a half-life of approximately 12 h and is predominantly metabolized in liver by CYP3A. Some of its metabolites have immunosuppressive activity. Most of the tacrolimus is excreted in feces, and a negligible amount (<1%) is excreted in urine without undergoing any metabolism.

Drug Interactions

As for cyclosporine, its blood concentration is impacted by drugs that act on the CYP3A system. Drugs that inhibit this enzyme will increase its blood concentration and drugs that enhance the activity of CYP3A will decrease the blood concentration of tacrolimus. In combination with cyclosporine, it produces additive renal toxicity.

Toxicity

The side effects associated with tacrolimus administration include nephro- and hepatotoxicity, hypertension, tremors, seizures, diabetes mellitus, neuropathy, blurred vision, depression, loss of appetite and confusion. Tacrolimus may cause opportunistic

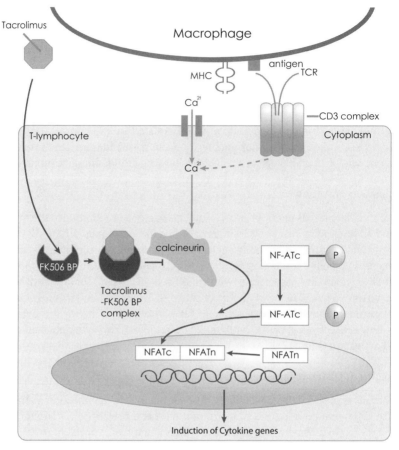

Fig. 4.2 Mechanism of action of tacrolimus. Tacrolimus readily diffuses into the cytoplasm of the target cell where it binds to immunophilins (FK506-BP). The tacrolimus–immunophilin complex stably associates with calcineurin and inhibits calcineurin activity. Calcineurin is a Ca^{2+}-dependent enzyme—serine/threonine phosphatase—which after activation by Ca^{2+}, dephosphorylates a cytosolic component of NFAT (NFATc, cytosolic factor of activated T cells). After dephosphorylation, NFATc migrates from the cytoplasm to the nucleus where it associates with NFATn and induces transcription of several cytokine genes including IL-2. Tacrolimus inhibits calcineurin activity after associating with immunophilins, resulting in the inhibition of IL-2 production and other cytokines (*see* Color Insert)

infection and could increase the severity of preexisting infections, including fungal or viral (herpes) infections. It has no effect on LDL cholesterol or uric acid. As was the case with cyclosporine, the administration of tacrolimus in combination with other immunosuppressive agents may increase the risk of tumors.

Clinical Uses

The immunosuppressive properties of tacrolimus are similar to those of cyclosporine, but tacrolimus is more potent than cyclosporine, so lower doses are required. Its primary use is for preventing rejection after allogeneic transplants to reduce the risk of organ rejection. Tacrolimus is also administered to patients who are showing signs of rejection despite treatment with cyclosporine. Furthermore, tacrolimus is used for the treatment of severe atopic dermatitis and for severe refractory uveitis.

Target of Rapamycin Inhibitors

Sirolimus

Sirolimus is a macrolide antibiotic (macrocyclic lactone) first isolated from soil samples of Eastern Island as a product of the bacterium *Streptomyces hygroscopicus* and shares a lot of its structure with tacrolimus.

Mechanism of Action

Sirolimus binds to the cytosolic protein FK-binding protein R (FKBP-12) but does not block calcineurin activity. It does not bind to cyclophilins, which are cytosolic receptors for cyclosporine. Unlike cyclosporine and tacrolimus, sirolimus does not inhibit the activation of NFAT responsive genes. After binding to its cytosolic receptors, sirolimus inhibits a protein kinase, the mammalian target of rapamycin (mTOR) pathway, via suppression of PP2-A. When mTOR is inhibited, the cells will not proceed to the S phase, and the cell cycle will be blocked (Fig. 4.3). As a result, sirolimus blocks T-cell proliferation but its effects are downstream of the IL-2 receptors. IL-2 binding to its receptors activates intracellular protein kinases that in turn activate gene transcription and T-cell proliferation.

Absorption, Distribution and Excretion

Sirolimus is rapidly absorbed after it is given orally, and in healthy individuals, peak blood levels are achieved about an hour after oral administration. However, it takes twice as long to reach peak blood levels in kidney transplant patients. Its systemic availability is about 15% and high-fat meals interfere with the bioavailability. Sirolimus is bound (40%) to proteins in plasma, and its elimination half-life is about 12–15 h for transplant patients, but may vary. It is predominantly metabolized by CYP2A4; the drug has a number of active metabolites and is excreted in feces.

Drug Interactions

Sirolimus is metabolized by CYP2A4 and is a substrate of the P-glycoprotein drug efflux pump; drugs like voriconazole, itraconazole, fluconazole and erythromycin increase its blood concentration. Conversely, the inducers of CYP3A4 will decrease blood levels of sirolimus. Cyclosporine increases the bioavailability of sirolimus, possibly due to P-GP inhibition and competition for CYP3A4. The bioavailability is more than 30–40% when the two drugs are administered 4 h apart and is more than

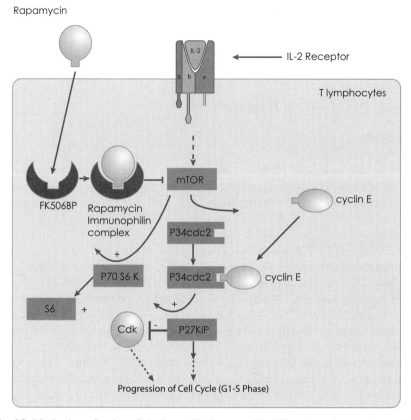

Rapamycin

Fig. 4.3 Mechanism of action of sirolimus. Sirolimus readily diffuses into the cytoplasm of the target cells where it binds to immunophilins (FK506-BP). The sirolimus–immunophilin complex does not inhibit calcineurin activity; instead it binds to the mTOR. The sirolimus–immunophilin–mTOR complex stops the cell cycle progression from G1 to S phase. The targets of sirolimus include the eukaryotic initiation factor (eIF-4F), 70-kDa S6 protein kinase (p70S6 K) and several cyclin-dependent kinases (cdk). As a consequence, it blocks downstream signaling pathway initiated after activation of IL-2 receptors, resulting in blockage of T-cell proliferation (*see* Color Insert)

100% when given at the same time. The combination of tacrolimus and sirolimus produced more renal toxicity than cyclosporine and sirolimus administered together.

Toxicity

The serious side effects of sirolimus may include an allergic reaction, increased risk of infection and lymphoma. Nephrotoxicity is not a concern when the drug is not used in combination with cyclosporine or tacrolimus. Less serious side effects associated with the administration of sirolimus include stomach upset, increased cholesterol and triglyceride levels, acne, insomnia, tremor, sore or weak muscles, water retention or swelling, anemia, leukopenia and thrombocytopenia. It could also impair wound healing.

Clinical Uses

Sirolimus is used for tissue transplantation where its major advantage over calcineurin inhibitors is that it is not nephrotoxic. Chronic renal failure in transplant patients who have taken calcineurin inhibitors for the long term can be prevented by the administration of sirolimus. Steroid-free immunosuppression can be achieved by administering sirolimus alone or in combination with mycophenolate mofetil and cyclosporine or tacrolimus. Since impaired wound healing is one of its potential side effects, some transplant centers use sirolimus only after several weeks of surgery.

Sirolimus has also been used in the production of sirolimus-eluting stents. These stents are used to treat obstructive coronary arteries. The rationale for the use of sirolimus in these stents is due to its antiproliferative activity. It has also received attention for cancer treatment due to its antiproliferative effects. In animal studies, sirolimus has shown some potential in the treatment of cancer where in combination with doxorubicin, a remission for Akt-positive lymphomas has been observed.

Everolimus

Everolimus is a derivative of sirolimus currently being used in Europe as an immunosuppressive agent. It inhibits P70S6 kinase (cyclin-dependent kinases, TOR). This kinase is involved in cellular proliferation signal, and its inhibition results in arresting the cell cycle at the G1–S phase. Everolimus has not received approval for clinical use in the United States. Although its mechanism of action is similar to sirolimus, it is more hydrophilic and has a distinct pharmacokinetic profile. Everolimus has a shorter half-life than sirolimus and works well in cardiac allograft vasculopathy or posttransplant lymphoproliferative disorders. In combination with calcineurin inhibitors, nephrotoxicity is enhanced, as was the case with sirolimus, and it also produces hyperlipidemia. Everolimus is also a substrate for CYP3A4, so the drugs affecting this enzyme system will alter its blood levels and clearance.

Sphingosine 1-Phosphate Receptor (S1P-R) Modulators

Fingolimod

Fingolimod is a synthetic analog of myriocin (ISP-1). It is derived from the culture filtrates of the fungus *Isaria sinclairii*. After phosphorylation, it binds to G protein-linked sphingosine-1 phosphate receptor 1 (S1P1) on lymphocytes and thymocytes with high affinity, resulting in the internalization of the receptors. As a result, the cells are not able to respond to serum lipid sphingosine 1-phosphate (S1P), and due to the lack of this signal, they are unable to egress from lymphoid organs.

Consequently, the recirculation of lymphocytes into graft site and peripheral inflammatory tissues is inhibited. Fingolimod has a novel immunosuppressive mechanism of action as it induces a reduction in peripheral blood lymphocyte count through emigration of blood lymphocytes to secondary lymphoid tissue as well as apoptotic T-cell death. The emigration of lymphocytes to secondary lymphoid tissue is achieved through the modulation of the lymphocyte chemotactic response to chemokines, which induces accelerated trafficking of T cells in the secondary lymphatic tissue. It also acts on endothelial cells where fingolimod preserves vascular integrity as a result of augmenting endothelial barrier function and adherens junction assembly. Fingolimod does not affect activation, proliferation and effector function of T and B cells. It has similar efficacy in the prevention of acute graft rejection as mycophenolate mofetil in new renal transplant patients. However, patients switched from fingolimod to mycophenolate mofetil exhibit significant improvement in arterial vasodilatory function if they were also receiving cyclosporine. It works in synergism with both the calcineurin inhibitors and the proliferation signal inhibitors to promote their immunosuppressive effects after tissue transplantation. Pharmacokinetics is characterized by a prolonged absorption phase, a large distribution volume and a long elimination half-life. The most common side effect associated with fingolimod is asymptomatic transient bradycardia. Initial trials have suggested that it does not produce hyperlipidemia, diabetes mellitus and renal toxicity, which are characteristic side effects of other immunosuppressive agents. The drugs have also shown some promising effects in phase II trials for the treatment of patients with relapsing MS.

Cytotoxic Agents

Mycophenolate Mofetil (CellCept)

Mycophenolic acid (MPA) was isolated from cultures of *Penicillium* spp. in 1896 and was purified in 1913. Initially the compound was studied for its antifungal and antibacterial effects and later for its antitumor effects. Many years later, its immunosuppressive activities were recognized and after further developmental work, an ester prodrug mycophenolate mofetil was developed, which was approved by the United States Food and Drug Administration for the prevention of acute renal allograft rejection in 1995 and for heart transplant recipients in 1998. Mycophenolate mofetil is a cytotoxic agent now used for immunosuppressive therapy and is the mofetil ester of MPA, which is the active immunosuppressive agent.

Mechanism of Action

Mycophenolate mofetil is a functionally selective cytotoxic agent for B and T lymphocytes, where it blocks the production of guanosine nucleotides required for DNA synthesis. For purine biosynthesis, B and T lymphocytes rely on de novo synthesis rather than on the salvage pathway. Lymphocytes have little or no salvage pathway as opposed to other blood marrow elements and parenchymal cells that depend on salvage pathways for the production of guanosine nucleotides. Mycophenolate mofetil is a selective, noncompetitive and reversible inhibitor of inosine monophosphate dehydrogenase (IMP-DH), which is associated with the de novo synthesis of guanosine nucleotides. As a result, it suppresses lymphocyte proliferation and antibody synthesis by B lymphocytes. Furthermore, mycophenolate mofetil may suppress the recruitment of leukocytes to the sites of inflammation by deglycosylating lymphocyte glycoproteins involved in adhesion and cell migration, which could be useful in organ transplantation. The synthesis of cytokines and their receptors and cytokine receptor-dependent signal transduction mechanisms are not specifically affected by mycophenolate mofetil.

Absorption, Distribution and Excretion

Mycophenolate mofetil is rapidly absorbed after oral administration, and the bioavailability of its oral dose is 94%. It is metabolized by esterases to free MPA, which is the active metabolite. The enterohepatic recirculation plays a crucial role in the serum levels of MPA. The active metabolite is further metabolized by glucuronyl transferase and is eliminated (90%) in urine as the MPA glucuronide (MPAG) as a result of the organic anion transport system in the proximal tubule. A small amount is excreted in feces.

Drug Interactions

The coadministration of mycophenolate mofetil with antacids results in decreased absorption. The plasma MPA concentration is significantly reduced by cholestyramine due to binding of the cholestyramine to MPAG in the intestine and interfering with the enterohepatic recirculation of the drug. The bioavailability of mycophenolate mofetil is higher when administered with tacrolimus as opposed to cyclosporine. The bioavailability of MPA is reduced by antibiotics including fluoroquinolones and metronidazole.

Toxicity

The side effects of mycophenolate mofetil include diarrhea, abdominal pain, constipation, nausea/vomiting, acne, dyspnea, cough, peripheral edema, increased risk of infections, drug-induced fever, dizziness, headaches, leukopenia and anemia.

Clinical Uses

Mycophenolate mofetil is used for tissue transplantation in combination with tacrolimus or cyclosporine or sirolimus plus glucocorticoids. It is used more than any other cytotoxic drug either at the time of the transplant or following the initiation of acute rejection. Mycophenolate mofetil is a prophylactic agent and cannot be used for chronic rejection or ongoing acute rejection.

Azathioprine

Originally developed for chemotherapy, azathioprine is used today mainly as an immunosuppressive agent and rarely as an antineoplastic drug. It was introduced as an immunosuppressive agent by a British pioneer of tissue transplantation, Roy Calne. Azathioprine was used to prevent rejection after tissue transplantation as a replacement for 6-mercaptopurine because it was less toxic. In addition to tissue transplantation, it is also used for rheumatoid arthritis and Crohn's disease. Azathioprine is a prodrug which in the body is converted to its active metabolites 6-mercaptopurine and 6-thioinosinic acid. Until the discovery of cyclosporine, azathioprine in combination with steroids was the standard treatment to prevent rejection after tissue transplantation.

Mechanism of Action

Azathioprine inhibits purine synthesis, which is necessary for the proliferation of cells, especially immunocompetent cells. It is converted to 6-mercaptopurine after it reacts with glutathione, and its metabolite, 6-mercaptopurine, is converted to additional metabolites, which inhibit de novo purine synthesis. This results from the synthesis of 6-thio IMP, 6-thio GMP and 6-thio GTP, and cell proliferation is inhibited after 6-thio GTP is inserted into host DNA.

Absorption, Distribution and Excretion

Azathioprine is orally absorbed with maximum blood levels attained within 1–2 h. The half-life of the parent drug is 10 min but its metabolite, 6-mercaptopurine, has a half-life of nearly 1 h, and some metabolites have an even longer half-life. The prodrug and the active metabolites both bind to plasma proteins with low affinity and are removed from tissues by oxidation or methylation.

Drug Interactions

The administration of purine analogs such as allopurinol and azathioprine together is not recommended. The enzyme thiopurine *S*-methyltransferase (TPMT) inhibits the activity of 6-mercaptopurine. Genetic polymorphisms of TPMT could increase azathioprine toxicity and therefore measuring levels of serum TPMT will be helpful in avoiding this toxic effect. Leukopenia, anemia and thrombocytopenia could develop when azathioprine is administered in conjunction with angiotensin-converting enzyme inhibitors or other drugs that cause myelosuppression.

Toxic Effects

Azathioprine is considered a human carcinogen although some of these studies have been suggested to be inconclusive. Individuals who have been previously treated with alkylating agents may be at a higher risk for cancer when treated with azathioprine. It is not known to cause fetal malformation. The short-time adverse effects associated with the use of azathioprine are myelosuppression including anemia, leukopenia and thrombocytopenia. Other side effects include increased risk of cancer and infection, GI disturbances, alopecia and hepatotoxicity.

Clinical Uses

Azathioprine is administered to patients who do not respond to calcineurin inhibitors, sirolimus and glucocorticoids. Daily doses of 3–10 mg/kg of azathioprine are administered 1 or 2 days before renal transplantation or on the day of surgery for prophylactic therapy. Mycophenolate mofetil is increasingly used in place of azathioprine for tissue transplantation since it is less myelotoxic and causes few opportunistic infections.

Cyclophosphamide

Cyclophosphamide disturbs the mechanisms associated with DNA synthesis and cell proliferation by alkylating DNA in proliferating and nonproliferating cells. Its mechanisms of immunosuppressive effects are similar to its antineoplastic actions. Cyclophosphamide affects both B and T cells, but it produces more toxicity to B cells because they recover slowly. It has some unpredictable effects on T-cell-mediated immunity where it actually augments some T-cell-mediated responses; however, the overall response is inhibitory.

Most of the preparative regimens used prior to allogeneic bone marrow transplantation to avoid the risk of rejection include cyclophosphamide in combination with another cytotoxic agent ± antithymocyte globulin or total lymphoid irradiation. The cytotoxic agents used in addition to cyclophosphamide are busulfan, fludarabine or treosulfan. A combination of fludarabine and cyclophosphamide ± antithymocyte globulin has also been used as preparative regimens for cord blood transplantation and allogeneic stem cell transplantation to avoid the risk of rejection.

Glucocorticoids

Glucocorticoids are steroids synthesized by the adrenal cortex and are made up of 21 carbon atoms. Although they possess multiple biological functions affecting numerous tissues and organ systems, only their immunosuppressive effects will be discussed here. Cortisone was the first immunosuppressant identified. Since the 1960s glucocorticoids have been used in tissue transplantation to prevent rejection. Today they are used in combination with other immunosuppressive agents to prevent rejection of the transplanted tissue. However, due to their serious effects, the focus has been on glucocorticoid withdrawal soon after tissue transplantation and switching to steroid-free immunosuppressive regimens.

Mechanism of Action

Glucocorticoids inhibit acquired or cell-mediated immunity. Their effects are mediated via inhibition of genes that code for various cytokines. The cytokines inhibited by glucocorticoids include IL-1, IL-2, IL-3, IL-4, IL-5, IL-6, IL-8 and IFN-γ. IL-2 inhibition by corticosteroids is the most crucial effect in immunosuppression, which results in the inhibition of T-cell proliferation and activation of cytolytic T cells. Glucocorticoids also slightly affect humoral immunity by inhibiting B-cell clonal expansion and antibody synthesis, and these effects are mediated via their ability to inhibit B cells' ability to express IL-2 and IL-2 receptors.

Glucocorticoids are potent inhibitors of all phases of the inflammatory process. This is accomplished by the induction of lipocortin (annexin-1) synthesis. Lipocortin binds to the cell membrane and prevents the access of phospholipase A_2 to its substrate arachidonic acid, resulting in diminished eicosanoid production. This effect is further pronounced by the inhibition of cyclooxygenase. Furthermore, lipocortin 1 escaping to the extracellular space is also stimulated by glucocorticoids, which subsequently cause inhibition of inflammation by binding to various leukocyte membrane receptors. The inflammatory processes affected by this mechanism include adhesion, chemotaxis, respiratory burst and phagocytosis. It also inhibits the release of various inflammatory mediators from mononuclear and polymorphonuclear phagocytes. Glucocorticoid–receptor complex inhibits NF-κB via increased expression of IκB, which results in increased apoptosis of activated cells. Overall, glucocorticoids cause a rapid, transient decrease in the number of circulating peripheral blood lymphocytes.

Absorption, Distribution and Excretion

Glucocorticoids are administered orally, intravenously or intramuscularly. They are absorbed locally at the site of administration including skin, eye and respiratory tract. Prolonged local exposure could eventually produce systemic effects. Most of the corticosteroids are bound to two plasma proteins, transcortin (corticosteroid-binding globulin, CBG) and albumin. At higher doses, when all the plasma-binding protein sites are saturated, corticosteroids could exist in unbound form. Their metabolism results in the formation of water-soluble derivatives. Their reduction can take place both in the liver and outside the liver, resulting in inactive compound, and subsequent reduction is only hepatic. These enzymatic reactions convert glucocorticoids either into glucuronide or sulfate form, which are then excreted in urine.

Side Effects

The most common side effects of glucocorticoids when used to prevent transplant rejection include hyperglycemia, increased risk of infection, poor wound healing, bone loss, ulcers, muscle weakness, water retention, skin irritation, excess facial hair growth and facial puffiness. In children their use could also cause growth retardation.

Clinical Uses

The glucocorticoids are employed in combination with other immunosuppressive agents to prevent transplant rejection. For acute transplant rejection, high doses of intravenous methylprednisolone are used. They are useful in suppressing allergic reactions to other immunosuppressive agents as well as the effects of cytokines associated with the use of anti-CD3. The glucocorticoids are used for bone marrow transplantation to prevent graft-versus-host disease. They are also used to treat autoimmune diseases, asthma, psoriasis, dermatomyositis and for the inflammatory manifestations of the eye.

Polyclonal Antibodies

Antithymocyte Globulin

Antithymocyte antibodies are antihuman thymocyte antibodies produced in rabbits and are used for immunosuppression. These polyclonal antibodies also possess cytotoxic antibodies that bind to various antigenic markers on T lymphocytes, resulting in their depletion from the circulation. Furthermore, they inhibit lymphocyte function by binding to important lymphocyte regulatory molecules on the lymphocyte cell surface. Antithymocyte globulins are used in combination with other immunosuppressive agents for the treatment of acute renal transplant rejection. They may be used initially in renal transplant patients instead of calcineurin inhibitors, which are nephrotoxic, to protect the transplanted tissue. Antithymocyte globulins are used at a dose of 1.5 mg/kg per day for 1–2 weeks for acute rejection of renal grafts. They have also been used for liver transplantation. The side effects of antithymocyte globulin include fever, chills, serum sickness, leukopenia, thrombocytopenia, increased risk of infection and malignancies when used in combination with other immunosuppressive agents.

Rho(D) Immune Globulin

Rho(D) immune globulin is one of the most specific and effective immunosuppressive treatments available. These IgG antibodies have high Rh(D)-specific titers. Administration of Rho(D) immune globulin prevents the response that develops in Rh⁻ mothers who were pregnant with an Rh⁺ fetus and consequently have become sensitized to the D antigen on fetal erythrocytes of the infant. In these Rh⁻ mothers, the antibody titers against Rh⁺ cells will continue to rise after each subsequent pregnancy, resulting in hemolytic disease of the newborn and erythroblastosis fetalis. This disease can be prevented by the administration of Rho(D) immunoglobulin to the Rh⁻ mother within 72 h of the birth of an Rh⁺ baby. The antibody could also be given after miscarriage, abortion or ectopic pregnancy. Rho(D) immune globulin is administered intramuscularly, and its half-life is about 3–4 weeks. The side effects include discomfort at the site of injection and mild fever. A very rare side effect is anaphylactic shock.

Monoclonal Antibodies

The monoclonal antibodies used as immunosuppressive agents in tissue transplantation include muromonoab-CD3, daclizumab and basiliximab. Muromonoab-CD3 binds to a specific site on CD3 receptors and interferes with the ability of the TCR to bind the antigen and also inhibits CD3 receptor-dependent signal transduction mechanisms, all of which result in immune suppression. Both daclizumab and basiliximab are monoclonal antibodies directed against IL-2 receptors and consequently inhibit IL-2-dependent responses after tissue transplantation, resulting in immune suppression. The monoclonal antibodies used as immunosuppressive agents are described in detail in Chapter 5.

Future Directions

The development of novel immunosuppressive agents in the past decade has resulted in a reduction in the incidence of acute rejection following tissue transplantation. However, the nonimmune side effects of the maintenance immunosuppressive regimens have hindered successful long-term outcomes. Consequently, a more selective group of immunosuppressive therapeutic agents that will not target ubiquitously expressed receptors and thus will have far fewer undesired side effects is required, which continues to be a major challenge for the available compounds. An alternate strategy that has drawn considerable interest in this regard is the development of pharmacological agents that inhibit T-cell activation by blocking the interaction between costimulatory receptor–ligand interactions. The positive costimulatory signals between CD40 and CD154 and between CD80/CD86 and CD28 receptors in T-cell stimulation have been well characterized and are considered potential therapeutic targets for intervention. Belatacept is one such specific immunosuppressive agent under development, which binds to CD80 and CD86 receptors on APCs. Full T-cell activation requires interaction of CD80 and CD86 molecules with CD28 costimulatory molecule on T cells. Belatacept is modified from abatacept, which is a human fusion protein combining the extracellular portion of CTLA-4 (cytotoxic T lymphocyte-associated antigen 4) with the Fc (constant region fragment) region of human IgG1. The difference between belatacept and abatacept is in two specific amino acid substitutions, making it more potent for binding to its ligands and a more efficacious T-cell activation inhibitor.

The reports from initial clinical trials suggest that the level of immunosuppression achieved by belatacept is equivalent to cyclosporine in renal transplant patients with less chronic allograft nephropathy, stable kidney function and improved cardiovascular and metabolic profiles.

Bibliography

Ballow M. 1997. Mechanism of action of intravenous serum immunoglobulin in autoimmune and inflammatory diseases. J All Clin Immunol. 100:151–157.

Boratynska M, Banasik M, Patrzalek D, Klinger M. 2006. Conversion from cyclosporine-based immunosuppression to tacrolimus/mycophenolate mofetil in patients with refractory and ongoing acute renal allograft rejection. Ann Transplant. 11:51–56.

Borel JF. 2002. History of the discovery of cyclosporin and its early pharmacologic development. Wien Klin Wochenschr. 114/12:433–437.

Bowman JM. 1990. RhD hemolytic disease of the newborn. NEJM. 339:1775–1777.

Brinkmann V, Pinschewer PD, Feng L, Chen S. 2001. FTY720: Altered lymphocyte traffic results in allograft protection. Transplantation. 72:764–769.

Brinkmann V, Lynch KR. 2002. FTY720: Targeting G-protein coupled receptors for sphingosine 1-phosphate in transplantation and autoimmunity. Curr Opin Immunol. 14:569–575.

Brinkmann V, Cyster JG, Hla T. 2004. FTY720: Sphingosine 1-phosphate receptor in the control of lymphocyte egress and endothelial barrier function. Amer J Transplant. 4:1019–1025.

Brochstein JA, Kernan NA, Groshen S, Cirricione C, et al. 1987. Allogenic bone marrow transplantation after hyperfractionated total body irradiation and cyclophosphamide in children with acute leukemia. NEJM. 317:1618–1624.

Brown EJ, Albers MW, Shin TB, Ichikawa K, et al. 1994. A mammalian protein targeted by G1-arresting rapamycin-receptor complex. Nature. 369:756–758.

Budde K, Schutz M, Glander P, Peters H, et al. 2006. FTY720 (fingolimod) in renal transplantation. Clin Transplant. 20:17–24.

Burke JF, Pirsch JD, Ramos EL, Salomon DR, et al. 1994. Long term safety and efficacy of cyclosporine in renal transplant recipients. NEJM. 331:358–363.

Busuttil RW, Lake JR. 2004. Role of tacrolimus in the evolution of liver transplantation. Transplantation. 77:S44–S51.

Calne RY, White DJ, Thiru S, Evans DB, et al. 1978. Cyclosporin A in patients receiving renal allografts from cadaver donors. Lancet. 2:1323–1327.

Chan M, pearson GJ. 2007. New advances in antirejection therapy. Curr Opin Cardiol. 22:117–122.

Christians U, Sewing KF. 1993. Cyclosporin metabolism in transplant patients. Pharmacol Ther. 57:291–345.

Citterio F. 2001. Steroid side effects and their impact on transplantation outcome. Transplantation. 72:S75–S80.

Contreas M, deSilva M. 1994. The prevention and management of the hemolytic disease of newborn. J Royal Soc Med. 87:256–258.

Crespo-Leiro MG. 2003. Tacrolimus in heart transplantation. Transplant Proc. 35:1981–1983.

DiPavoda FE. 1990. Pharmacology of cyclosporine (Sandimmune). V. Pharmacological effects on immune function in vitro studies. Pharmacol Rev. 41:373–405.

Eisen HJ, Tuzcu EM, Dorent R, Kobashigawa J. 2003. Everolimus for the prevention of allograft rejection and vasculopathy in cardiac transplant recipients. NEJM. 349:847–858.

Fahr A. 1993. Cyclosporin clinical pharmacokinetics. Clin Pharmacokin. 24:472–495.

Ferry C, Socie G. 2003. Busulfan-cyclophosphamide versus total body irradiation-cyclophosphamide as preparative regimen before allogeneic hematopoietic stem cell transplantation for acute myeloid leukemia: What have we learned? Exp Hematol. 31:1182–1186.

Formica RN, Lorber KM, Friedman AL, Bia MJ, et al. 2004. The evolving experience using everolimus in clinical transplantation. Transplant Proc. 36:495S–499S.

Giebel S, Wojnar J, Krawczyk-Kulis M, Markiewicz M, et al. 2006. Treosulfan, cyclophosphamide and antithymocyte globulin for allogeneic hematopoietic cell transplantation in acquired severe aplastic anemia. Ann Transplant. 11:23–27.

Grinyo JM, Cruzado JM. 2006. Mycophenolate mofetil and sirolimus combination in renal transplantation. Am J Transplant. 6:1991–1999.

Guessner RW, Kandaswamy R, Humar A, Gruessner AC, et al. 2005. Calcineurin inhibitor – and steroid free immune suppression in pancreas-kidney and solitary pancreas transplantation. Transplant. 79:1184–1189.

Gummert JF, Ikonen T, Morris RE. 1999. New immunosuppressive drugs. J Am Soc Nephrol. 10:1366–1380.

Halloran PF. 2004. Immunosuppressive drugs for kidney transplantation. NEJM. 351:2715–2729.

Henry ML. 1999. Cyclosporine and tacrolimus: A comparison of efficacy and safety profiles. Clin Transplant. 13:209–220.

Hong JC, Kahan BD. 2000. Sirolimus-induced thrombocytopenia and leukopenia in renal transplant recipients: Risk factors, incidence, progression and management. Transplantation. 69:2085–2090.

Hosenpud JD. 2005. Immunosuppression in cardiac transplantation. NEJM. 352:2749–2750.

Iacono AT, Johnson BA, Grgurich WF, Yossef JG, et al. 2006. A randomized trial of inhaled cyclosporine in lung-transplant recipients. 354:141–150.

Ingelfinger JR, Schwartz RS. 2005. Immunosuppression – the promise of specificity. 353:836–839.

Kahan BD. 1999. The potential role of rapamycin in pediatric transplantation as observed from adult studies. Pediat Transplant. 3:175–180.

Kahan BD. 2004. FTY720: From bench to bedside. Transplant Proc. 36:531S–543S.

Kappos L, Antel J, Comi G, Montalban X, et al. 2006. Oral fingolimod (FYT720) for relapsing multiple sclerosis. NEJM. 355:1124–1140.

Keogh A. 2005. Long term benefits of mycophenolate mofetil after heart transplantation. Transplantation. 79:S45–S46.

Kino T, Hatanaka H, Hashimoto M, Nishiyama M, et al. 1987. FK506, a novel immunosuppressant isolated from a streptomyces. I. Fermentation, isolation, and physico-chemical and biological characteristics. J Antibiot. 40:1249–1255.

Kreis H. 1992. Antilymphocytic globulins in kidney transplantation. Kidney Int. 42:S188–S192.

Krensky AM, Strom TB, Bluestone JA. 2005. Immunomodulators: Immunosuppressive agents, tolerogens and Immunostimulants. In: Brunton LL, Ed. Goodman and Gilman's, the Pharmacological basis of Therapeutics, 11th edition, The McGraw-Hill Companies.

Larsen CP, Knechtle SJ, Adams A, Pearson T, et al. 2006. A new look at blockade of T cell costimulation. A therapeutic strategy for long term maintenance immunosuppression. Am J Transplant. 6:876–883.

Liu J, Farmer J, Lane W, Friedman J, et al. 1991. Calcineurin is a common target of cyclophilin-cyclosporin A and FKBP-FK506 complexes. Cell. 66:807–815.

Losa Garcia JE, Mateos RF, Jimenez A, Salgado MJ, et al. 1998. Effect of cyclosporin A on inflammatory cytokine production by human alveolar macrophages. Resp Med. 92:722–728.

MacDonald AS. 2003. Rapamycin in combination with cyclosporine or tacrolimus in liver, pancreas and kidney transplantation. Transplant Proc. 35:2015–2085.

Mizutami E, Narimatsu H, Murata M, Tomita A, et al. 2007. Successful second cord blood transplantation using fludarabine and cyclophosphamide as a preparative regimen for graft rejection following reduced intensity cord blood transplantation. Bone Marrow Transplant. 40:85–87.

Neumayer HH. 2005. Introducing everolimus (Certican) in organ transplantation: An overview of preclinical and early clinical developments. Transplantation. 79:S72–75.

Perkins JD. 2006. Steroid use in liver transplantation: None, perioperative or full course. Liver Transplant. 12:1294–1295.

Raught B, Gingras AC, Sonenberg N. 2001. The target of rapamycin (TOR) proteins. Proc Natl Acad Sci USA. 98:7037–7044.

Reichenspurner H. 2005. Overview of tacrolimus based immunosuppression after heart or lung transplantation. J Heart Lung Transplant. 24:119–130.

Rhen T, Cidlowski JA. 2005. Antiinflammatory action of glucocorticoids – new mechanisms for old drugs. NEJM. 353:1711–1723.

Rippin SJ, Serra AL, Marti HP, Wuthrich RP. 2007. Six year followup of azathioprine and mycophenolate mofetil use during the first six months of renal transplantation. Clin Nephrol. 67:374–380.

Sanchez T, Estrada-Hernandez T, Paik JH, Wu MT, et al. 2003. Phosphorylation and action of the immunomodulator FTY720 inhibits vascular endothelial cell growth factor-induced vascular permeability. J Biol Chem. 278:47281–47290.

Schmeding M, Newmann UP, Neuhaus R, Neuhaus P. 2006. Mycophenolate mofetil in liver transplantation – Is monotherapy safe? Clin Transplant. 20:75–79.

Schulak JA. 2004. The steroid immunosuppression in kidney transplantation: A passing era. J Surg Res. 117:154–162.

Sehgal SN. 2006. Rapamune (RAPA, rapamycin, sirolimus) mechanism of action immunosuppressive effect results from blockade of signal transduction and inhibition of cell cycle progression. Clin Biochem. 39:484–489.

Socie G, Clift RA, Blaise D, Devergie A, et al. 2001. Busulfan plus cyclophosphamide compared with total-body irradiation plus cyclophosphamide before marrow transplantation for myeloid leukemia: Long term followup of four randomized studies. Blood. 98:3569–3574.

Starzl TE, Klintmalm GB, Porter KA, Iwatsuki S. 1981. Liver transplantation with use of cyclosporin a and prednisone. NEJM. 305:266–269.

Sykes A. 2006. Inhaled cyclosporine may increase survival after lung transplantation. Thorax. 61:305–305.

Tanabe K, Tokumoto T, Ishida H, Ishikawa N, et al. 2004. Excellent outcome of ABO-incompatible living kidney transplantation under pre transplantation immunosuppression with tacrolimus, mycophenolate mofetil and steroid. Transpl Proc. 36:2175–2177.

Teraok S, Sato S, Sekijima M, Iwado K, et al. 2005. Comparative study of clinical outcome in kidney transplantation between early steroid withdrawal protocol using basiliximab, calcineurin inhibitor, and mycophenolate mofetil and triple regimen consisting of calcineurin inhibitor, mycophenolate and steroid. Transplant Proc. 37:791–794.

Vincenti F, Larsen C, Durrbach A, Wekerle T, et al. 2005. Costimulation blockade with belatacept in renal transplantation. NEJM. 353:770–781.

Vincenti F. 2007. Costimulation blockade – what will the future bring? Nephrol Dial Transplant 22:1293–1296.

Webster AC, Lee VW, Chapman JR, Craig JC. 2006. Target of rapamycin inhibitors (sirolimus and everolimus) for primary immune suppression of kidney transplant recipients: A systemic view and meta-analysis of randomized trials. Transplantation. 81:1234–1248.

Wente MN, Sauer P, Mehrabi A, Weitz J, et al. 2006. Review of the clinical experience with a modified release form of tacrolimus [FK506E(MR4)] in transplantation. Clin Transplant. 17:80–84.

Westhoff TH, Schmidt S, Glander P, Liefeld L, et al. 2007. The impact of FTY720 (fingolimod) on vasodilatory function and arterial elasticity in renal transplant patients. Nephrol Dial Transplant. 22:2354–2358.

Chapter 5
Monoclonal Antibodies as Therapeutic Agents

Introduction

In 1895, Hericourt and Richet reported the first clinical trials testing the principle of antibody production. They injected cancer cells into animals to obtain antiserum to treat cancer patients; this was the first time several patients with cancer were administered tailor-made serum for treatment. Several patients showed improvement, which was encouraging, but none of the patients were completely cured. These trials were repeated in the early 1900s but the results were not consistent. The problems included the variability of the antisera and the side effects of polyclonal antibodies – some of which were directed against self.

Paul Ehrlich, at the beginning of the twentieth century, first proposed the concept of "magic bullets." He reasoned that the development of compounds that selectively target a disease-causing organism could potentially deliver toxins to that organism. The specificity of antibodies, which are extremely specific and bind to and attack one particular antigen, provided a powerful tool to advance the idea envisioned by Ehrlich. As a result, the idea that antibodies could be used as therapeutic agents was not seen as far-fetched. His vision is being pursued today since antibodies not only are specific with high affinity but can also recruit effector functions of immune response. These effector mechanisms include antibody-mediated cytolysis, complement-mediated cytolysis and antibody-dependent cell-mediated cytotoxicity. Furthermore, antibodies can be used to deliver radiation, chemotherapeutic agents and/or toxins, which make them extremely useful in treating infectious diseases, cancer, tissue rejection, graft-versus-host disease, auto-immune and inflammatory diseases.

Antibodies have been used for years to detect small levels of antigens in biological fluids. The focus of interest included enzymes, hormones, microorganisms, toxins, drugs and other proteins. Their specificity provided a distinct advantage in targeting the intended antigen. However, their use in the diagnostic laboratory was limited due to severe problems with their propagation in large amounts. The specific sera could be propagated only by conventional immunization, which was further complicated by other issues such as inability to predict immune response, heterogeneity of specific antibodies and immunogenicity of minor contaminants. This technique produced only polyclonal antibodies that possessed many different

M.M. Khan, *Immunopharmacology*, DOI: 10.1007/978-0-387-77976-8_5,
© Springer Science+Business Media, LLC 2008

variable regions. Furthermore, it was almost impossible to produce antibodies against weak and/or impure antigens.

Historically, antibodies have been obtained from the serum of animals. The serum contains a mixture of polyclonal antibodies. In 1890, Emil Behring immunized rabbits and mice against tetanus and diphtheria and reported that the antitoxin serum could protect against a lethal dose of the toxin. Since then, antisera have been used to protect from pathogens and toxins, but serum sickness was a major drawback for their clinical use. Antisera may produce immune responses, which could cause severe allergic reactions, and may even lead to anaphylactic shock and death.

Production of Monoclonal Antibodies

In the 1970s, the B-cell cancer myeloma was studied, and it was established that these cells produced a single type of antibody. A revolution started in 1975 in the area of generating antibodies when Kohler and Milstein succeeded in fusing murine

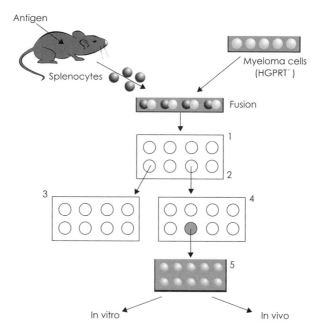

Fig. 5.1 Technique to produce monoclonal antibodies. After mice are injected with the antigen, the splenocytes are isolated and fused with myeloma cells. The fused cells are cultured (1) and the supernatants are assayed for the presence of desired antibodies (2). The cells from antibody-positive wells are cloned (3) and supernatants are tested for the desired antibodies (4) followed by expansion of the positive clones (5) and in vitro or in vivo propagation to produce the needed amount of monoclonal antibodies (*see* Color Insert)

myeloma cells with sheep red blood cell-immunized splenocytes. This technique resulted in the production of purified antibodies and was named the hybridoma technology. This turned out to be an ideal technique to produce antibodies since each hybrid cell is made up of genetic material from a single myeloma cell, and a single B cell produces only one type of antibody called monoclonal antibody. This fused cell could be potentially propagated indefinitely, overcoming the problem of limited production of antibodies and their heterogeneity.

To produce monoclonal antibodies by hybridoma technology, mice (or other animals) are injected with the antigen. After 72 h, spleens are removed and splenocyte suspensions are prepared; the splenocytes are then mixed with mouse myeloma cells grown in cultures. A chemical reagent or a virus is added to the mixture, which causes the fusion of the cell membranes of splenocytes and myeloma cells. This results in the formation of random fused cells containing splenocytes–myeloma cells, splenocytes–splenocytes and myeloma cells–myeloma cells. Selection is performed via culturing the newly fused hybridoma cells in special media containing hypoxanthine aminopterin thymidine (HAT) medium for 10–14 days. Aminopterin in the medium blocks the de novo pathway of purine biosynthesis, which causes the death of unfused myeloma cells, since they are unable to produce nucleotides by de novo or salvage pathway. Under these conditions, only the hybrids of splenocyte–myeloma cells can survive and propagate, and the unfused myeloma cells and splenocytes will die. The surviving hybridomas are then cultured in small colonies (Fig. 5.1).

Screening and Cloning

The hybridomas are allowed to propagate for 2–3 weeks. The screening of the desired hybridomas, which are the hybrids producing the desired antibody, is done by radioimmunoassay, enzyme-linked immunosorbent assay or flow cytometry. The desired hybrids are then cloned to ascertain the hybrids producing the desired antibody, are kept separate and the rest are discarded. Various hybridomas in the culture may secrete a mixture of antibodies directed against various epitopes on the antigen. The cloning is done by limited dilution where hybridomas are diluted and one cell is pipetted into a microtiter plate well. The single cell produces a colony of identical cells. A large-scale production of monoclonal antibodies is done by one of many techniques, including tissue culture, injection into peritoneal cavities of mice and industrial bioreactors.

Murine Antibodies as Therapeutic Agents

Following the development of hybridoma technology, the potential of using these antibodies appeared endless. However, when the mouse antibodies were employed clinically, they were found to be of rather limited use. These murine antibodies were rapidly inactivated by the human immune response, that is, the production of antibodies against the murine monoclonal antibodies. This reaction by the human immune response against mouse antibodies produces what are called human anti-mouse antibodies (HAMA). This response not only causes flu-like symptoms, allergic reaction and in extreme cases systemic shock or death, but also results in the rapid neutralization and clearance of mouse monoclonal antibodies being adminis-tered as therapeutic agent.

There have been advances not only in the area of hybridoma and monoclonal antibody production but also in recombinant DNA technology. New information emerged about the organization and expression of immunoglobulin genes in B cells as well as the rearrangement and mutation of germline immunoglobulin genes, which form the repertoire of functional immunoglobulin genes. This resulted in combining hybridoma and recombinant DNA technologies to produce antibodies possessing more human components. As a consequence, recombinant antibodies have become important therapeutic agents, since they are designer antibodies that are less immunogenic, smaller and have greater affinity for the target. Furthermore, they also possess the potential to carry more complex molecules including radiation, toxins, chemotherapeutic agents and enzymes.

Chimeric Antibodies

The word *chimeric* comes from *chimera*, a mythical beast with the head of a lion, the body of a goat and the tail of a dragon. To overcome the problems associated with the HAMA response, the idea of chimeric antibodies was an important alternative. The rationale was based on the ability of the Fc fragment of the antibody to conduct signal transduction mechanisms, where the fusion of the Fc fragment of human with the Fab fragment from another species may result in minimizing the devel-opment of HAMA. By using the recombinant DNA technology, parts of the Fab fragment of murine antibody were added to the Fc fragment of the human antibody and murine variable region genes were cloned into a mammalian expression system. This contained human heavy and light chain constant region gene components and

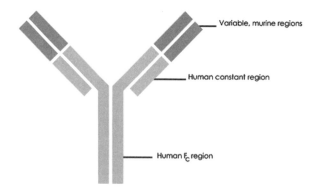

Fig. 5.2 Depiction of a chimeric monoclonal antibody (*see* Color Insert)

the desired biological function led the way in choosing the Fc segment. As a result, this antibody has the binding characteristics of the mouse antibody and the signal transduction capabilities of the human antibody. The chimeric antibodies are 30% mouse and 70% human, and therefore they could still produce the HAMA response, referred to as human antichimeric antibody (HACA) responses. Human IgG1 is the antibody of choice for chimeric antibodies when effector function is required as well. Otherwise, if the effector function of IgG1 is not desired, other subclasses of IgG could be employed (Fig. 5.2).

Complementarity-Determining Region-Grafted Antibodies

Complementarity-determining region (CDR)-grafted antibodies are produced by recombinant DNA technology. This technique was developed by Greg Winter and his colleagues in 1986 whereby only a part of the mouse antibody that binds to its target is inserted into human antibody. Also termed second-generation or hyper-chimeric antibodies, this technique employed CDR of the mouse antibody and is grafted into the human antibody. The DNA encoding the CDR from the heavy and light chains is used. This technique produces an antibody that is more human than the chimeric antibody described in the previous section. The method was further improved by modifying any framework or "scaffold" residues, which are important for the antibody-binding site. These antibodies are now called humanized antibodies, which are 90–95% human. However, human immune response could still react to the mouse region of the antibody.

Therapeutic Uses of Monoclonal Antibodies

Tissue Transplantation

Muromonoab-CD3

Muromonoab-CD3 binds to the CD3 glycoprotein Σ chain that is in close proximity to the antigen recognition complex. The components of the antigen recognition complex are involved in the recognition of the antigen, signal transduction and resultant proliferation of T lymphocytes. When muromonoab-CD3 binds to the Σ chain of CD3, the antigen is blocked and is unable to bind to the TCR. The inability of the TCR to bind to the antigen also results from its internalization following treatment with the antibody. It also binds to T cells resulting in their depletion from the blood and peripheral lymphoid organs, rapidly after its administration. Two to seven days following the initiation of therapy, the numbers of CD4$^+$ and CD8$^+$ begin to increase in the peripheral blood without any consequence on the immunosuppressive effects of the antibody. The number of CD3$^+$ cells returns to normal within a week after the end of therapy with the antibody. During the second week of therapy, there is production of antibodies directed against the drug. The antibodies compete with the drug for binding to its receptors, and the antibody titer ($\geq 1{:}1000$) must be monitored to decide if the treatment should be continued. The antibody also inhibits the secretion of a number of cytokines except IL-4 and IL-10. The effects of IL-2 secretion are particularly pronounced.

Muromonoab-CD3 is used for the treatment of acute organ transplant rejection. It is effective in preventing graft rejection after kidney, heart or liver transplantation. Muromonoab-CD3 is effective in patients who after acute cardiac or liver allograft rejection do not respond to steroid therapy. It is administered intravenously and with a dose of 5 mg/day, a general concentration range of 400–1500 ng/ml can be achieved. A serum concentration of 600–1150 ng/ml in renal transplant patients produces desirable immunosuppressive effects. The levels of CD3 expression, their production and antibodies to the drug determine its rate of clearance. In the absence of antibodies to muromonoab-CD3, its half-life is about 18 h.

A major side effect, a cytokine-release syndrome, is observed within 30–60 min after the administration of muromonoab-CD3. The side effect could last for several days, and is the result of increased production of several cytokines including IL-2, IL-6, TNF-α and IFN-γ. TNF-α appears to be responsible for most of the discomfort, and the symptoms are more pronounced after the first administration of muromonoab-CD3.

Daclizumab (Zenapax)

Daclizumab is an IL-2 receptor antagonist that binds to the TAC unit (CD25) of the IL-2 receptor with high affinity, thus preventing the binding of IL-2 to its receptors. Daclizumab is a CDR-grafted (humanized) monoclonal antibody, which is 90% human IgG and 10% murine antibody, and the murine part binds to the CD25 receptor.

Since the TAC subunit of IL-2 receptors is expressed only on activated and not on resting T cells, its effects are restricted to activated T cells and may not alter the entire immune response. In allograft rejection, IL-2-activated T cells play a critical role and consequently inhibition of IL-2-dependent responses by daclizumab makes it an important drug for preventing tissue rejection. It also interferes with the development of immune response to antigenic challenges and is suggested to inhibit the expression of the IL-2α and -β chains and to cause a decrease in circulating lymphocytes.

Daclizumab is used for the prophylaxis of acute rejection in patients receiving kidney transplants. A dose of 1 mg/kg is sufficient to completely block all the IL-2 receptors. It is administered in five doses at a 2-week interval where its elimination half-life is about 20 days. A combination of several other immunosuppressive agents including cyclosporine (or tacrolimus, rapamycin), mycophenolate mofetil and corticosteroids can be used with daclizumab. When it is used in combination with tacrolimus, the doses of tacrolimus are reduced. After tissue transplantation, the addition of daclizumab to the standard immunosuppressive regimen produces reduction in tissue rejection up to 50%. Daclizumab can cause hypersensitivity reactions, but it does not cause cytokine-release syndrome. There is a low incidence of opportunistic infections and lymphoproliferative disorders with daclizumab. When administered in combination with other immunosuppressive agents, it does not alter the side effects of other pharmacological agents. However, increased incidence of wound infection and cellulitis have been reported when daclizumab is used in combination with other immunosuppressive agents.

Basiliximab (Simulect)

Basiliximab, a glycoprotein produced by recombinant DNA technology, is a chimeric (human 60%, murine 40%) monoclonal antibody (IgG1κ) that binds to and blocks the IL-2 receptor-α chain (CD25), which is also known as TAC and is expressed only on activated T cells. It has higher affinity for IL-2 receptors than daclizumab. After binding to the IL-2 receptor-α chain, basiliximab causes the inhibition of IL-2-induced T-cell activation, which is critical in the etiology and pathogenesis of tissue rejection pathway. As is the case for daclizumab, it also inhibits antigen-induced immune response; whether the clearance of the drug results in the restoration of the immune response has not been established.

Basiliximab is used for the prophylaxis of acute rejection for patients undergoing kidney transplantation where it is employed in combination with other standard immunosuppressive therapy regimens. After tissue transplantation, its addition to the standard immunosuppressive regimen results in inhibiting tissue rejection up to approximately 30%. Both daclizumab and basiliximab have similar effects on the expression of IL-2α and -β chains.

Basiliximab is contraindicated in patients with known hypersensitivity to this antibody. There is a slightly increased risk of GI discomfort compared with placebo after treatment with basiliximab. However, there is no difference in the incidence of malignancies and infections between patients receiving basiliximab versus placebo. Furthermore, it does not increase the adverse effects resulting from the administration of a standard immunosuppressive regimen and other medications.

Psoriasis

Psoriasis is a disease of the immune system that involves T lymphocytes. The etiology and pathogenesis of psoriasis results from complex communications that cause activation of T lymphocytes and trafficking to the skin. Further reactivation causes inflammation and overproduction of skin, resulting in lesions and plaques. In psoriatic skin, there is an upregulation of intracellular adhesion molecule-1 (ICAM-1) on endothelium and keratinocytes.

Efalizumab (Raptiva)

Efalizumab is a humanized IgG1κ antibody produced by recombinant DNA technology. It exhibits immunosuppressive function. It binds to CD11a, which is the α-subunit of leukocyte function antigen (LFA)-1. Efalizumab decreases the cell surface expression of CD11a, which is expressed on all leukocytes. It interferes with the adhesion of LFA-1 to ICAM-1, which is required for the initiation of a number of immune and inflammatory processes. These processes include T-cell activation and migration of T cells to the site of inflammation after their adhesion to endothelial cells. The plaque formation in psoriatic skin results from helper T-cell activation and trafficking to the skin. The net effect of efalizumab is the inhibition of T-cell proliferation, cytokine release, migration and interaction with tissue-specific cells.

Efalizumab is administered weekly by subcutaneous injections to treat psoriasis, where a steady state is reached after 4 weeks. Its side effects include serious infections, thrombocytopenia, hemolytic anemia and the probability of malignancies due to its immunosuppressive effects. Other common side effects produced by efalizumab within 2 weeks of administration are nausea, fever, chills and headache. Furthermore, it can produce symptoms associated with hypersensitivity reaction.

Rheumatoid Arthritis

Infliximab (Remicade)

Infliximab is a chimeric monoclonal antibody (IgG1κ) produced by recombinant DNA technology and is directed against TNF-α. It is composed of human constant and mouse variable regions. Infliximab binds to the soluble and the membrane-bound form of TNF-α, resulting in the neutralization of its biological activity. This is achieved via the inability of TNF-α to bind to its receptors in the presence of infliximab. The prominent effects of TNF-α include enhanced production of IL-1 and IL-6, induction of adhesion molecules and inflammatory cell traffic and stimulation of acute-phase response and tissue-degrading enzymes. Infliximab suppresses the expression of adhesion molecules, decreases IL-6 levels, inhibits acute-phase response and may alter other immune cells. It lyses the cells expressing transmembrane TNF-α receptors.

Infliximab is used to reduce signs, symptoms and progression of rheumatoid arthritis. It is indicated for patients with moderate to severe active rheumatoid arthritis, where infliximab reduces infiltration of inflammatory cells into the inflamed areas of joints. Infliximab is also indicated in moderate to severely active Crohn's disease. It is used to reduce signs and symptoms and to maintain clinical remission. The number of draining enterocutaneous and rectovaginal fistulas is reduced by infliximab. It helps to maintain fistula closure in patients with fistulizing Crohn's disease.

Infliximab is administered in combination with methotrexate for rheumatoid arthritis. A dose of 3 mg/kg is administered via intravenous infusion and is repeated after 2 and 6 weeks followed by the maintenance dose every 8 weeks. The recommended dose for Crohn's disease is 5 mg/kg. The side effects associated with the administration of infliximab include acute infusion reactions (fever, chills, chest pain, hypotension and rare anaphylaxis), increased risk of infection, production of autoantibodies (lupus-like syndrome), increased risk of malignancies, immuno-genicity, cough, back pain, nausea, vomiting and several leukocyte disorders.

Adalimumab (Humira)

Adalimumab, a human monoclonal antibody (IgG1) directed against TNF-α, is produced by phage display technology. It binds to TNF-α resulting in the inhibi-tion of its ability to bind to its p55 and p75 cell surface receptors. Adalimumab modulates biological responses induced or regulated by TNF including expression of adhesion molecules, acute-phase response, IL-1 and IL-6 levels and metallopro-teinases. It neutralizes the elevated levels of TNF found in the synovial fluid of patients with rheumatoid arthritis and psoriatic arthritis.

Adalimumab is used to reduce signs and symptoms and progression of rheuma-toid arthritis and psoriatic arthritis. It could be administered alone or in combination with methotrexate. The recommended dose for adults is 40 mg by subcutaneous injection, administered every other week. The most common side effect associated with the administration of adalimumab is injection site reaction. The most serious side effects resulting from treatment with this antibody include increased risk of infections and malignancies and neurological disorders. Additional side effects are the production of autoantibodies, immunogenicity and GI disorders.

Thrombosis

Abciximab (ReoPro)

Abciximab is a Fab fragment (7E3) of a chimeric monoclonal antibody that is directed against GPIIb/IIIa receptors, members of the integrin family of adhesion molecules. These receptors are located on platelets where they are involved in platelet aggregation. Abciximab inhibits platelet aggregation by blocking GPIIb/IIIa receptors, thus preventing the binding of fibrinogen, von Willebrand factor and other molecules promoting adhesion to the receptors on platelets. It increases bleeding and activated clotting times and reduces the response of platelets to adenosine diphos-phate. The Fab fragment of abciximab does not directly interact with the arginine-glycine-aspartic acid (RGD) binding site of GPIIb/IIIa but appears to be involved in steric hindrance, resulting in interference for other larger molecules.

The antibody fragment also binds to vitronectin receptors (av integrin) that are present on a number of tissues including vessel wall endothelial cells, smooth muscle cells and platelets. These receptors are involved in the proliferative prop-erties of endothelial and smooth muscle cells and the procoagulant properties of the platelets.

Abciximab is indicated in patients with heart disease caused by poor blood flow in the arteries of the heart (ischemic cardiac complications). It is also used in patients with unstable angina not responding to conventional treatment and for precutaneous coronary intervention. It can reduce the incidence of abrupt closure and restenosis resulting from percutaneous transluminal coronary angioplasty. Abciximab may also be used in unstable angina and acute therapy of myocardial infarctions.

An initial bolus dose of 0.25 mg/kg of abciximab is administered intravenously followed by continuous infusion of 5 or 10 mg/min for 12–96 h. Following an IV administration, its initial half-life is several minutes followed by a second-phase half-life of about 30 min, which is attributed to rapid binding to its receptors on the platelets. Within 10 min after administration, abciximab exhibits its inhibitory effects on platelet aggregation after a single bolus injection and remains in circulation for several days bound to platelets. The side effects of abciximab include irritation at the injection site, nausea, vomiting, dizziness and bleeding. Occasional breathing trouble, rapid or abnormal heartbeat, chest pain, swelling of the feet or ankles may occur after its administration. Human antichimeric antibodies may be produced in response to abciximab.

Treatment of Cancer

Rituximab (Rituxan)

Rituximab is a chimeric (murine/human) monoclonal antibody (IgG1κ) specific for cells possessing a CD20 antigen, which is a human B lymphocyte-restricted differentiation antigen also termed Bp35 or B1. It is distributed on pre-B cells and with a higher density on mature B cells but is not present on hematopoietic stem cells, pro-B cells and plasma cells. CD20 is a transmembrane phosphoprotein, which is also expressed in more than 90% of B-cell non-Hodgkin's lymphomas (NHLs). It does not shed from its cell surface membrane-bound form and also does not internalize after binding to its antibody. CD20 antigen is not expressed in most tissues and regulates cell cycle initiation and differentiation. Rituximab contains human constant region and murine light and heavy chain variable region, and its Fab domain binds to CD20 antigen on B cells and its Fc domain recruits immune effector cells and mediators for B-cell lysis. The lysis can be achieved by ADCC and apoptosis.

Rituximab is used for the treatment of relapsed or refractory, low-grade or follicular B-cell lymphoma (CD20+) and NHL. In combination with CHOP or other standard chemotherapeutic regimens, rituximab is the first-line treatment for diffuse large B-cell, CD20+, NHL. The suggested dose is 375 mg/m^2 intravenous infusion and generally four to eight doses are administered once a week. Fifty percent of patients have responded to the treatment with full or partial remission.

Most side effects are felt after the first treatment with rituximab and attention should be given to the rate of infusion. The most common immediate side effects of rituximab are fever, chills and respiratory symptoms, but these effects are much milder than the traditional chemotherapy. Other infusion reactions include nausea, angioedema, headache, hypotension, puritus, utricaria, rash and vomiting. The adverse effects decrease with each subsequent administration of the drug. Other side effects associated with rituximab include B-cell depletion, cytopenia, immunogenicity and multiple pulmonary events.

Trastuzumab (Herceptin)

Trastuzumab is a CDR-grafted (humanized) monoclonal antibody (IgG1κ) produced by recombinant DNA technology. It binds to human epidermal growth factor (EGF) type 2 receptor (HER-2). Trastuzumab is composed of two antigen-specific sites that bind to the extracellular domain of the HER-2 receptors resulting in inhibition of the induction of its intracellular tyrosine kinase. The rest of the antibody is human IgG, which has a conserved Fc portion. The proposed mechanisms by which trastuzumab may inhibit signal transduction involve antagonism of HER-2 receptor dimerization, activation of immune response, suppression of the shedding of the extracellular domain and/or enhanced endocytic breakdown of the receptor. HER-1, HER-2, HER-3 and HER-4 (also called epidermal growth factor receptors (EGFRs) Erb B-1, Erb B-2, Erb B-3 and Erb B-4, respectively) are transmembrane tyrosine kinase receptors, which are partially homologous. These receptors regulate various biological functions including cell growth, differentiation, migration, adhesion, survival and their responses. The cell proliferation via HER-2 signaling is mediated through the RAS–MAPK pathway, and cell death is inhibited via the phosphatidylinositol 3-kinase–AKT–mTOR pathway. In 20–30% of women with invasive breast carcinomas, there is a genetic alteration, which results in the overexpression of the HER-2 gene. This causes an increased amount of the growth factor receptor protein on the tumor cell surface and also results in elevated levels of circulating shed fragments of its extracellular domain. The HER-2 overexpression is associated with more aggressive metastatic breast cancer. Trastuzumab inhibits the proliferation of tumor cells overexpressing HER-2. It causes lysis via antibody-dependent cell-mediated cytotoxicity.

Trastuzumab is indicated as a single agent for the treatment of patients whose breast cancer has an overexpression of the HER-2 protein and has been previously treated by standard chemotherapeutic regimen. Trastuzumab is administered in combination with paclitaxel in patients who have not been previously treated with chemotherapy and their tumors overexpress HER-2. The initial dose for trastuzumab is 4 mg/kg given as a 90-min infusion, and the weekly maintenance dose is 2 mg/kg administered as a 30-min infusion. Its half-life is 1.7 and 12 days when a dose of 10 and 500 mg is administered, and the volume of distribution is 44 ml/kg, which is serum volume on average.

Trastuzumab causes the development of ventricular dysfunction and congestive heart failure. Patients treated with trastuzumab have exhibited symptoms of cardiac dysfunction including dyspnea, increased cough, peripheral edema and reduced ejection fraction. Congestive heart failure could also be severe. Left ventricular function must be evaluated in all patients before and during trastuzumab therapy. Patients receiving trastuzumab in combination with anthracyclin and cyclophosphamide exhibit more severe cardiomyopathy. Other side effects associated with trastuzumab when administered in combination with chemotherapeutic agents include anemia and leukopenia and increased risk of upper respiratory infections, but when given alone, it does not produce frequent hematological toxicity. Trastuzumab may also cause severe hypersensitivity reactions including anaphylaxis and pulmonary events. Infusion-associated symptoms including chills and fevers have been reported in about 40% of patients during the first infusion of trastuzumab.

Alemtuzumab (Campath)

Alemtuzumab is a CDR-grafted (humanized) monoclonal antibody (IgG1κ) produced by recombinant DNA technology. It binds to CD52, which is a 21–28-kDa cell surface glycoprotein. This receptor is present on the cell surface of normal and malignant T and B cells, NK cells and monocytes/macrophages. It mediates its effects via antibody-dependent lysis of target cells after binding to CD52 receptors.

Alemtuzumab is prescribed for the treatment of B-cell chronic lymphocytic leukemia in patients who did not respond to the standard chemotherapeutic regimen. It has also been used for refractory celiac disease. Alemtuzumab is administered daily by a 2-h IV infusion with an initial dose of 3 mg, and the dose is then increased to 10 mg in patients who do not encounter serious side effects. If tolerating a dose of 10 mg/day is not a problem, a maintenance dose of 30 mg/day is administered on alternate days (three times/week), which is generally initiated in most patients in 3–7 days. The average half-life of the drug over the dosing interval is about 12 days, and a steady-state level is achieved in about 6 weeks after initiation of the 30-mg dose. The side effects associated with alemtuzumab include immunosuppression/opportunistic infections, myelosuppression and infusion-related adverse events, resulting in fever, fatigue, pain, anorexia and edema. It is not administered to patients with active systemic infection, AIDS or patients hypersensitive to alemtuzumab.

Gemtuzumab–Ozogamicin (Mylotarg)

Gemtuzumab–ozogamicin is a conjugated CDR-grafted (humanized) monoclonal antibody (IgG4κ) that is conjugated to calicheamicin, a cytotoxic chemotherapeutic agent. The antibody is directed against CD33 antigen that is expressed on the cell surface of leukemic blast cells and normal cells of myelomonocytic lineage, but is not found on normal hematopoietic stem cells. Gemtuzumab–ozogamicin is indicated for the treatment of CD33-positive acute myeloid leukemia (AML). More than 80% of patients with AML express CD33 antigen on the cell surface of the leukemic blast cells. Gemtuzumab binds to CD33 antigen via its antibody fragment, resulting in a complex that is subsequently internalized. The internalization causes the release of calicheamicin derivative inside the lysosomes where the cytotoxic agent then binds to DNA in the minor groove and causes double strand breaks in the DNA and cell death.

Gemtuzumab–ozogamicin is administered as a dose of 9 mg/m^2 and is infused over a 2-h period; prior to its infusion, reduction of the leukocyte count in patients to below 30,000/ml is recommended. A second dose is administered after a 14-day interval. The elimination half-life of total calicheamicin is about 41 h after the first dose, 64 h after the second dose, and in its unconjugated form it is 143 h. The most common side effects associated with gemtuzumab–ozogamicin are fever and chills.

Other adverse effects resulting from the administration of gemtuzumab–ozogamicin include myelosuppression, thrombocytopenia, neutropenia, increased risk of infection, abdominal pain, vomiting and headache.

Ibritumomab (Zevalin)

Ibritumomab, a murine monoclonal antibody (IgG1κ), is first conjugated to a linker–chelator tiuxetan. Tiuxetan is responsible for a high-affinity and conformationally restricted chelation to one of the two radiation sources indium-111 or yttrium-90. The antibody is directed against CD20 antigen that is not shed from the cell surface and is not internalized after binding to the antibody. The antibody for treatment is supplied in two vials, one containing antibody conjugated to indium-111 and the other to yttrium-90. After binding to cells expressing CD20 antigen, ibritumomab delivers radiation, which enhances the killing effect of the antibody.

Ibritumomab along with radioisotopes, which constitutes its therapeutic regimen, is indicated for the treatment of relapsed or refractory low-grade follicular, or transformed B-cell NHL. It is also used in patients with rituximab refractory follicular NHL. Ibritumomab therapy involves two steps. The patient first receives a single infusion of rituximab ($250\,mg/m^2$), which is administered initially to clear the majority of B cells and to limit the toxicity from radiation. The immunoconjugate is linked to indium-111 for the first transfusion, $5.0\,mCi$ of which is administered intravenously. Seven to nine days after the first infusion, a second infusion of rituximab ($250\,mg/m^2$) is given followed by the administration of the immunoconjugate linked to yttrium-90 ($0.4\,mCi/kg$). Ibritumomab induces apoptosis in cells expressing CD20 antigen and the beta emission from yttrium-90 causes damage to the cells by the formation of free radicals. Since CD20 antigen is not expressed on B-cell precursors, the B cells recover usually in about 9 months after treatment with ibritumomab regimen. The side effects associated with ibritumomab therapeutic regimen include anemia, thrombocytopenia, neutropenia, increased risk of infections predominantly bacterial in nature, hemorrhage, allergic reactions including the bronchospasm and angioedema, GI discomfort and increased cough. There is also some risk (2%) of secondary malignancies after treatment with ibritumomab regimen.

Tositumomab (Bexxar)

The tositumomab therapeutic regimen is also composed of a monoclonal antibody and radioisotope iodine-131. Tositumomab is a murine monoclonal antibody (IgG2aλ) specific for the CD20 antigen. The covalent linkage of iodine-131 to tositumomab is used to produce iodine-I131 tositumomab. The therapeutic regimen of tositumomab causes lysis of the target cells by various mechanisms including apoptosis, ADCC and complement-dependent cytotoxicity. The cell death also results from the ionizing radiation of iodine-131.

This therapeutic regimen is indicated for the treatment of relapsed or refractory, low-grade, follicular or transformed NHL, all expressing CD20 antigen. It is not an initial treatment of choice but is used in patients who have not successfully responded to standard chemotherapeutic regimen or the combination of chemotherapy and rituximab. Tositumomab therapeutic regimen is used as a single course of treatment and is not used in combination with other chemotherapeutic or radiation treatments. It is administered to patients in two steps, termed dosimetric and therapeutic, and each step has two components. During the first component of the dosimetric step, intravenous administration of 450 mg of tositumomab is followed by the intravenous administration of the second component (iodine-131 tositumomab at a dose of 5.0 mCi iodine-131 and 35 mg tositumomab). The therapeutic step is then followed 7–14 days later, which is composed of an initial administration of 450 mg of tositumomab followed by the intravenous infusion of iodine-131 tositumomab. The median half-life of tositumomab is 67 h, ranging from 28 to 115 h. Iodine-131 is eliminated by decay and excreted in urine. Sixty-seven percent of the injected dose is cleared in 5 days with 98% of the clearance in the urine.

The adverse effects associated with tositumomab therapeutic regimen include anemia, thrombocytopenia, neutropenia, infections, hemorrhage and allergic reactions. Increased risk of secondary neoplasia and myelodysplasia has also been reported with this regimen. Other side effects produced by tositumomab therapy include pneumonia, pleural effusion and dehydration, GI discomfort and infusional toxicity. Delayed adverse reactions include hypothyroidism and HAMA.

Cetuximab (Erbitux)

Cetuximab is a chimeric monoclonal antibody produced by hybridoma technology and is composed of murine Fv region and human IgG1 (heavy and kappa light chain constant regions) molecule. The murine region is specific for human epidermal growth factor receptor (EGFR, HER-1, c-ERB-1). EGFR belong to the tyrosine kinase type 1 receptor subfamily, which also includes HER-2, HER-3 and HER-4. EGFR a transmembrane glycoprotein, is distributed on epithelial tissue as well as on tumors of the colon, rectum, head and neck.

Cetuximab is an antagonist of EGF and its other ligands including transforming growth factor-α and causes the inhibition of EGF-mediated signal transduction events including the phosphorylation and activation of various kinases associated with EGFR. The inhibition of the binding of EGF to its receptors suppresses cell growth and decreases the production of matrix metalloproteinase and vascular EGF (VEGF). Cetuximab after binding to EGFR induces apoptosis.

Cetuximab is indicated for the treatment of head, neck and colorectal cancers, and in combination with radiation therapy it is used for the treatment of squamous cell carcinoma (locally or regionally advanced) of the head and neck. However, in patients with recurrent or metastatic squamous cell carcinoma of the head and neck whose response to the previous platinum-based therapy has not been positive, it is not given in combination with radiation therapy and is administered alone. Generally, a dose of 400 mg/m^2 is initially administered in combination with radiation therapy followed by a maintenance dose of 250 mg/m^2 and is administered 1 h before the radiation therapy. Similar doses are used when it is employed as a single agent.

Cetuximab is used for treatment of metastatic colorectal carcinoma, which expresses EGFR, in combination with irinotecan, provided that these patients are refractory to irinotecan-based standard chemotherapeutic regimen. However, in patients who could not tolerate irinotecan-based chemotherapeutic regimen, it is used as a single treatment for EGFR-expressing metastatic colorectal cancer. The doses are similar to the doses used for squamous cell carcinoma of the head and neck.

Whether administered in combination or as a single therapy for cancers of the head, neck and colon, cetuximab exhibits similar pharmacokinetic characteristics. After a 2-h infusion of 400 mg/m^2, the half-life is 97 h, ranging from 41 to 213 h, and after initial and subsequent maintenance doses, the half-life is about 112 h, ranging from 63 to 230 h. The adverse effects associated with cetuximab include immunogenicity, electrolyte depletion (hypomagnesemia) and infusion reactions. Infusion reactions involve airway obstruction, urticaria and hypotension.

Bevacizumab (Avastin)

Bevacizumab is a recombinant CDR-grafted (humanized) monoclonal antibody (IgG1) that is directed against human VEGF. Its antibody-binding region is murine-specific for VEGF. VEGF promotes the proliferation of endothelial cells and the formation of new blood vessels. The receptors (Flt-1 and KDR) for VEGF are distributed on the endothelial cells. Bevacizumab binds to VEGF and inhibits its binding to VEGF receptors on endothelial cells.

In combination with a standard chemotherapeutic regimen (5-fluouracil-based), bevacizumab is used for the treatment of patients with malignant colon or rectal cancer, where it is indicated for the first- or second-line treatment. Bevacizumab in combination with bolus IFL is administered at a dose of 5 mg/kg and 10 mg/kg in combination with FOLFOX4. Bevacizumab is administered as an intravenous infusion. Its average half-life is approximately 20 days (ranging from 11 to 50 days) and the steady state is reached in 100 days.

The serious adverse effects associated with bevacizumab include GI perforation, hemorrhage, hypertension, complications in wound healing, nephritic syndrome, congestive heart failure and arterial thromboembolic events. Patients receiving bevacizumab commonly experience pain, asthenia, headache, abdominal pain, nausea, vomiting, anorexia, upper respiratory infection and exfoliative dermatitis.

Bibliography

Albanell J, Codony J, Rovira A, Mellado B, et al. 2003. Mechanism of action of anti-HER2 monoclonal antibodies: Scientific update on trastuzumab and 2C4. Adv Exp Med Biol. 532: 253–268.

Baca M, Presta LG, O'Connor SJ, Wells JA. 1997. Antibody humanization using monovalent phage display. J Biol Chem. 272:10678–10684.

Bartelds GM, Wijbrandts CA, Nurmohamed MT, Stapel S, et al. 2007. Clinical response to adalimumab: Relationship to anti-adalimumab antibodies and serum adalimumab concentrations in rheumatoid arthritis. Ann Rheum Dis. 66:921–926.

Bennett JM, Kaminski MS, Leonard JP, Vase JM, et al. 2005. Assessment of treatment-related myelodysplastic syndromes and acute myeloid leukemia in patients with non-Hodgkin lymphoma treated with tositumomab and Iodine I 131 tositumomab. Blood. 1052: 4576–4582.

Bielekova B, Richert N, Howard T, Blevins G, et al. 2004. Humanized anti-CD25 (daclizumab) inhibits disease activity in multiple sclerosis patients failing to respond to interferon beta. Proc Natl Acad Sci. 101:8705–8708.

Binder M, Otto F, Mertelsmann R, Veelken H, et al. 2006. The epitope recognized by rituximab. Blood. 108:1975–1978.

Bonner JA, Harari PM, Giralt J, Azarmia N, et al. 2006. Radiotherapy plus centuximab for squamous-cell carcinoma of the head and neck. NEJM. 354:567–578.

Bordigoni P, Dimicoli S, Clement L, Baumann C. 2006. Daclizumab, an efficient treatment for steroid-refractory acute graft-versus-host disease. Br J Hematol. 135:382–385.

Borro JM, De-la-Torre M, Miguelez C, Fernandez R, et al. 2005. Comparative study of basiliximab treatment in lung transplantation. Transplant Proc. 37:3996–3998.

Boulianne GL, Hozumi N, Shulman MJ. 1984. Production of functional chimeric mouse/human antibody. Nature. 312:643–646.

Boyne J, Elter T, Engert A. 2003. An overview of the current clinical use of the anti-CD20 monoclonal antibody rituximab. 14:520–535.

Braendstrup P, Bjerrum OW, Nielsen OJ, Jensen BA, et al. 2005. Rituximab chimeric anti-CD20 monoclonal antibody treatment for adult refractory idiopathic thrombocytopenic purpura. Am J Hematol. 78:275–280.

Brennan DC, Daller JA, Lake KD, Cibrik D. 2006. Rabbit antithymocyte globulin versus basiliximab in renal transplantation. NEJM. 355:1967–1977.

Bruggemann M, Witner G, Waldmann H, Neuberger MS. 1989. The immunogenicity of chimeric antibodies. J Exp Med. 170:2153–2157.

Burton C, Kaczmarski R, Jan-Mohamed R. 2003. Interstitial pneumonitis related to rituximab therapy. NEJM. 348:2690–2691.

Carter P, Presta L, Gorman CM, Ridgway JB, et al. 1992. Humanization of anti-p185HER2 antibody for human cancer therapy. Proc Natl Acad Sci USA. 89:4285–4289.

Carter PJ. 2006. Potent antibody therapeutics by design. Nat Rev Immunol. 6:343–357.

Cartron G, Watier H, Golay J, Solal-Celigny P. 2004. From bench to the bedside: Ways to improve rituximab efficacy. Blood. 104:2635–2642.

Chothia C, Lesk AM, Tramontano A, Levitt M, et al. 1989. Conformations of immunoglobulin hypervariable regions. Nature. 342:877–883.

Choy EH, Pnaayi GS. 2001. Cytokine pathways and joint inflammation in rheumatoid arthritis. NEJM. 344:907–916.

Clark M. 2000. Antibody humanization: A case of the "emperor's new clothes?" Immunol. Today. 21:397–402.

Coiffier B, Lepage E, Briere J, Herbrecht R, et al. 2002. CHOP chemotherapy plus rituximab compared with CHOP alone in elderly patients with diffuse large-B-cell lymphoma. NEJM. 346:235–242.

Connors JM. 2005. Radioimmunotherapy – hot new treatment for lymphoma. NEJM. 352: 496–498.

Cosimi AB, Burton RD, Colvin RB, Goldstein G. 1981. Treatment of cute renal allograft rejection with OKT3 monoclonal antibody. Transplantation. 32:535–539.

Cunningham D, Humblet Y, ,Siena S, Khayat D, et al. 2004. Centuximab monotherapy and centuximab plus irinotecan in irinotecan-refractory metastatic colorectal cancer. NEJM. 351: 337–345.

Davies AJ, Rohatiner AZ, Howell S, Britton KE, et al. 2004. Tositumomab and Iodine I 131 tositumomab for recurrent indolent and transformed B cell non-Hodgkin's lymphoma. J Clin Oncol. 22:1469–1479.

Davis TA, White CA, Grillo-Lopez AJ, Velasquez WS, et al. 1999. Single-agent monoclonal antibody efficacy in bulky non-Hodgkin's lymphoma: Results of a phase II trial of Rituximab. J Clin Onc. 17:1851–1857.

Dedrick RL, Walicke P, Garovoy M. 2002. Anti-adhesion antibodies efalizumab, a humanized anti-CD11a monoclonal antibody. Transplant Immunol. 9:181–186.

Edwards J, Szczepanski L, Szechinski J, Filipowicz-Sosnowaska A, et al. 2004. Efficacy of B-cell-targeted therapy with rituximab in patients with rheumatoid arthritis. NEJM. 350:2572–2581.

Ettenger RB, Marik J, Rosenthal JT, Fine RN, et al. 1988. OKT3 for rejection reversal in pediatric renal transplantation. Clin Transplant. 2:180–184.

Foote J, Winter G. 1992. Antibody framework residues affecting the conformation of the hypervariable loops. J Mol biol. 224:487–499.

Francis RJ, Sharma SK, Springer C, Green AJ, et al. 2002. A phase I trial of antibody directed enzyme prodrug therapy (ADEPT) in patients with advanced colorectal carcinoma or other CEA producing tumours. Br J Cancer. 87:600–607.

Gaston RS, Deierhoi MH, Patterson T, Prasthofer E, et al. 1991. OKT3 first-dose reaction: Association with T cell subsets and cytokine release. Kid Internat. 39:141–148.

Goldberg RM. 2006. Therapy for metastatic colorectal cancer. The Oncologist. 11:981–987.

Goldman M, Abramowicz D, DePauw L, Alelgre M, et al. 1989. OKT3-induced cytokine released attenuation by high-dose methylprednisolone. Lancet. 2:802–803.

Goldstein G, Fuccello AJ, Norman DJ, Sheild CF, et al. 1986. OKT3 monoclonal antibody plasma levels during therapy and the subsequent development of host antibodies to OKT3. Transplantation. 42:507–511.

Goncalves LF, Ribeiro AR, Berdichevski R, Joelsons G. 2007. Basiliximab improves graft survival in renal transplant recipients with delayed graft function. Transplant Proc. 39:437–438.

Graves JE, Nunley K, Heffernan MP. 2007. Off-label uses of biologics in dermatology: Rituximab, omalizumab, infliximab, etanercept, adalimumab, efalizumab and alefacept. J Am Acad Dermatol. 56:e55–e79.

Griggen JG. 2007. How I treat indolent lymphoma. Blood. 109:4617–4626.

Hale G, Dyer MJS, Clark MR, Phillips JM, et al. 1988. Remission induction in non-Hodgkin lymphoma with reshaped human monoclonal antibody CAMPATH-1 H. Lancet. 2:1394–1399.

Hericourt J, Richet CH. 1895. de la serotherapie dans la traitement du cancer, Physologie Pathologique. 120:567–569.

Hershberger RE, Randall SC, Eisen HJ, Bergh C-H, et al. 2005. Daclizumab to prevent rejection after cardiac transplantation. NEJM. 352:2705–2713.

Himmelweit F. 1960. The Collected Papers of Paul Erlich, Vol. 3,59, Pergmon, London.

Hoogenboom HR. 2005. Selecting and screening recombinant antibody libraries. Nat Biotech. 23:1117–1125.

Hooks MA, Wade CS, Millikan WJ. 1991. Muromonab CD-3: A review of its pharmacology, pharmacokinetics, and clinical use in transplantation. Pharmacotherapy. 11:26–37.

Hortobagyi GN. 2005. Trastuzumab in the treatment of breast cancer. NEJM. 353:1734–1736.

Hosenpud JD. 2005. Immunosuppression in cardiac transplantation. NEJM. 352:2749–2750.

Hudis CA. 2007. Trastuzumab-mechanism of action and use in clinical practice. NEJM. 357: 39–51.

Hudson PJ, Souriau C. 1992. Engineered antibodies. Nat Med. 9:129–134.

Hurwitz H, Fehrenbacher L, Novotny W, Cartwright T, et al. 2004. Bevacizumab plus irinotecan, fluorouracil, and leucovorin for metastatic colorectal cancer. NEJM. 350:2335–2342.

Janeway C, Travers P, Walport M, Schlomchik M. 2001. Immunobiology, Fifth Edition, Garland Science, New York and London.

Jazirehi AR, Bonavida B. 2005. Cellular and molecular signal transduction pathways modulated by rituximab (Rituxan, anti-CD20 mab) in non-Hodgkin's lymphoma: Implications in characterization of therapeutic intervention. Oncogene. 24:2121–2143.

Jensen M, Klehr M, Vogel A, Schmitz S, et al. 2007. One step generation of fully chimeric antibodies using C gamma 1 and C kappa mutant mice. J Immunother. 30:338–349.

Jones PT, Dear PH, Foote J, Neuberger MS, et al. 1986. Replacing the complementarity-determining regions in a human antibody with those from a mouse. Nature. 321:522–525.

Kaminski MS, Tuck M, Estes J, Kolstad A, et al. 2005. ^{131}I-Tositumomab therapy as initial treatment for follicular lymphoma. NEJM. 352:441–449.

Kapic E, Becic F, Kusturica J. 2004. Basiliximab, mechanism of action and pharmacological properties. Med Arch. 58:373–376.

King DJ, Adair JR, Angal S, Low DC, et al. 1992. Expression, purification and characterization of a mouse-human chimeric antibody and chimeric Fab fragment. Biochem J. 281:317–323.

Knight DM, Trinh H, Le J, Siegel S, et al. 1993. Construction and initial characterization of a mouse-human chimeric anti-TNF antibody. Mol Immunol. 30:1443–1453.

Kohler G, Milstein C. 1975. Continuous culture of fused cells secreting antibody of predefined specificity. Nature. 256:495–497.

Krauss WC, park JW, Kirpotin DB, Hong K, Benz CC. 2002. Emerging antibody-based HER2 (ErbB-2/neu) therapeutics. Breast Dis. 11:113–124.

Leahy MF, Seymour JF, Hicks RJ, Turner JH. 2006. Multicenter phase II clinical study of Iodine-131-Rituximab radiotherapy in relapsed or refractory indolent non-Hodgkin's lymphoma. J Clin Oncol. 24:4418–4425.

Lind P, Lechner P, Hausmann B. 1991. Development of human anti-mouse antibodies (HAMA) after single and repeated diagnostic application of intact murine monoclonal antibodies. Antibody Immunoconjug Radiopharm. 4:811–818.

Littlejohns P. 2006. Trastuzumab for early breast cancer: Evolution or revolution? Lancet Oncol. 7:22–23.

Lo-Coco F, Cimino G, Breccia M, Noguera NI, et al. 2004. Gemtuzumab ozogamicin (Mylotarg) as a single agent for molecularly relapsed acute promyelocytic leukemia. Blood. 104: 1995–1999.

Mascelli MA, Zhou H, Sweet R, Getsy J, et al. 2007. Molecular, biologic and pharmacokinetic properties of monoclonal antibodies: Impact of these parameters on early clinical development. J Clin Pharmacol. 47:553–565.

McArthur HL, Chia S. 2007. Cardiotoxicity of trastuzumab in clinical practice. NEJM. 357:94–95.

McCune SL, Gockerman JP, Rizzieri DA. 2001. Monoclonal antibody therapy in the treatment of non-Hodgkin's lymphoma. JAMA. 286:1149–1152.

McLaughlin P, Grillo-Lopez AJ, Link BK, Levy R, et al. 1998. Rituximab chimeric anti-CD20 monoclonal antibody therapy for relapsed indolent lymphoma: Half of patients respond to a four-dose treatment program. J Clin Oncol. 16:2825–2833.

Menter A, Leonardi CL, Sterry W, Bos JD. 2006. Long-term management of plaque psoriasis with continuous efalizumab therapy. J Am Acad Dermatol. 54:S182–S188.

Morris JA, Hanson JE, Steffen BJ, Chu AH, et al. 2005. Daclizumab is associated with decreased rejection and improved patient survival in renal transplant patients. Clin Transplant. 19: 340–345.

Morrison SL, Johnson MJ, Herzenberg LA, Oi VT. 1984. Chimeric human antibody molecules: Mouse antigen binding domains with human constant region domains. Proc Natl Acad Sci USA. 81:6851–6855.

Nadler LM, Stashenko P, Hardy R, Kaplan WB, et al. 1980. Serotherapy of a patient with a monoclonal antibody directed against a human lymphoma associated antigen. Cancer Res. 40:3147–3154.

O'Dell JR. 2004. Therapeutic strategies for rheumatoid arthritis. NEJM. 350:2591–2602.

Okamoto H. 2006. Effect of rituximab in refractory SLE: Inhibition of Th1. Rheumatology. 45:121–122.

Olsen NJ, Stein CM 2004. New drugs for rheumatoid arthritis. NEJM. 350:2167–2179.

Ortho Multicenter Transplant Study Group. 1985. A randomized clinical trial of OKT3 monoclonal antibody for acute rejection of cadaveric renal transplants. NEJM. 313:337–342.

Pegram M, Hsu S, Lewis G, Pietras R. 1999. Inhibitory effects of combination of HER-2/neu antibody and chemotherapeutic agents used for treatment of human breast cancer. Oncogene. 18:2241–2251.

Pendley C, Schantz A, Wagner C. 2003. Immunogenicity of therapeutic monoclonal antibodies. Curr Opin Mol Ther. 5:172–179.

Penichet ML, Morrison SL. 2004. Designing and engineering human forms of monoclonal antibodies. Drug Deve Res. 61:121–136.

Piccart-Gebhart MJ, Procter M, Leyland-Jones B, Goldhirsch A, et al. 2005. Trastuzumab after adjuvant chemotherapy in HER2-positive breast cancer. NEJM. 353:1659–1672.

Piccoli R, Olson KA, Vallee BL, Fett JW. 1998. Chimeric anti-angiogenin antibody cAb 26-2F inhibits the formation of human breast cancer xenograft in athymic mice. PNAS USA. 95:4579–4583.

Posner MR, Wirth LJ. 2006. Centuximab and radiotherapy for head and neck cancer. NEJM. 354:634–636.

Present D, Rutgeerts P, Targan S, Hanauer S, et al. 1999. Infliximab for the treatment of fistulas in patients with Crohn's disease. NEJM. 340:1398–1405.

Presta LG, Lahr SJ, Shields RI, Porter JP, et al. 1993. Humanization of an antibody directed against IgE J Immunol. 151:2623–2632.

Przepiorka D, Kernan NA, Ipppoliti C, Papandopoulas EB, et al. 2000. Daclizumab, a humanized anti-interleukin-2 receptor alpha chain antibody, for the treatment of acute graft-versus-host disease Blood. 2000:83–89.

Queen C, Schneider WP, Selick HE, Payne PW, et al. 1989. A humanized antibody that binds to the interleukin 2 receptors. PNAS USA. 86:10029–10033.

Rai KR, Freter CD, Mercier RJ, Cooper MR. 2002. Alemtuzumab in previously treated chronic lymphocytic leukemia patients who also had received Fludarabine. J Clin Oncol. 20:3891–3897.

Ramirez CB, Marino IR. 2007. The role of basiliximab induction therapy in organ transplantation. Expert Opin Biol Ther. 7:137–148.

Rawstron AC, Kennedy B, Moreton P, Dickinson AJ, et al. 2004. Early prediction of outcome and response to alemtuzumab therapy in chronic lymphocytic leukemia. Blood. 103:2027–2031.

Riechmann L, Clark M, Waldmann H, Winter G. 1988. Reshaping human antibodies for therapy. Nature. 332:323–327.

Ritz J, Schlossman SF. 1982. Utilization of monoclonal antibodies in the treatment of leukemia and lymphoma. Blood. 59:1–11.

Roskos LK, Davis CG, Schwag GM. 2004. The clinical pharmacology of therapeutic monoclonal antibodies. Drug Dev Res. 61:108–120.

Rutgeerts P, Sanborn WJ, Feagan BG, Reinisch W, et al. 2005. Infliximab for induction and maintenance therapy for ulcerative colitis. NEJM. 353:2462–2476.

St. Clair EW. 2002. Infliximab treatment for rheumatic disease. Clinical and radiological efficacy. Ann Rehum Dis. 61:ii67–ii69.

Sands B, Anderson F, Bernstein C, Chey W, et al. 2004. Infliximab maintenance therapy for fistulizing Crohn's disease. NEJM. 350:876–885.

Scheinberg DA, Chapman BA. 1995. In: Birch JR, Lennox ES, Eds. Monoclonal Antibodies: Principles and Applications, Wiley-Liss, New York, pp. 45–105.

Schmidt-Hieber M, Fietz T, Knauf W, Uharek L. 2005. Efficacy of the interleukin-2 receptor antagonist basiliximab in steroid refractory graft-versus-host disease. Br J Hemtaol. 130:568–574.

Schmitz U, Versmold A, Kaufmann P, Frank HG. 2000. Phage display: A molecular tool for the generation of antibodies – a review. Placenta. 21:S106–S112.

Schrag D. 2004. The price tag on progress – chemotherapy for colorectal cancer. NEJM. 351:317–319.

Schroeder TJ, Michael AT, First MR, Hariharan S, et al. 1994. Variations in serum OKT3 concentration based upon age, sex, transplanted organ, treatment regimen and anti-OKT3 status. Ther Drug Monit. 16:361–367.

Sfikakis PP. 2002. Behcet's disease: A new target for anti-tumour necrosis factor treatment. Ann Rheum Dis. 61:ii51–ii56.

Sharkey RM, GHoodenberg DM. 2006. Targeted therapy of cancer: New prospects for antibodies and immuno conjugates. CA Cancer J Clin. 56:226–243.

Shields CJ, Winter DC, Becker JM, Prushik SG, et al. 2006. Infliximab for ulcerative colitis. NEJM. 354:1424–1426.

Shimoni A, Zwas ST, Oksman Y, Hardan I, et al. 2007. Yttrium-90-ibritumomab tiuxetan (Zevalin) combined with high-dose BEAM chemotherapy and autologous stem cell transplantation for chemo-refractory aggressive non-Hodgkin's lymphoma. Exp Hematol. 35:534–540.

Smith I, Procter M, Gelber RD, Guillaume S, et al. 2007. 2-year follow-up of trastuzumab after adjuvant chemotherapy in HER2-positive breast cancer: A randomized controlled trial. Lancet. 369:29–36.

Song H, Du Y, Sgouros G, Prideaux A, et al. 2007. Therapeutic potential of ^{90}Y and ^{131}I-labeled anti-CD20 monoclonal antibody in treating non-Hodgkin's lymphoma with pulmonary involvement. JNM. 48:150–157.

Stern M, Herrmann R. 2005. Overview of monoclonal antibodies in cancer therapy: Present and promise. Crit Rev Oncol Hematol. 54:11–29.

Tan P, Mitchell DA, Buss TN, Holmes MA, et al. 2002. Super humanized antibodies: Reduction of immunogenic potential by complementarity-determining region grafting with human germline sequences: Application to an anti-CD28. J Immunol. 169:1119–1125.

Topol EJ, Moliterno DJ, Hermann HC, Powers ER, et al. 2001. Comparison of two platelet glyco-proteins IIb/IIIa inhibitors, Tirofiban and abciximab, for the prevention of ischemic events with precutaneous coronary revascularization. NEJM. 344:1888–1894.

Valabrega G, Montemurro F, Aglietta M. 2007. Trastuzumab: Mechanism of action, resistance and future perspectives in HER2-overexpressing breast cancer. Ann Oncol. 18:977–984.

Verbeek WHM, Mulder CJJ, Zweegman S, Vivas S, et al. 2006. Alemtuzumab for refractory celiac disease. NEJM. 355:1396–1397.

Verhoeyen M, Milstein C, Winter G. 1988. Reshaping human antibodies: Grafting an antilysozyme activity. Science. 239:1534–1536.

Vincenti F, Kirkman R, Light S, Bumgardner G, et al. 1998. Interleuin-2 receptor blockade with daclizumab to prevent acute rejection in renal transplantation. NEJM. 338:161–165.

Vivas S, Morales JM, Ramos F, Suarez-Vilela D. 2006. Alemtuzumab for refractory celiac disease in a patient at risk for enteropathy-associated T-cell lymphoma. NEJM. 354:23–24.

Waldmann T.A. 1991. Monoclonal antibodies in diagnosis and treatment. Science. 252:1657–1662.

Waldmann TA, Levy R, Coller BS. 2000. Emerging therapies: Spectrum of applications of mono-clonal antibody therapy. Hematology. 2000:394–408.

Waldmann TA. 2003. Immunotherapy: Past, present and future. Nat Med. 9:269–277.

Waldmann TA. 2007. Anti-Tac (daclizumab, Zenapax) in the treatment of leukemia, autoimmune diseases and the prevention of allograft rejection: A 25-year personal odyssey. J Clin Immunol. 27:1–18.

Wellington K, Perry CM. 2005. Efalizumab. Am J clin Dermatol. 6:113–118.

White CA. 2004. Radioimmunotherapy in non-Hodgkin's lymphoma: Focus on 90Y-ibritumonomab tiuxetan (Zevalin). J Exp Ther Oncol. 4:305–316.

Wiseman GA, Witzig, TE. 2005. Yttrium-90 (90Y) ibritumonomab tiuxetan (Zevalin) induces long-term durable responses in patients with relapsed or refractory B-cell non-Hodgkin's lymphoma. Cancer Biother Radiopharm. 20:185–188.

Chapter 6
Allergic Disease

LIVERPOOL
JOHN MOORES UNIVERSITY
AVRIL ROBARTS LRC
TITHEBARN STREET
LIVERPOOL L2 2ER
TEL. 0151 231 4022

Introduction

The term *allergy* (from Greek *allos*, meaning "other," and *ergon*, meaning "work") was introduced in 1906 by von Priquet. *Atopy* (from Greek, meaning "out of place") is also used to describe allergic disease. In 1919, a physician first reported allergy-related symptoms of transient asthma following blood transfusion. In the study, a response to horse dander suggested that an allergic reaction could be mediated by a factor in the blood. The passive transfer of a positive skin test performed by Prausnitz and Kustner in an experiment in 1921 led to a search to discover the substances responsible for allergic reaction. Until 1960 the focus was on several labile complexes in the serum rather than on one molecule, and it was assumed that antibodies were not involved in allergic reaction.

In 1965, Bennick and Johansson in Sweden reported a new class of immunoglobulin called IgND, and it was subsequently found that the levels of IgND were severalfold higher in allergic asthmatics than in normal subjects. A new immunoglobulin was also identified by Ishizaka's group in the United States in 1966–1967 when they found that an antiserum could suppress the allergic reaction; it was termed "γE-globulin" although they were unable to isolate it due to its very low levels in the serum. In 1967, it was determined that IgND and γE-globulin were the same molecules and were given a new name, IgE.

Hypersensitivity Disease

The human immune response protects the host by maintaining homeostasis. The immune system is activated as a result of exposure to antigen, allergen, infection and/or neoplasm, resulting in the alteration of homeostasis. In some instances, an immune response generated against a foreign invader is harmful to the host itself, which is referred to as hypersensitivity disease and has been divided into five subtypes:

1. *Type I Reactions*: These reactions result from the production of IgE; they are also termed immediate hypersensitivity disease. These are allergic reactions that

M.M. Khan, *Immunopharmacology*, DOI: 10.1007/978-0-387-77976-8_6,
© Springer Science+Business Media, LLC 2008

emanate from reexposure to allergens where their binding to IgE results in the degranulation and secretion of endogenous mediators including histamine, leukotrienes, prostaglandins and cytokines from sensitized mast cells (Fig. 6.1). These products cause vasodilation and smooth-muscle contraction. These reactions vary from local to systematic producing symptoms from mild allergic reaction to rare anaphylactic shock resulting in death. The clinical examples of type I reactions include allergic rhinitis, allergic asthma, atopic eczema, anaphylaxis and urticaria.

2. *Type II Reactions*: These reactions result from the recognition of cells to which antigens (extrinsic or intrinsic) are bound by macrophages or dendritic cells,

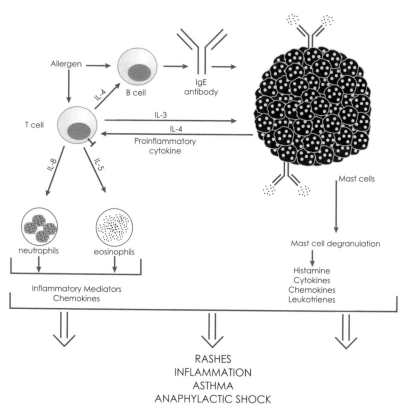

Fig. 6.1 Immediate hypersensitivity reaction. These reactions are the result of the production of IgE antibody in response to an allergen. IgE binds to the mast cells via Fc receptors and its reexposure to the allergen causes degranulation and secretion of endogenous mediators. In allergic responses, TH_2 cells are important in recognizing allergens in the context of MHC molecules and secrete IL-4, IL-5 and IL-13. IL-4 induces isotope switching from IgG to IgE, and IL-5 is involved in eosinophil recruitment. IL-8 serves as a chemical signal to attract neutrophils at the site of inflammation. The collective effects of endogenous mediators include rashes, inflammation, smooth-muscle contraction, bronchospasm, asthma and severe anaphylactic shock, which may even cause death (*see* Color Insert)

resulting in a B-cell response and the production of antibodies against foreign antigens. They also involve ADCC and complement plays a role in this process as well. The reactions result from the hemolytic disease of the newborn as well as adverse effects to some drugs, specifically those causing hemolytic anemia. Other clinical examples include transfusion reactions, autoimmune hemolytic anemia, immune thrombocytopenia, pernicious anemia and rheumatic fever.

3. *Type III Reactions*: These reactions involve the presence of antigen–antibody complexes, particularly those formed as a result of the production of autoantibodies. These complexes deposit in various tissues and involve inflammatory cells as well as complement, resulting in tissue damage due to the production of proteolytic enzymes by polymorphonuclear leukocytes and macrophages. A number of autoimmune diseases result from these reactions. Some clinical examples include systemic lupus erythematosus, rheumatoid arthritis, immune complex glomerulonephritis, Arthus reaction and serum sickness.

4. *Type IV Reactions*: Also termed delay-type hypersensitivity reaction, these take 48–72 h to develop and are not antibody-mediated. Antigens are recognized by $CD4^+$ and/or $CD8^+$ cells in the context of MHC class restrictions on APCs. These reactions are T-cell-mediated where activated T cells release cytokines, resulting in the development of granulomas from macrophages. These mechanisms are responsible for symptoms that may include transplant rejection, contact dermatitis, leprosy, tuberculosis and sarcoidosis.

5. *Type V Reactions*: This subtype is occasionally used as separate from type II reactions. In this case, antibodies bind to the cell surface receptors instead of cell surface components, resulting in the impairment of cell signal either via augmentation or suppression. Some clinical examples include myasthenia gravis and Graves' disease.

IgE-Mediated Responses

IgE, one of the five isotypes of antibodies made by humans, was designed as an antiparasitic immunoglobulin. Helminths result in a strong IgE response, which also includes the production of parasite-specific IgE antibody. The beneficial effects of IgEs include roles of early recognition of invading organisms and allergens and stimulation of immunity due to better antigen presentation. IgE could expel allergenic materials from the body since typical allergic reaction produces sneezing, coughing, mucus, tears, bronchoconstriction, inflammation and even vomiting. Nonetheless, the production of IgE instead of IgG in response to an antigen contributes to the etiology and pathogenesis of allergic disease. IgE binds to cells via high-affinity or low-affinity Fc receptors. Mast cells primarily express high- affinity receptors, whereas eosinophils, lymphocytes and various other cell types express low-affinity receptors. An allergic reaction has three fundamental features:

(1) Synthesis of IgE specific to an allergen;
(2) Initial binding of allergen-specific IgE to high-affinity Fc receptors on mast cells followed by second exposure to the same antigen, which causes cross-linking of pairs of IgE molecules, resulting in the activation and degranulation of mast cells, which produces the signs and symptoms of immediate hypersensitivity reaction; and
(3) Allergic inflammation—resulting from the recruitment and activation of eosinophils.

These processes depend on chemotactic activation of the effector cells and altered cellular migration to promote an allergic reaction. During these events, there is also an upregulation of the adhesion molecules, both on the effector cells and on high endothelial venules in blood vessels, lungs and other tissues.

In atopic/allergic subjects, IgE is produced following contact with low levels of environmental allergens. The immune response begins with sensitization after an allergen such as pollen, animal dander or house dust mite is inhaled, and the allergen is processed by APCs such as Langerhans cells. These cells, which are present in the epithelium linings of the lungs and nose airways uptake, process and present the processed antigen to the TCR in the context of HLA molecules. Activated T cells secrete cytokines that along with other cell–cell interactions transform B cells to plasma cells, which are the antibody-secreting cells. In allergic disease, plasma cells produce IgE instead of IgG, which could bind to its specific allergen via its Fab region. The IgE binds to the Fc receptors on the cells, and this phase of immune response is called the sensitization phase.

After reexposure to the same allergen, a robust and rapid memory response is initiated. Signal transduction emanating from the cross-linking of IgE antibodies bound to mast cells by allergen results in the degranulation and release of mast cell mediators. Mast cells also regulate the expression of IgE receptors in order to maintain a given number of unoccupied receptors. It is feasible that this phenomenon is regulated by the levels of circulating IgE.

We are all (atopic or nonatopic) exposed to aeroallergens, which include pollens, house dust mites and dander. The nonatopic individuals mount a weak immune response to the allergens by producing antigen-specific IgG1 and IgG4 antibodies, whereas the atopic individuals produce allergen-specific IgE antibodies. T cells from atopic individuals respond to the aeroallergens against which they were generated by elaborating TH_2 cytokines including IL-4, IL-5 and IL-13. The infiltration of TH_2 cells into the affected tissues is the hallmark of the allergic disease and asthma.

The common environmental allergens that cross the placenta prime the T cells of the fetus and all newborns have predominantly TH_2 cells. As the infant grows, this composition shifts to TH_1 cells and TH_1-mediated responses against aeroallergens. However, in atopic infants, there is further development of TH_2 cells and TH_2-mediated responses. Epidemiological studies on families, particularly on twins, have suggested that IgE production is due to genetic predisposition. This is based on their ability to produce more TH_2 cells and subsequently IL-4-dependent immune response to aeroallergens.

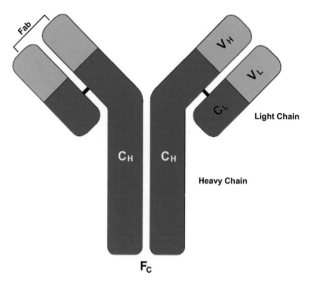

Fig. 1.1 The immunoglobulin molecule; each immunoglobulin molecule is composed of two identical heavy (CH + VH) and two identical light chains (CL + VL). The antigen binds to the Fab region, which varies according to the specificity of the antibody. The rest of the domains (blue) are constant. The classes of the antibody molecules differ based on the Fc region of the heavy chain

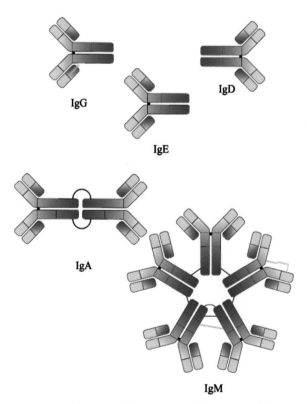

Fig. 1.2 Classes of immunoglobulins; this figure depicts five classes of immunoglobulins, IgG, IgE, IgD, IgA, IgM. IgA can be present as a monomer or a dimer molecule whereas IgM exists as a pentamer

LIVERPOOL JOHN MOORES UNIVERSITY
LEARNING & INFORMATION SERVICES

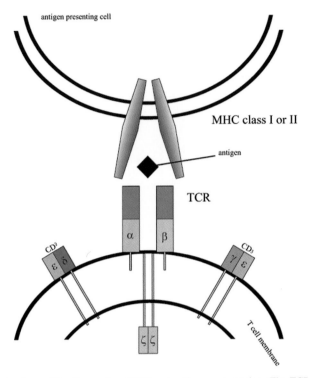

antigen presenting cell

MHC class I or II

antigen

TCR

CD3

CD3

α β

ε δ

γ ε

ζ ζ

T cell membrane

Fig. 1.3 The structure of T-cell receptor (TCR) and antigen presentation. The TCR (TCR-2) is a heterodimer composed of two transmembrane glycoprotein chains α and β, which are disulfide-linked polypeptides. TCR-1 is structurally similar but composed of γ and σ polypeptides. The TCR is also associated with the CD3 complex. The latter is composed of six polypeptide glycoprotein chains known as CD3γ, CD3σ, CD3ε and another protein known as ζ. When the TCR recognizes antigen, the CD3 complex is involved in signal transduction. The antigen is recognized by the TCR only in context with MHC molecules after it is taken up and processed by the antigen-presenting cells (APCs) [A part of this diagram is based on one published by Janeway and Travers, 1996]

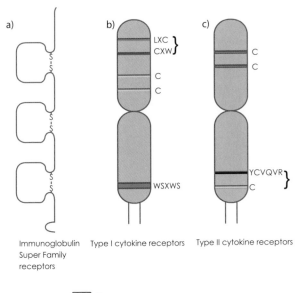

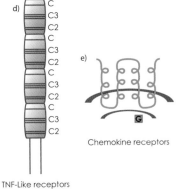

Fig. 3.1 Families of cytokine receptors; the cytokine receptors are classified into five major families: immunoglobulin superfamily receptors, type I cytokine receptors, type II cytokine receptors, TNF-like receptors and chemokine receptors. The drawings illustrate their general biochemical structure

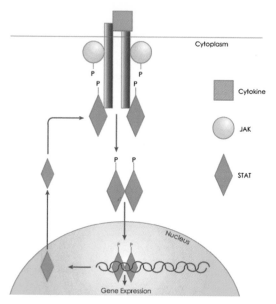

Fig. 3.2 The JAK/STAT signaling pathway; after cytokine binds to its receptors, the associated Janus kinase (JAK) is induced. This results in the phosphorylation of the receptor's cytoplasmic domain. STAT is recruited after the phosphorylation of the receptor's cytoplasmic domain, which after phosphorylation dimerizes and migrates into the nucleus. In the nucleus, STAT binds to its niche in the DNA and induces gene expression

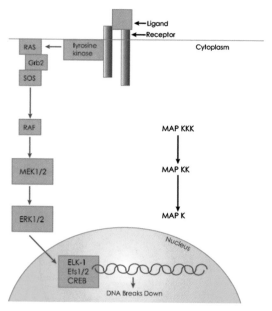

Fig. 3.3 The ERK/MAP kinase signaling pathway; cytokines and growth factors activate tyrosine kinase to which the adaptor protein Grb2 binds. This localizes SOS to plasma membrane. RAS is then activated by SOS. Activated RAS then binds to RAF, which forms a transient membrane-anchoring signal. Active RAF kinase phosphorylates MEK. The activated MEK phosphorylates ERK1/ERK2, which also migrates to the nucleus to phosphorylate ELK-1, Ets1/2 and CREB, resulting in the activation and expression of respective genes

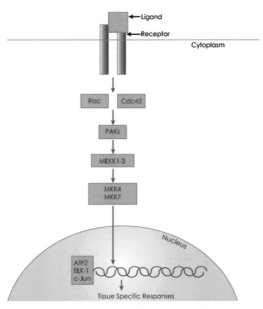

Fig. 3.4 The JNK/SAPK signaling pathway; various signals including cytokines activate the MAP kinases. The JNK/SAPK cascade is activated in response to inflammatory cytokines, heat shock or ultraviolet radiation. Two small G proteins, Rac and cdc42, mediate the activation of the MAP kinases. After activation, cdc42 binds to and activates PAK65 protein kinase. This results in the activation of MEKK, which eventually phosphorylates JNK/SAPK that migrates to the nucleus and activates the expression of several genes specifically the phosphorylation of c-Jun

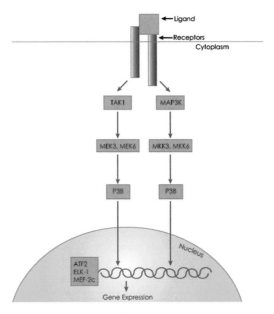

Fig. 3.5 The p38 kinase signaling pathway; inflammatory cytokines, osmotic stress and endotoxins activate this signaling pathway. The maximum activation of p38 requires two MAP2Ks, MKK3 and MKK6. Following activation, p38 translocates to the nucleus and phosphorylates ATF2, ELK1 and MEF-2C. p38 can also be activated independent of MAP2Ks by TAK1-binding protein via TAB1

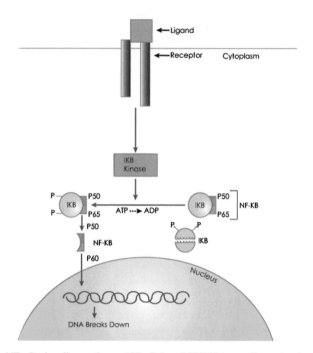

Fig. 3.6 The NF-κB signaling pathway. NF-κB is a P50/P65 heterodimer that is present in the cytoplasm in an inactive form as it is noncovalently associated with IκB. IκB is an inhibitor of NF-κB and prevents its migration and gene expression signal in the nucleus. Selected cytokines after binding to their specific receptors phosphorylate IκB resulting in its degradation, which removes the inactivity barrier from NF-κB. This is followed by the migration of NF-κB into the nucleus where it binds to specific genes and results in their induction and gene expression

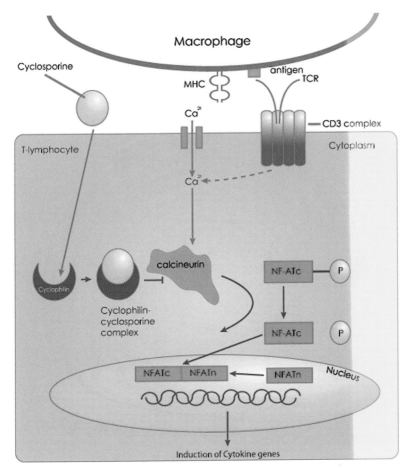

Fig. 4.1 Mechanism of action of cyclosporine. Cyclosporine readily diffuses into the cytoplasm of the target cells where it binds to cyclophilins. The cyclosporine–cyclophilin complex stably associates with calcineurin and inhibits calcineurin activity. Calcineurin is a Ca^{2+}-dependent enzyme— serine/threonine phosphatase— which after activation by Ca^{2+}, dephosphorylates a cytosolic component of NFAT (NFATc, cytosolic factor of activated T cells). After dephosphorylation, NFATc migrates from the cytoplasm to the nucleus where it associates with NFATn and induces transcription of several cytokine genes including IL-2. Cyclosporine inhibits calcineurin activity after associating with cyclophilins, resulting in the inhibition of IL-2 production and other cytokines

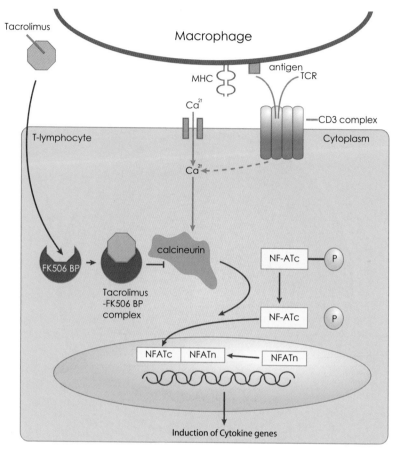

Fig. 4.2 Mechanism of action of tacrolimus. Tacrolimus readily diffuses into the cytoplasm of the target cell where it binds to immunophilins (FK506-BP). The tacrolimus–immunophilin complex stably associates with calcineurin and inhibits calcineurin activity. Calcineurin is a Ca^{2+}-dependent enzyme—serine/threonine phosphatase—which after activation by Ca^{2+}, dephosphorylates a cytosolic component of NFAT (NFATc, cytosolic factor of activated T cells). After dephosphorylation, NFATc migrates from the cytoplasm to the nucleus where it associates with NFATn and induces transcription of several cytokine genes including IL-2. Tacrolimus inhibits calcineurin activity after associating with immunophilins, resulting in the inhibition of IL-2 production and other cytokines

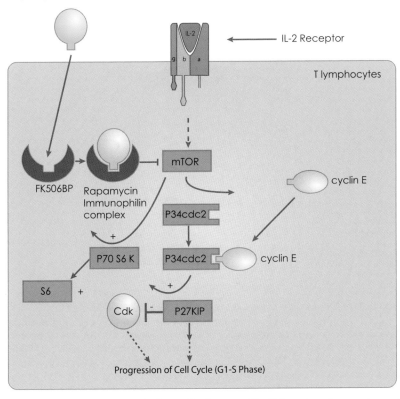

Fig. 4.3 Mechanism of action of sirolimus. Sirolimus readily diffuses into the cytoplasm of the target cells where it binds to immunophilins (FK506-BP). The sirolimus–immunophilin complex does not inhibit calcineurin activity; instead it binds to the mTOR. The sirolimus–immunophilin–mTOR complex stops the cell cycle progression from G1 to S phase. The targets of sirolimus include the eukaryotic initiation factor (eIF-4F), 70-kDa S6 protein kinase (p70S6 K) and several cyclin-dependent kinases (cdk). As a consequence, it blocks downstream signaling pathway initiated after activation of IL-2 receptors, resulting in blockage of T-cell proliferation

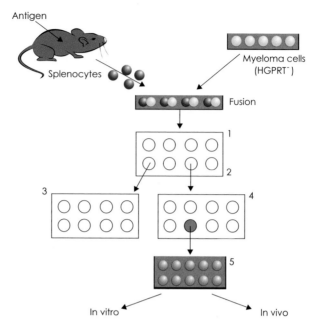

Fig. 5.1 Technique to produce monoclonal antibodies. After mice are injected with the antigen, the splenocytes are isolated and fused with myeloma cells. The fused cells are cultured (1) and the supernatants are assayed for the presence of desired antibodies (2). The cells from antibody-positive wells are cloned (3) and supernatants are tested for the desired antibodies (4) followed by expansion of the positive clones (5) and in vitro or in vivo propagation to produce the needed amount of monoclonal antibodies

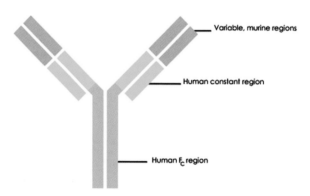

Fig. 5.2 Depiction of a chimeric monoclonal antibody

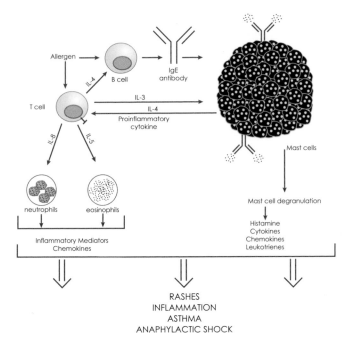

Fig. 6.1 Immediate hypersensitivity reaction. These reactions are the result of the production of IgE antibody in response to an allergen. IgE binds to the mast cells via Fc receptors and its reexposure to the allergen causes degranulation and secretion of endogenous mediators. In allergic responses, TH$_2$ cells are important in recognizing allergens in the context of MHC molecules and secrete IL-4, IL-5 and IL-13. IL-4 induces isotope switching from IgG to IgE, and IL-5 is involved in eosinophil recruitment. IL-8 serves as a chemical signal to attract neutrophils at the site of inflammation. The collective effects of endogenous mediators include rashes, inflammation, smooth-muscle contraction, bronchospasm, asthma and severe anaphylactic shock, which may even cause death

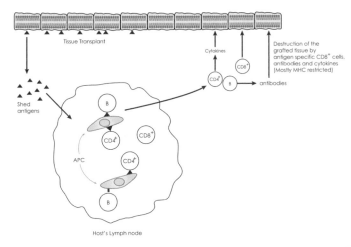

Fig. 7.1 Phases of tissue transplant rejection. The transplanted tissue sheds antigens. These antigens undergo uptake, processing and presentation to the T cells in the secondary lymphoid tissue by APCs, which include macrophages, B cells, Langerhans cells or dendritic cells. This phase results in the production of antibodies and antigen-specific TH and Tc cells. The antibodies and effector cells then migrate to the grafted tissue where TH cells secrete cytokines and which in combination with the antibodies and Tc cells destroy the grafted tissue

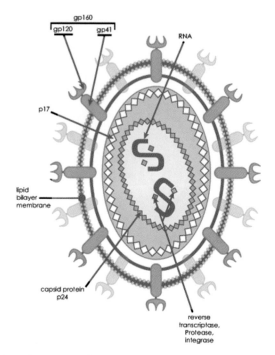

Fig. 8.1 HIV, a schematic representation

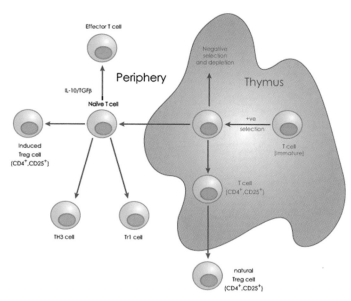

Fig. 9.1 Development of CD4$^+$ regulatory T cells. Natural CD4$^+$CD25$^+$ Treg cells develop in the thymus as a result of positive selection between TCR and host antigens. The thymus-derived Treg cells are specific for antigens seen in the thymus. The autoreactive T cells undergo negative selection and are depleted by apoptosis. The acquired Treg cells develop in the periphery from naïve precursors, and their specificity lies in antigens other than the ones that come in contact with the thymus. Tr1 cells are induced in the periphery when naïve T cells are exposed to an antigen in the presence of IL-10. They are not identified by a particular cell surface marker. TH3 cells are generated from naïve CD4$^+$ T cells as a result of low doses of antigen via the oral route, and they secrete TGF-β. (This diagram is based on one published by Lohret et al, 2006)

Regulation of IgE Synthesis

The IgE production between atopic and nonatopic subjects is critically different. Nonatopic individuals normally synthesize IgM and IgG and only small amounts of IgE, whereas atopic individuals produce high levels of IgE in response to aeroallergens. The synthesis of IgE is regulated by four interacting factors: heredity, the natural history of antigen exposure, the nature of the antigen and T helper cells, and their cytokines. It has been reported that atopy runs in families and atopic individuals have high levels of IgE. MHC class II molecules play a role in antigen presentation. In one example, different class II molecules may serve to present ragweed antigen. Atopic reaction develops as a result of repeated exposure to a particulate antigen. Consequently, a change in geographical location often benefits patients with allergic rhinitis or asthma. The influence of natural history of exposure to an antigen seen with bee sting is the most dramatic example. First exposure does not do anything but second exposure could be potentially fatal.

Allergens that produce strong immediate hypersensitivity reactions are proteins or chemicals bound to proteins. It has not been established why some antigens produce strong allergic reactions while others do not. It is feasible that the property of being allergenic may be due to the nature of the antigen itself, and the epitopes seen by T cells may contribute to this effect. Alternatively, some protein antigens interact naturally with adjuvants that favor IgE synthesis. A number of allergens induce proteolysis, since they are enzymes in their natural state and their enzymatic properties render allergenic activity. Aerodynamic properties also result in allergenic activity depending on the size of the particles. In the United States, the major allergens include Der p1 and Der p2 (house dust mites), Fel d1 (cat), Amb a1, 2, 3, 5 and 6 (short ragweed), Amb t5 (giant ragweed), Phl p1 and Phl p5 (timothy grasses) and Bet v1 (birch tree). Other serious allergens include ara H 1, 2, and 3 (peanuts) and Hev b 1, 2, 3, 4, 5, 6, 7 (latex).

Immediate Hypersensitivity Reaction

Allergen-dependent immune response constitutes two phases. The initial response, which occurs within 15 min of recognizing the allergen, is called the early-phase reaction, or immediate hypersensitivity reaction. Four to six hours after the termination of the symptoms of the first phase, the second or late phase starts, which may persist for days or up to weeks. The mast cells release autacoids including histamine, leukotrienes, prostaglandins and thromboxane during the early-phase reaction. The local responses produced by autacoids include sneezing, itching, runny nose, edema, mucus secretion, nasal congestion and bronchoconstriction. The symptoms of the late-phase reaction include cellular infiltration, fibrin deposition and tissue destruction in lungs. With continued allergic responses, this results in allergic hyperreactivity response, including edema and serious inflammatory response.

The classical example of immediate hypersensitivity in humans is the "wheal and flare" reaction. Challenging a sensitized individual with an intradermal injection of an allergen results in redness at the injection site. This is due to locally dilated blood vessels engorged with red blood cells. During the second phase, the leakage of plasma from the venules results in rapid swelling of the injection site. The soft swelling is called a wheal and the area of the skin involved in this soft swelling could be as large as several centimeters in diameter. In the third phase, there is dilation of the blood vessels at the margins of the wheal. The blood vessels become engorged with red blood cells, which produce a flare, the characteristic red rim. It takes about 5–10 min after the administration of the allergen for the development of a full wheal and flare reaction, and it lasts less than an hour. There is a slight separation of the endothelial cells in the venules of the wheal as observed by electron microscopy. This separation results in the escape of macromolecules and fluid, and there is a release of preformed mediators by mast cells in the area of wheal and flare.

In nonatopic infants, according to the hygiene hypothesis, the main stimuli for TH_1-dependent immune response are the microbes. Macrophages not only secrete IL-12, but also engulf the microbes, and IL-12 induces TH_1 cells and NK cells. The stimulated TH_1 cells and NK cells secrete IFN-γ, which results in a nonatopic state where an individual is protected from the allergic disease. It has been suggested that the lifestyle in the West may be responsible for an increase in the prevalence of allergic disease in the developed industrial states. This could be due to an absence of microbial antigens that stimulate TH_1 cells. The contributing factors could be a comparatively clean environment and a widespread use of antibiotics in early life.

The hygiene hypothesis was initially proposed by David Strachan in 1989. He observed that allergic disease was less common in children from larger families than in children from families with only one child. This was attributed to exposure to more infectious agents among children belonging to larger families. According to the hygiene hypothesis, in families with only one child, insufficient stimulation of TH_1 responses that are elicited by bacterial and/or viral infections results in a robust TH_2 response secreting high levels of IL-4, IL-5 and IL-13 and consequently leading to allergic disease. The atopic status depends on the environment, beginning in early childhood. Children living on farms or with multiple pets, which exposes them to high levels of endotoxins, have lower incidence of allergic disease. The exposure to animal sheds, hay lofts and the consumption of unpasteurized cow's milk play an important role in this regard. Furthermore, the timing of exposure is critical since protection was the best when the exposure took place during the first years of life as opposed to the later years. The high exposure to endotoxins promotes the development of IFN-γ-secreting TH_1 cells, which inhibits the development of TH_2 responses. Three aspects of the hygiene hypothesis have been proposed to overcome the criticism of this hypothesis, particularly the observations that there has also been an increase in the incidence of autoimmune diseases such as MS, Crohn's disease, Type 1 diabetes and inflammatory bowel disease, which are associated with overactive TH_1 response in the same patients with increased allergic disease. These three possibilities include overt and unapparent bacterial and viral infections, noninvasive exposure to microbes in the environment and the effects of infections on innate and acquired immune responses including the development of regulatory T cells.

The mechanisms are complex since timing of the exposure to allergen, state of development and prenatal factors acting either in utero or modulating subsequent development may all play some role in the development of allergic disease. New data suggest that the interactions between the environment and genes may be the result of multiple genes responding to the environment at several stages of development rather than a single gene ultimately leading to the etiology and pathogenesis of allergic disease.

Some observations have suggested that presence of microbial compounds in the environment may affect hosts' immune response. For example, the children of farmers who are exposed to high levels of endotoxins exhibit increased expression of TLR2 and CD14 compared to children with less exposure. This suggests a modulation of innate immune response in the absence of an infection. How this observation corresponds to a lower incidence in allergic disease needs to be determined. Activation of TLR induces inflammatory and acquired immune response, and inhibition of one of these pathways (MyD88) results in decreased TH_1 responses and enhanced IgE production.

To respond to the criticism of an increased prevalence of both autoimmune and allergic disease, an alternative mechanism has been proposed. According to this hypothesis, a lack of stimuli from infectious agents during development results in poor development of T regulatory cells. A poor T regulatory cell function increases the risk of the development of an autoimmune response due to the inability to suppress TH_1 function, which may also lead to an enhanced TH_2 response.

The microbes by producing an environment rich in IL-12 could drive TH_1-mediated responses. The production of IgE is induced by IL-4 and IL-13, which initiate transcription of the gene for a particular region of the IgE heavy chain ϵ class of the constant region. STAT-6 and NF-κB, two transcription factors, are also required for the production of IgE. IL-4 activates STAT-6, and NF-κB pathway needs costimulatory molecule CD40 and its ligand CD154. Furthermore, the presence of the transcription factors GATA-3 and C-maf favors TH_2 cell-mediated allergen-specific responses. The TH_2 paradigm suggests that the etiology and pathogenesis of allergic disease involves a complex interaction of T cells, B cells and APCs, resulting in higher production of IL-4 and IL-13, lower production of IFN-γ and elevation of IgE.

Although the mechanisms related to TH_1/TH_2 balance may be pivotal in explaining the etiology of allergic disease, they may not be sufficient. For example, serum levels of IgE are also dependent on the factors regulating terminal differentiation of class-switched B cells and the rate of IgE secretion, in addition to isotype switching alone. This process is regulated by IL-6. Furthermore, regulatory T cells and their interaction with dendritic cells both play an important role in the regulation of the development of allergic disease.

In addition to hygiene and allergic paradigms, epithelial and viral paradigms have also been proposed to explain the increases in atopic disorders. It seems that in addition to CD4$^+$ cells, a number of other cell types including the CD4$^+$CD25$^+$, dendritic cells and others may be involved in the pathogenesis of allergic disease. Nonetheless, it is difficult to imagine a strong allergen-driven allergic/asthmatic response if T helper cell activation is shut down. Consequently, developmental dysregulation of T helper cells (CD4$^+$) in childhood is important in the pathophysiology of allergic/asthmatic disease.

Allergic Inflammation

The immediate hypersensitivity reaction is followed by the late-phase reaction, which is manifested by red, edematous and indurated swelling in skin, blockage in nose and wheezing in lungs. Acute allergic reactions are the result of immediate hypersensitivity, and the mediators released by the mast cells produce their clinical symptoms including anaphylaxis, rhinoconjunctivitis and urticaria. Three cysteinyl leukotrienes C_4, D_4 and E_4 are produced by mast cells, and their pathological effects include smooth muscle contraction, vasodilation, hypersecretion of mucus and increased vascular permeability. The granules of mast cells also contain a protease, tryptase, which activates receptors on endothelial and epithelial cells resulting in a number of processes that also include an enhanced expression of adhesion molecules. The role of histamine in asthmatic disease has not been fully elucidated although the new generation of antihistamines has shown some benefits in childhood asthma.

The activation of mast cells and T cells results in late-phase reactions. Both immediate hypersensitivity and late-phase reactions are evident in the skin of atopic as well as nonatopic individuals after the cross-linking of IgE-bound mast cells with an antibody against IgE. The atopic asthmatic could develop late-phase reaction even in the absence of mast cell-related immediate hypersensitivity reaction, which is mast cell-independent and HLA-dependent, suggesting the role of T cells by themselves in causing asthma symptoms in atopic asthmatics.

APCs, specifically cutaneous Langerhans cells and dendritic cells are important in atopic eczema and asthma, respectively. These APCs are responsible for presenting the processed allergen to the T cells in an HLA-restricted manner. This process is further stimulated by the presence of GM-CSF in the airways of atopic asthmatics. GM-CSF also serves as a stimulus for macrophage production, which presents antigen to T helper cells with a resultant increase in the development of TH_2 cells and the elaboration of their cytokines in atopic asthmatic patients. The most important cytokines associated with allergic inflammation and released by TH_2 cells include IL-4, IL-5, IL-9 and IL-13. The stimulation of IgE production is the most important role of IL-4 and IL-13, which also augment the expression of VCAM-1. The development of mast cells is regulated by IL-4 and IL-9, whereas the production and differentiation of eosinophils is regulated by IL-5 and IL-9. The AHR is dependent on IL-9 and IL-13, and the production of mucus is regulated by IL-4, IL-9 and IL-13.

Cysteinyl leukotrienes and platelet-activating factors released by eosinophils injure mucosal surfaces of the airways. Furthermore, they damage M_2 muscarinic receptors and as a result the cholinergic response is unchecked in the absence of inhibitory M_2 muscarinic receptors. The production of eosinophils is regulated by IL-5, which also regulates their terminal differentiation. The preferential accumulation of eosinophils in the airways of the atopic asthmatics results from a combination of interaction between adhesion receptors, chemokines and cytokines. The interacting molecules include $\alpha_4\beta_1$ integrin, vascular adhesion molecules, eotaxin-1, eotaxin-2, eotaxin-3, monocyte chemotactic protein-3 and -4, RANTES, GM-CSF, IL-3 and IL-5. Allergic inflammation also results from several neuropeptides including substance P, neurokinin A and calcitonin gene-related peptide. Furthermore, thymus and activation-related chemokine (TARC), a ligand for the foregoing TH_2 cells, is upregulated in bronchial epithelial cells in allergen-challenged asthmatics. Consequently, TARC represents a new molecule involved in allergen-induced asthma. Production of IL-1 and TNF-α leads to the induction of transcription factor NF-κB and TARC expression. TARC expression results in the recruitment of TH_2 cells, where it is a ligand for the chemokine receptor CR4 that is expressed on TH_2 cells.

Tissue-specific transcription factors, T-bet, GATA-3, CMAF, regulate T helper cell differentiation, and genetic defects in these factors may alter the $TH_1:TH_2$ balance. However, paradigms other than the $TH_1:TH_2$ imbalance have also been proposed for the etiology of allergic/asthmatic disease. Allergic inflammation may also occur through innate immune stimulation, which is not directly related to $TH_1:TH_2$ dichotomy. TNF-α may be involved in the inflammatory response in adults with asthmatic disease and as a consequence, inhibition of TNF-α may be useful in severe asthma. IL-10 also plays an interesting role in this process; while being produced by both T regulatory and TH_2 cells, it may downregulate early life inflammation, however, and in other situations it could serve as a TH_2 proinflammatory cytokine depending on its cellular origin.

It has been suggested that the increased risk of allergic inflammation may result from T-cell dysregulation. This could result either from a global lack of suppression of proinflammatory cytokine production or by the upregulation of TH_2 response. Regulatory T cells participate in the induction of immune tolerance. $CD4^+CD25^+$ regulatory T cells play a specific role in protection against allergic disease and asthma.

Receptors Associated with Allergic Disease and Asthma

IL-1 receptors, TLRs and CCR8 receptors are linked to the pathogenesis of asthmatic disease. IL-1 induces the proliferation of TH_2 cells, specific antibody responses, and expression of exotoxin as well as eosinophil chemoattractants. It may play a role in T-cell priming through induction of CD40L and OX40 receptors. The role of TLRs in allergic disease and asthma is being increasingly elucidated. It has been proposed that since bacterial infections mount TH_1 responses through TLR, it is feasible that a reduction in TLR ligands on bacteria coupled to allergen exposure would cause an inhibition from TH_2 to TH_1 responses. This further suggests a dual role of TLR and allergen in inducing TH_2 responses. Activation of TLR7 provides protective effects against the development of experimental asthma, and both IL-10 and IL-12 are required for the mediation of these effects. TH_2 and TH_1 cells exhibit differential receptor expression. In neutrophilic asthma, there is an upregulation of the innate immune response and increased expression of TLR2, TLR4, SPA, IL-8 and IL-1β.

Genetic Predisposition

Atopy runs in families, and various genetic and phenotypic polymorphisms have been correlated in population studies with the onset and development of allergic disease/asthma. There is a correlation of atopic sensitization in atopic children who have asthma at age 7 compared with children without asthma. The chromosomal regions that are linked to allergic disease/asthma include 2q33, 5p15, 11p15, 17q11.1, 19q13 and 21q21. Furthermore, chromosomal region 14q contains several polymorphic markers and TAP-1 (transporters associated with antigen processing) – Acc 1 allele polymorphism is associated with atopic bronchial asthma. TAP is found in chromosomal regions DQB1 and DRB1 and helps translocate peptides to MHC I surface glycoproteins.

The Role of T Cells in Allergic Disease and Asthma

Different types of T cells including $CD4^+$, $CD8^+$ and NKT cells respond to allergens. In allergic responses, TH_2 cells are vital in recognizing the allergens in the context of MHC class II molecules as they produce their effector function via elaboration of IL-4, IL-5 and IL-13. The isotype switching from IgG to IgE production is induced by IL-4, and as IL-5 recruits eosinophils, both lead to the etiology and pathogenesis of allergic disease and asthma. Allergic inflammation and its clinical symptoms emanate from various immune pathways orchestrated by multiple cytokines produced by TH_2 cells. It has not yet been established why allergens preferentially promote TH_2 responses in atopic individuals. The patients with seasonal allergies possess allergen-specific memory T cells. One distinguishing characteristic of memory T cells is the expression of the CD45RO isoform as opposed to CD45RA that is expressed on naïve T cells, and they are further classified into effector memory T cells ($CCR7^-$) and central memory T cells ($CCR7^+$) both of which have different functions. However, it is not clear where memory TH_2 cells fit into this classification. CCR4 has been suggested to be a marker for TH_2 cells but its distribution is also diverse. Selective expression of a gene encoding chemoattractant receptor-homologous molecules expressed on TH_2 cells ($CRTH_2$) has also received particular attention in order to search for markers exclusively expressed on TH_2 cells. $CRTH_2$ is a prostaglandin D2 receptor present in approximately 0.5% of memory $CD4^+$ T cells ($CD45RO^+$). These cells specifically secrete IL-4, IL-5 and IL-13 in response to stimulation of the TCR. Allergen-specific recall responses to these cells require priming of dendritic cells by thymic stromal lymphopoietin (TSLP), which is an IL-7-like cytokine.

Although the role of TH_2 cells has been well established in allergic disease, the contribution of other types of T cells in this process is under investigation. The presence of gamma–delta T cells plays a role in allergic disease as the mice lacking these cells exhibit decrease in specific IgE, IgG1 and pulmonary IL-5 levels even after the development of pulmonary allergic inflammation. A prominent role for cytolytic T cells ($CD8^+$) in asthmatic disease has also emerged. A series of diverse observations suggest the involvement of $CD8^+$ cells in allergic disease. These findings include the following:

(1) The peripheral blood of individuals with atopic dermatitis has higher numbers of cutaneous lymphocyte-associated antigens $(CLA)^+CD8^+$ T cells, which produce TH_2 cytokines.
(2) During infancy, wheezing is also associated with the sequestering of $CD8^+$ cells in the airways during an acute asthma attack, and asthma deaths are associated with activated $CD8^+$ T-cell infiltration into peribronchial tissue.
(3) The bronchial lavage of atopic subjects exhibits increased TH_2 cytokine (IL-4, IL-5, IL-13) mRNA levels in both $CD4^+$ and $CD8^+$ T cells.
(4) A significant infiltration of $CD8^+$ T cells is found in late asthmatic reactions induced by Fel d 1-derived peptides.
(5) Allergen-specific $CD8^+$ T cells produce significantly less IL-10 in subjects with severe disease compared to those with mild disease. Consequently, $CD8^+$ T cells also play an important role in the pathogenesis of allergic disease.

The Role of Signal Transducers and Activators of Transcription in Allergic Disease

Signal transducers and activators of transcription are pivotal in transmitting signals for many cytokines that play an important role in the etiology and pathogenesis of allergic disease. They also impact other cell types including epithelial cells, lymphocytes, mast cells, eosinophils and dendritic cells by their signaling, which has an impact on allergic disease. One role that STATs play in allergic disease is via their effects on regulatory T cells. STAT3 and STAT5 play a role in upregulation of Foxp3 expression by IL-2, and the development of T regulatory cells via IL-2 requires STAT5-dependent signaling mechanisms.

Allergic inflammation is produced as a result of dysregulation in STAT pathways. Allergen-induced airway inflammation is mediated via STAT6, which also plays an important role in AHR, mucus production, chemokine expression, airway eosinophilia and cell trafficking of TH_2 cells. Several genes, including eotaxin-1, eotaxin-3, P-selectin and arginase 1, which are involved in allergic inflammation, are induced by STAT6. IL-13 and IL-4 are the primary inducers of STAT6 activation but STAT6-independent pathways are also involved in allergic inflammation. Furthermore, STAT1, STAT3, STAT4 and STAT5 also play a role in allergic inflammation, either directly or indirectly. STAT5 is crucial for the activation of mast cells by IgE, and STAT4 is a modulator of AHR and allergen-induced chemokine production.

Neural Pathways in Allergic Disease and Asthma

The pathophysiological changes in asthma and allergic disease may be modulated by sensory nervous receptors. These sensory nervous receptors are divided into three groups: C-fiber receptors, the rapidly adapting receptors and deltanociceptive receptors. These receptors are activated by inflammatory and immunological changes. The axon reflex neurogenic inflammation–bronchoconstriction and mucus secretion may be mediated by the activation of C-fiber receptors. The activation of receptors causes reflexes, which include bronchoconstriction, mucus secretion and mucosal vasodilation. Three basic reflex pathways are responsible for the cough reflex.

Adrenergic, cholinergic and nonadrenergic–noncholinergic (NANC) nerves regulate the function of airways. Nervous system-mediated bronchoconstriction of the airways is produced by the cholinergic system. This is attributed to the neurotransmitter acetylcholine, which mediates its effects via nicotinic and muscarinic receptors, and these actions are short-lived. Acetylcholine also modulates its own release from postganglionic cholinergic nerves. The postganglionic nerves are home to M2 autoreceptors at prejunctions, which control the release of acetylcholine after their stimulation. Allergens, viruses and ozone inhibit the activation of neuronal muscarinic M2 autoreceptors. The neuronal bronchodilatory mechanisms constitute inhibitory NANC (i-NANC) mechanisms, which utilize nitric oxide and vasoactive intestinal peptide (VIP) as their neurotransmitters. The excitatory part of the NANC employs substance P and neurokinin A as neurotransmitters, and smooth muscles of both large and small airways express neurokinin A-2 receptors. Tachykinins produce bronchoconstrictor effects by interacting with NK2 receptors, whereas NK1 receptors are involved in exhibiting the proinflammatory effects of substance P.

Role of Neuropeptides in Airway Inflammation

The stimulation of C fibers by capsaicin causes a subset of sensory airway neurons to release several neuropeptides, which include tachykinin, substance P and neurokinin A. In addition to capsaicin, other endogenous mediators including histamine, prostaglandins and bradykinins can also result in their release. These neuropeptides are responsible for neurogenic inflammation, which is characterized by vasodilation, mucus secretion, plasma protein extravasation, increased expression of the adhesion molecules and bronchoconstriction.

Neuropeptides are degraded by enzymes called neutral endopeptidases, which are present adjacent to neuropeptide receptors. Neutral endopeptidases regulate the neuropeptide-induced responses by modulating their levels. Inhibitors of these enzymes are being studied as a potential therapeutic agent for asthmatic disease. In addition to these inhibitors, the role of tachykinin receptor antagonists is also being investigated.

Besides neuropeptides, nitric oxide is an inflammatory mediator in the airways, which is also a vasodilator and a neurotransmitter. Nitric oxide is produced by the enzymatic action of nitric oxide synthetase on L-arginine. Airways contain this enzyme in three different forms, two of which termed neuronal and endothelial nitric oxide synthetase are constitutive whereas the third form called inducible nitric oxide synthetase is inducible. The inflammatory cytokines including IL-1 and TNF-α augment the expression of inducible nitric oxide synthetase in human airway epithelial cells. Nitric oxide causes bronchodilation as a result of the relaxation of bronchial smooth muscles. It has also been suggested that nitric oxide is the neurotransmitter of the inhibitory NANC bronchodilation. The detrimental effects of nitric oxide include airway inflammation and vasodilation. It causes airway edema by increasing the erudition of plasma due to increased blood flow to post-capillary venules. The increased blood flow may also contribute to an increased mucus secretion. The role of nitric oxide in inflammatory responses has not yet been established.

Nitric oxide may suppress TH_1 subset and its high levels may increase the expression of TH_2 subset. The action of inducible nitric oxide synthetase may result in increased exhaled nitric oxide in asthmatics. The airway epithelial cells are the predominant source of increased nitric oxide in asthmatic subjects. Since nitric oxide has both beneficial and adverse effects on bronchial airways, its precise contribution to the etiology and pathogenesis of asthmatic disease requires additional investigation.

Neurotrophins and Allergic Disease and Asthma

The role of neurotrophins in the pathogenesis of allergic disease and asthma has been appreciated only recently. Neurotrophins are a family of growth factors that have common receptors and physiologic effects and are the principal regulators of neuronal activity, differentiation and maintenance. In humans, there are at least four defined neurotrophins also termed NGFs. These polypeptides include NGF, neurotrophin 3 (NT-3), neurotrophin 4/5 (NT-4/5) and brain-derived neurotrophic factor (BDNF). They all act on a common group of tropomyosin-related tyrosine kinase (Trk) receptors. Neurotrophin receptors are expressed in the neurons of both the central and peripheral nervous systems.

Airway epithelia and alveolar macrophages constitutively express neurotrophins, and under normal conditions, low levels of neurotrophins are produced. However, in allergic inflammation, these levels rise considerably in atopic patients. After inhalation of an allergen by asthmatic patients, there are high concentrations of NGF, NT-3 and BDNF. The cellular sources of the increased neurotrophins include invading leukocytes and resident lung cells. Inflammatory cytokines, IL-1 and TNF-α stimulate epithelial neurotrophin expression. The increased production of neurotrophins during allergic inflammation suggests a role of neurotrophins in allergic disease and asthma.

The allergic airway inflammation affects the neural reflex pathway and the function of afferent sensory nerves and synaptic transmission. Coughing and sneezing could result from protective axonal responses after activation of sensory airway nerves due to allergens. Inflammatory mediators also regulate the changes in airway sensory innervations. The neurotrophins affect allergic hyperresponsiveness by enhancing peptidergic sensory airway nerve function.

Allergic asthma patients have higher blood levels of NGF. The influx of inflammatory cells in lungs is observed as the neurotrophin expression is augmented. Multiple targets may be affected by neurotrophins as they play a role in allergic inflammation, which include recruitment, maintenance and activation of mast cells and eosinophils and facilitation of TH_2 response. Whether neurotrophins can alter TH_1/TH_2 balance in humans has not yet been established.

The neurotrophins play a role as the mediators of neurogenic inflammation. This is mediated by their effects on both the immune and nervous systems. The function of sensory neurons is regulated by neurotrophins. Furthermore, NGF regulates the expression of neuropeptide genes in adult sensory neurons. The release of neuropeptide substance P from neurons is augmented in allergic inflammation with contribution from neurotrophins, which are overproduced in this process. In parallel, the neurotrophin receptors are expressed on leukocytes and consequently neurotrophins modulate neurogenic inflammation by at least two mechanisms. One mechanism of regulation is via controlling the extent and intensity of local immune response directly by activation of the neurotrophin receptors. The other possible mechanism of regulation of neurogenic inflammation by neurotrophins is via modulating the synthesis of neuropeptides in local neurons. The effects of neurotrophins are brief, self-limited and in the vicinity of the area of their local synthesis. They possess a very short half-life but could be produced continuously when the symptoms of allergic inflammation are manifested.

Gene for Allergic Disease

A protein, Ndfip1, that protects mice from developing a severe allergic disease, which could be fatal, has been identified. Ndfip1 binds to Nedd4, a family of proteins known as ubiquitin ligases. Ndfip1 blocks the activated T cells to stimulate IL-4 production.

Specific Immunotherapy

Clinical tolerance to allergens may be achieved by the administration of allergen extracts, which is called specific immunotherapy. In addition to its therapeutic value, it has been suggested that this treatment may also alter the progression of disease. Specific allergen immunotherapy (SIT) is used to alleviate the condition associated with atopy by administering allergen extracts. Immunotherapy for allergic disease was first developed at the end of the nineteenth century in England, which was based on the principles described by Noon and Freeman. This is initiated by subcutaneously injecting small but increasing quantities of specific allergens over a period of weeks or months, which is followed by administering maintenance injections every 4–6 weeks for a period of 2–4 years. Two other protocols referred to as the semirush protocol and the rush protocol have also been used for the treatment of allergic disease by immunotherapy. These are different from the regular updosing phase, which is a series of weekly injections. The semirush protocol entails the administration of several doses on each day every week, whereas the rush protocol constitutes injections of a series of increasing quantities of allergens in one day.

A number of theories have been outlined to explain the mechanism of action of immunotherapy, which includes induction of IgG, inhibition of allergen-specific IgE, inhibition of the recruitment of effector cells, shift from TH_2 to TH_1 cytokines, induction of T regulatory cells and T-cell anergy. Originally, it was believed that immunotherapy abolishes atopic responses as a result of inhibiting allergen-specific antibodies. However, during the initial phase of immunotherapy, allergen-specific IgE levels rise, and this is followed by their return to the original levels during the maintenance therapy. It also appears that the systemic beneficial effects of immunotherapy are much more pronounced than the limited observed effects during the skin test. No late-phase skin responses are observed if the patient responds well to therapy. Furthermore, it does not appear that the immunotherapy is beneficial because of the ability of IgG to neutralize the allergens and to inhibit the allergic response since the rise of IgG is observed after the beneficial effects of the therapy, and the allergen first encounters IgE before it is recognized by IgG. The immunotherapy modifies allergen-specific T-cell responses, which include a reduction in eosinophil and T-cell recruitment in response to allergen and increased expression of TH_1 cells. The effects on regulatory T cells may be mediated via inducing the ability of allergen-specific B cells to produce IgG4. The effectiveness of immunotherapy in allergic rhinitis and specifically to aeroallergens may last up to 6 years and is more beneficial for seasonal rhinitis than for perennial rhinitis. The possible new technologies that are under development for immunotherapy include recombinant allergens, T-cell peptide vaccines, agents that stimulate TH_1 cells, antibodies to IgE, allergen–immunostimulant complexes and hypoallergenic allergens.

Sublingual Immunotherapy

Some patients experience discomfort and frequent swelling at the site of injection. As a consequence, sublingual immunotherapy has been explored in Europe. An extract of allergens is placed under the tongue for a couple of minutes and is then swallowed. The dose for sublingual immunotherapy is 3–300 times greater than that for subcutaneous immunotherapy. This form of therapy is more effective for seasonal allergens than for perennial allergens in both adults and children, and its long-term efficacy is similar to subcutaneous immunotherapy.

Ragweed Toll-Like Receptor 9 Agonist Vaccine for Immunotherapy

A novel approach for immunotherapy in allergic disease involves the conjugation of immunostimulatory sequences of DNA to specific allergens. In a recent clinical study, a vaccine that consisted of a ragweed-pollen antigen (Amb α1) conjugated to phosphorothioate oligo deoxyribonucleotide immunostimulatory sequence of DNA was tested in subjects with allergic rhinitis to ragweed. This protocol has advantages over the regular immunotherapy methods due to shorter regimens and a low risk of serious adverse effects. In this particular study, the patients were administered six injections of the conjugated allergen once a week, and the results demonstrated a substantial reduction in the symptoms of allergic rhinitis. The mechanisms responsible for immune tolerance to the allergen are not fully understood at this time.

Probiotics for the Treatment of Allergic Disease

Probiotics are the first compounds tested to treat allergic disease, which are based on the concept of the hygiene hypothesis. The preliminary results have shown that the use of probiotics reduces the risk of the development of atopic eczema; however, additional clinical trials are needed before their approval for use in a wider population. Other immunomodulatory compounds derived from bacteria (CpGs), mycobacteria and helminths are also being tested to prevent allergic disease.

Bibliography

Aberg N, Sundell J, Eriksson B, Hesselmar B. 1996. Prevalence of allergic disease in school children in relation to family history, upper respiratory tract infections and residential characteristics. Allergy. 51:232–237.
Akdis CA, Blasker K. 1999. IL-10-induced anergy in peripheral T cells and reactivation by micro environmental cytokines: Two key steps in specific immunotherapy. FASEB J. 13:603–609.
Akdis CA, Blaser K. 2003. Histamine in the immune regulation of allergic inflammation. J All Clin Immunol. 112:15–22.
Akdis M, Blaser K, Akdis CA. 2005. T regulatory cells in allergy: Novel concepts in the pathogenesis, prevention and treatment of allergic diseases. J All Clin Immunol. 116:9961–969.
Alam R, Kampern G. 2002. Cytokines in allergic inflammation. J Clin All Immunol. 16:255–274.
Annesi I, Oryszczyn M, Frette C, Neukirch F, et al. 1992. Total circulating IgE and FEV1 in adult men: An epidemiologic longitudinal study. Chest. 101:642–648.
Annunziato F, Cosmi L, Galli G, Beltrame C, et al. 1999. Assessment of chemokine receptor expression by human Th1 and Th2 cells in vitro and in vivo. J Leukoc Biol. 65:691–699.
Bach J-F. 2002. The effects of infections on susceptibility to autoimmune and allergic diseases. NEJM. 347:911–920.
Barnes PJ. 1991. Neuropeptides and asthma. Am Rev Respir Dis. 143:S28–S32.

Barnes PJ. 1999. Anti-IgE antibody therapy for asthma. NEJM. 341:2006–2008.

Barnes PJ. 2000. New directions in allergic diseases: Mechanism based anti-inflammatory therapies. J All Clin Immunol. 106:5–16.

Beasley R, Burgess R, Crane J, Pearce N, et al. 1993. Pathology of asthma and its clinical implications. J All Clin Immunol. 92:148–154.

Bisset LR, Schmid-Grendelmeier P. 2005. Chemokines and their receptors in the pathogenesis of allergic asthma: Progress and perspective. Curr Opin Pulm Med. 11:35–42.

Boulet LP, Chapman KR, Cote J, Kalra S, et al. 1997. Inhibitory effects of an anti-IgE antibody E25 on allergen-induced early asthmatic response. Am J Respir Crit Care Med. 155: 1835–1840.

Bousquet J, Jeffery PK, Busse WW, Johnson M, et al. 2000. Asthma: From bronchoconstriction to airway inflammation and remodeling. Am J Respir Crit Care Med. 161:1720–1745.

Braun-Fahrlander C, Riedler J, Herz U, Eder W, et al. 2002. Environmental exposure to endotoxin and its relation to asthma in school-age children. NEJM. 347:869–877.

Burrows B, Martinez FD, Halonen M, Barbee RA, et al. 1989. Association of asthma with serum IgE levels and skin-test reactivity to allergens. NEJM. 320:271–277.

Busse WW, Coffman RL, Gelfand EW, Kay AB, et al. 1995. Mechanisms of persistent airway inflammation in asthma. A role for T cells and T-cell products. Am J Respir Crit Care Med. 152:388–393.

Busse WW, Fahy JV, Rick RB. 1998. A pilot study of the effects of an anti-IgE antibody (E25) on airway inflammation in moderate-severe asthma. Am J Respir Crit Care Med. 157:456A.

Busse WW. 2000. Mechanisms and advances in allergic diseases. J All Clin Immunol. 105: S593–S598.

Busse W, Kraft M. 2005. Cysteinyl leukotrienes in allergic inflammation: Strategic target for therapy. Chest. 127:1312–1326.

Chen W, Hershey GKK. 2007. Signal transducer and activator of transcriptional signals in allergic disease. J All Clin Immunol. 119:529–541.

Cherwinski HM, Schumaker JH, Brown KD, Mosmann TR. 1987. Two types of mouse helper T cell clone. III. Further differences in lymphokine synthesis between Th1 and Th2 clones revealed by RNA hybridization, functionally monospecific bioassays, and monoclonal antibodies. J Exp Med. 166:1299–1244.

Chung F. 2001. Anti-inflammatory cytokines in asthma and allergy: Interleukin 10, interleukin 12, interferon gamma. Mediators Inflamm. 10:51–59.

Cohn L, Homer RJ, Niu N, Bottomly K. 1999. T helper 1 cells and interferon regulate allergic airway inflammation and mucus production. J Exp Med. 190:1309–1317.

Contopoulos-Loannidis DG, Kouri LN, Loannidis JPA. 2007. Genetic predisposition to asthma and atopy. Respiration. 74:8–12.

Corren J, Diaz-Sanchez D, Reimann J, ,Saxon A, Adelman D. 1998. Effects of anti-IgE antibody therapy on nasal reactivity to allergen and IgE synthesis in vivo. J All Clin Immunol. 101:S105.

Cosmi L, Annunziato F, Iwasaki M, Galli G, et al. 2000. CRTH2 is the most reliable marker for detection of circulating human type 2 Th2 and type 2 T cytotoxic cells in health and disease. Eur J Immunol. 30:2972–2979.

Creticos PS, Schroeder JT, Hamilton RG, Balcer-Whaley SL, et al. 2006. Immunotherapy with a ragweed-toll like receptor 9 agonist vaccine for allergic rhinitis. NEJM. 355:1445–1455.

Cui ZH, Joetham M, Aydintug MK, Hahn YS, et al. 2003. Reversal of allergic airway hyperactivity after long term allergen challenge depends on (gamma) (delta) T cells. Am J Respir Crit Care Med. 168:1324–1332.

Davis P. 2005. Carl Prausnitz and the mysteries of allergy. Biomed Sci. December:1247.

Demoly P, Bousquet J. 1997. Anti-IgE therapy for asthma. Am J Respir Crit Care Med. 155: 1825–1827.

Denburg JA. (Ed.) 1998. Allergy and Allergic Disease: The New Mechanisms and Therapeutics. Humana Press: New Jersey. ISBN: 0-89603-404-6.

Durham SR, Walker SM, Varga EM, Jacobsen MR, et al. 1999. Long-term clinical efficacy of grass pollen immunotherapy. NEJM. 341:468–475.

Elias JA. 2000. Airway remodeling in asthma. Am J Respir Crit Care Med. 161:S168–S171.

Emanuel MB. 1988. Hay fever, a post industrial revolution epidemic: A History of its growth during the 19th century. Clin Allergy. 18:295–304.

Erpenbeck VJ, Hohlfeld JM, Discher M, Krentel H, et al. 2003. Increased messenger RNA expression of C-maf and GATA-3 after segmental allergen challenge in allergic asthmatics. Chest. 123:370S–371S.

Fahey JV, Fleming HE, Wong HH, Liu J, et al. 1997. The effect of an anti-IgE monoclonal antibody on the early-and late-phase responses to allergen inhalation in asthmatic subjects. Am J Respir Crit Care Med. 155:1828–1834.

Fireman P. 2003. Understanding asthma pathophysiology. Allergy Asthma Proc. 24:79–83.

Freeman J. 1911. Further observations on the treatment of hay fever by hypodermic inoculations of pollen vaccine. Lancet. 2:814–817.

Freeman J. 1914. Vaccination against hay fever: Report of results during the first three years. Lancet. 1:1178–1180.

Frew AJ, Corrigan CJ, Maestrelli P, Tsai JJ, et al. 1989. T lymphocytes in allergen-induced late-phase reaction and asthma. Int Arch Allergy Appl Immunol. 88:63–67.

Frew AJ, Kay AB. 1990. Eosinophils and T-lymphocytes in late phase allergic reactions. J All Clin Immunol. 85:533–539.

Galli SJ. 1993. New concepts about the mast cells. NEJM. 328:257–265.

Gangur V, Oppenheim JJ. 2000. Are chemokines essential or secondary participants in allergic responses? Ann All Asthma Immunol. 84:569–581.

Gell PGH, Coombs RRA. (Eds.). 1963. Clinical Aspects of Immunology. 1st Ed., Blackwell, Oxford, England.

Gould HJ, Sutton BJ, Beavil AJ, Beavil RL, et al. 2003. The biology of IgE and the basis of allergic disease. Ann Rev Immunol. 21:579–628.

Guarner F, Bourdet-Sicard R, Brandtzaeg P, Gill HS, et al. 2006. Mechanisms of disease: The hygiene hypothesis revisited. Nat Clin Prct Gastroenterol Hepatol. 3:275–284.

Hang LW, Hsia TC, Chen WC, Chen HY, et al. 2003. Tap 1 Gene Acc1 polymorphism is associated with atopic bronchial asthma. J Clin Lab Anal. 17:57–60.

Hauber HP, Bergeron C, Hamid Q. 2004. IL-9 in allergic inflammation. Int Arch All Immunol. 134:79–87.

Hawrylowicz CM, O'Garra A. 2005. Potential role of interleukin-10 secreting regulatory T cells in allergy and asthma. Nat Rev Immunol. 5:271–283.

Hennino A, Vocanson M, Toussiant Y, Rodet K, et al. 2007. Skin-infiltrating CD8[+] T cells initiate atopic dermatitis lesions. J Immunol. 178:5571–5577.

Hollingsworth JW, Whitehead GS, Lin K, Nakano H, et al. 2006. TLR4 signaling attenuates ongoing allergic inflammation. J Immunol. 176:5856–5862.

Howarth PH, Salagean M, Dokic D. 2000. Allergic rhinitis: Not purely a histamine-related disease. Allergy, 55: 7–16.

Illig T, Wjst M. 2002. Genetics of asthma and related phenotypes. Paed. Respir Rev. 3:47–51.

Ishizaka K, Ishizaka T. 1967. Identification of gamma-E-antibodies as carrier of reaginic activity. J Immunol. 99:1187–1198.

Ishizaka K, Ishizaka T, Terry WD. 1967. Antigenic structure of gamma-E-globulin and reaginic antibody. J Immunol. 99:849–858.

Johansson SGO. 1967. Raised levels of a new immunoglobulin class (IgND) in asthma. Lancet. 2:951–953.

Johansson SGO, Bennich H, Wide L. 1968. A new class of immunoglobulins in human serum. Immunology,. 14: 265–272.

Joos GF, DeSwert KO, Schelfhout V, Pauwels RA. 2003. The role of neural inflammation in asthma and chronic obstructive pulmonary disease. Ann NY Acad Sci. 992:218–230.

Kaliner M, Lemanske R. 1984. Inflammatory responses to mast cell granules. Fed Proc. 43: 2846–2851.

Kalliomaki M, Salminen S, Poussa T, Arvilommi H, et al. 2003. Probiotics and prevention of atopic disease: 4 year follow up of a randomized placebo controlled trial. Lancet. 361: 1869–1871.

Karlsson MR, Rugtveit J, Brandtzaeg P. 2004. Allergen responsive CD4$^+$ CD25$^+$ regulatory T cells in children who have outgrown cow's milk allergy. J Exp Med. 12:1679–1688.

Kaufman HS, Frick OL. 1976. The development of allergy in infants of allergic parents: A prospective study concerning the role of heredity. Ann Allergy. 37:410–415.

Kay AB. 2001a. Allergy and allergic diseases – First of two parts. NEJM. 344:30–37.

Kay AB. 2001b. Allergy and allergic diseases – Second of two parts. NEJM. 344:109–113.

Kozyrskyj AL, Ernst P, Becker AB. 2007. Increased risk of childhood asthma from antibiotic use in early life. Chest. 131:1753–1759.

Lasley MV. 1999. Allergic disease prevention and risk factor identification. Immunol. All. Clin North Am. 19:149–159.

Lauener RP, birchler T, Adamski J, Braun-Fahrlander C, et al. 2002. Expression of CD14 and Toll-like receptor 2 in farmer's children. Lancet. 36:465–466.

Lawlor GJ, Fischer TJ, Adelman DC, eds. 1995. Manual of Allergy and Immunology. 3rd Ed. Lippincott-Raven, Philadelphia, PA.

Lichtenstein LM, Ishizaka K, Norman PS, Sobotka AD, et al. 1973. IgE antibody measurements in ragweed hay fever relationship to clinical severity and the results of immunotherapy. J Clin Inv. 52:472–482.

Luckacs NW. 2001. Role of chemokines in the pathogenesis of asthma. Nat Rev Immunol. 1: 108–116.

Martinez FD. 2007. Gene-environment interaction in asthma: With apologies t William of Ockham. Proc Am Thorac Soc. 4:26–31.

Miyahara N, Takeda K, Kodama T, Joetham A, et al. 2004. Contribution of antigen-primed CD8$^+$ T cells to the development of airway hyper-responsiveness and inflammation is associated with IL-13. J Immunol. 172:2549–2558.

Mosmann TR, Coffman R. 1989. TH$_1$ and TH$_2$ cells: Different patterns of lymphokine secretion lead to different functional properties. Ann Rev Immunol. 7:145–173.

Nagata K, Tanaka K, Ogawa K, Kemmostu K, et al. 1999. Selective expression of a novel surface molecule by human Th$_2$ cells in vivo. J Immunol. 162:1278–1286.

Nakajima H, Takatsu K. 2007. Role of cytokines in allergic airway inflammation. Int Arch All Immunol. 142:265–273.

Nassenstein C, Braun A, Erpenbeck A, Lommatzsch M, et al. 2003. The neurotrophins nerve growth factors, brain derived neurotrophic factor, neurotrophin 3 and neurotrophin 4 are survival and activation factors for eosinophils in patients with allergic bronchial asthma. J Exp Med. 198:455–467.

Nassenstein C, Braun A, Nockher WA, Renz H. 2005. Neurotrophin effects of eosinophils in allergic inflammation. Curr All Asthma Rep. 5:204–211.

Ngoc PL, Gold DR, Tzianbos AO, Weiss ST, et al. 2005. Cytokines, allergy and asthma. Curr Opin All Clin Immunol. 5:161–166.

Nigo YI, Yamashita M, Hirahara K, Shinnakasu R. 2006. Regulation of allergic airway inflammation through the Toll-like receptor 4-mediated modification of mast cell function. PNAS USA. 103:2286–2291.

Nockher WA, Renz H. 2006. Neurotrophins and asthma: Novel insight into neuroimmune interaction. J All Clin Immunol. 117:67–71.

Noon L. 1911. Prophylactic inoculation against hayfever. Lancet. 1:1572–1573.

Norman PS, Ohman JL, Long AA, Creticos PS, et al. 1996. Treatment of cat allergy with T-cell reactive peptides. Am J Respir. Crit Care Med. 154:1623–1628.

O'Byrne PM. 2006. Cytokines and their antagonists for the treatment of asthma. Chest. 130: 244–250.

Oliver PM, Cao X, Worthen GS, Shi P, et al. 2006. Ndfip1 protein promotes the function of itch ubiquitin ligase to prevent T cell activation and T helper 2 cell-mediated inflammation. Immunity. 25:929–940.

Ono S. 2000. Molecular genetics of allergic diseases. Ann Rev Immunol. 18:347–366.

Panina-Bordignon P, Papi A, Mariani M, DiLucia P. 2001. The C–C chemokine receptors and CCR4 and CCR8 identify airway T cells of allergen-challenged atopic asthmatics. J Clin Inv. 107:1357–1364.

Passalacqua G. 2007. Allergic rhinitis and its impact on asthma update. Allergen immunotherapy. J All Clin Immunol. 119:881–891.

Pepys J. 1994. "Atopy;" a study in definition. Allergy. 49:397–399.

Pernis AB, Rothman PB. 2002. JAK-STAT signaling in asthma. J Clin Inv. 109:1279–1283.

Platts-Mills TA. 2001. The role of immunoglobulin E in allergy and asthma. Am J Respir Crit Care Med. 164:S1–S5.

Prussin C, Metcalfe DD. 2003. IgE, mast cells, basophils and eosinophils. J All Clin Immunol. 111:5486–5494.

Racila DM, Kline JN. 2005. Perspectives in asthma: Molecular use of microbial products in asthma prevention and treatment. J All Clin Immunol. 116:1202–1205.

Rajan TV. 2003. The Gell-Coombs classification of hypersensitivity reactions: A reinterpretation. Trends Immunol. 24:376–379.

Ray A, Cohn L. 1999. TH₂ cells and GATA-3 in asthma: New insights into the regulation of airway inflammation. J Clin Inv. 104:995–1006.

Reynolds C, Ozerovitch L, Wilsen R, Altmann D, et al. 2007. Toll-like receptors 2 and 4 and innate immunity in neutrophilic asthma and idiopathic bronchiectasis. Thorax. 62:279.

Robertson DG, Kerigan AT, Hargreave FE, Chalmers R, et al. 1974. Late asthmatic responses induced by ragweed pollen allergen. J All Clin Immunol. 54:244–254.

Rochlitzer S, Nassenstein C, Braun A. 2006. The contribution of neurotrophins to the pathogenesis of allergic asthma. Biochem Soc Trans. 34:594–599.

Romagnani S. 1994. Lymphokine production by human T cells in disease states. Ann Rev Immunol. 12:227–257.

Romagnani S. 2000. The roles of lymphocytes in allergic disease. J All Clin Immunol. 105:399–408.

Romagnani S. 2002. Cytokines and chemoattractants in allergic inflammation. Mol Immunol. 38:881–885.

Rothenberg ME. 1998. Eosinophilia. NEJM. 338:1592–1600.

Schaub Bm, Lauener R, Von ME. 2006. The many faces of the hygiene hypothesis. J All Clin Immunol. 117:969–977.

Schmitz N, Kurrer M, Kopf M. 2003. The IL-1 receptor 1 is critical for TH2 cell type airway immune responses in a mild but not in more severe asthma model. Eur J Immunol. 33:991–1000.

Schnare M, Barton GM, Holt AC, Taked K, et al. 2001. Toll-like receptors control activation of adaptive immune responses. Nat Immunol. 2:947–950.

Schoenwetter WF. 2000. Allergic rhinitis: Epidemiology and natural history. Allergy Asthma Proc. 21:1–6.

Sel S. Wegmann M. Sel S, Bauer S, et al. 2007. Immunomodulatory effects of viral TLR ligands on experimental asthma depend on the additive effects of IL-12 and Il-10. J Immunol. 178:7805–7813.

Seneviratne SL, Jones L, King AS, Black A, et al. 2002. Allergen specific CD8⁺ T cells in atopic disease. J Clin Inv. 110:P1283–1291.

Serafin WE, Austen KF. 1987. Mediators of immediate hypersensitivity reactions. NEJM. 317:30–34.

Simpson JL, Grissell TV, Douwes J, Scott RJ, et al. 2007. Innate immune activation in neutrophilic asthma and bronchiectasis. Thorax. 62:211–218.

Solley GO, Gleich GJ, Jordon RE, Schroeter AL. 1976. The late phase of the immediate wheal and flare skin reaction. Its dependence upon IgE antibodies. J Clin Inv. 58:408–420.

Spitz E, Gelfand EW, Sheffer AL, Austen KF. 1972. Serum IgE in clinical immunology and allergy. J All Clin Immunol. 49:337–347.

Strachan DP. 1989. Hay fever, hygiene and household size. Brit Med J. 299:1259–1260.

Strachan DP. 2000a. Family size, infection and atopy: The first decade of the "hygiene hypothesis." Thorax. 55:S2–S10.

Strachan DP. 2000b. Family size, infection and atopy: The first decade of the "hygiene hypothesis." Thorax. 1:S2–S10.

Strachan DP, Wong HJ, Spector TD. 2001. Concordance and interrelationship of atopic diseases and markers of allergic sensitization among adult female twins. J All Clin Immunol. 108: 901–907.

Takhar P, Smurthwaite L, Coker HA, Fear DJ, et al. 2005. Allergen derives class switching to IgE in the nasal mucosa in allergic rhinitis. J Immunol. 174:5024–5032.

Umetuse DT, DeKruff RH. 2006. The regulation of allergy and asthma. Immunol Rev. 212: 238–255.

Undem BJ, Carr MJ. 2002. The role of nerves in asthma. Curr All Asthma Rep. 2:159–165.

Van Hove CL, Maes T, Joos GF, Tournoy KG. 2007. Prolonged inhaled allergen exposure can induce persistent tolerance. Am J Respir Cell Mol Biol. 36:573–584.

Varney VA, Gaga M, Frew AJ, Eber VR, et al. 1991. Usefulness of immunotherapy in patients with severe summer hayfever uncontrolled by anti-allergic drugs. Br Med J. 302:265–269.

Von Mutius E. 2002. Environmental factors influencing the development of progression of pediatric asthma. J All Clin Immunol. 109:525–532.

Wagner R. 1968. Clemens Von Priquet: His Life and Work. The John Hopkins Press, Baltimore, MD.

Wang YH, Ito T, Wang YH, Homey B, et al. 2006. Maintenance and polarization of human Th2 central memory T cells by thymic stromal lymphopoietin activated dendritic cells. Immunity. 24:827–838.

Weber RW. 1997. Immunotherapy with allergens. JAMA. 278:1881–1887.

Weiss ST. 2002. Eat dirt – The hygiene hypothesis and allergic diseases. 347:930–931.

Widdicombe JG. 2003. Overview of neural pathways in allergy and asthma. Pulm Pharmacol Ther. 16:23–30.

Wills-Karp M. 2001. IL-12/IL-13 axis in allergic asthma. J All Clin Immunol. 107:9–18.

Woodfolk JA. 2007. T-cell responses to allergens. J All Clin Immunol. 119:280–294.

Zuany-Amorim C, Ruffie C, Haile S, Vargaftig BB, et al. 1998. Requirement of γδ T cells in allergic airway inflammation. Science. 280:1265–1267.

Chapter 7
Tissue Transplantation

Introduction

The concept of transplantation may go back to the stories of chimeras including sphinxes and mermaids. According to a third-century legend of Saints Cosmas and Damian, these two brother physicians replaced the cancerous leg of a church sacristan using the limb of a dead Moor. It is suggested that about 2000 years ago, transplantation surgery was done for nose reconstruction by Vaidya, the Ayurvedic physicians. Gaspare Tagliacozzi in the late sixteenth century performed reconstructive rhinoplasty by using skin from the arms of the patients. The allografts of skin always failed and no further progress was reported until the twentieth century.

The original experiments of tissue transplantation were done by using skin transplants in inbred strains of mice. In 1908, Alexis Carrel first reported kidney transplantation in cats, but all the cats died, and only some exhibited a functional urinary output. George Schone published in 1912 the difficulties associated with tissue transplantation, which were not related to surgery, asepsis and anesthesia. He wrote six laws of tissue transplantation, which to date are accepted:

(1) Transplanting a tissue to a different species fails;
(2) Transplants in unrelated species members fail;
(3) Autografts succeed;
(4) First allograft is initially accepted but is rejected later;
(5) Second allograft is rejected rapidly; and
(6) Close relationship between the donor and the recipient aids the acceptance of the graft.

In 1935, a Russian surgeon performed the first human kidney transplant but the patient was not able to have kidney function and died. According to reports in 1954, the first successful kidney transplant was performed in the United States, where the donated organ was from an identical twin. Today tissue transplantation is a form of medical therapy where a failing or defective organ is replaced. The development of modern surgical techniques has helped overcome many obstacles facing this unique form of therapy. The other obstacle, the availability of donor organs, remains more in some cases than others.

M.M. Khan, *Immunopharmacology*, DOI: 10.1007/978-0-387-77976-8_7,
© Springer Science+Business Media, LLC 2008

However, the most serious obstacle to tissue transplantation is the immune response, specifically the acquired immunity. Blood transfusion provided the early understanding of the difficulties associated with this therapy. Considering the fact that there are only a few blood types, the complexity of the tissue type results in a major immune response against the grafted tissue. Both the blood and tissue types are important for matching an organ for the transplant. Matching of blood type is important because of the expression of blood group antigens on vascular endothelium.

Several basic rules of tissue transplantation were established using skin transplantation in inbred strains of mice. Autografts, the transplantation of skin from one part of the body to another or between genetically identical animals or individuals, survive. Allogeneic grafts, the transplantation of skin between nonidentical animals or individuals, are rejected. This rejection is T cell-dependent and is not exhibited by nude mice. The rejection after the first grafting of the tissue is called first set rejection, and if the same recipient receives a second graft from the same donor, the rejection process is more rapid and is called second set rejection. This rapid rejection is not observed when the graft is from a different donor.

The above-mentioned observations were also described in humans by P.B. Medawar in the 1940s. Based on his observations in humans, and later in animal experiments, he concluded that the graft rejection was a result of immune response, which eventually led to the discovery of MHC. The development of acquired immune response to transplanted tissue results in the rejection of the tissue where cytotoxic T cells, inflammatory T cells and antibodies play a major role in the rejection process.

Tissue Typing in Transplantation

After tissue transplantation, the severity and the period of rejection depend on the tissue type, and this process involves the specificity and memory components of the immune response. Avrion Mitchison in the 1950s observed that allograft immunity could be transferred by the components of the cellular immune response, and antibodies present in the serum that were part of the humoral response were not associated with this process. Future studies delineated the role of T lymphocytes in the allograft rejection process, and the role of both CD4$^+$ and CD8$^+$ cells was established.

The differences in HLA are responsible for the initiation of an immune response against the nonself HLA molecules. Tissues with similar HLA types are designated as histocompatible, whereas histoincompatibility induces immune responses resulting in allograft rejection. Consequently, the matching of the HLA types between donor and recipient is instrumental in controlling the rejection but it does not prevent it, which could be attributed to the polymorphism and complexity of the HLA. HLA-identical unrelated individuals mostly do not possess identical MHC genotypes. As expected, this will not be an issue with HLA-identical siblings, but still the grafts between HLA-identical siblings are slowly rejected with the exception of identical twins.

The HLA tissue typing plays an important role before the suitability of a donor and recipient could be determined. The first test is the matching of the blood groups followed by HLA typing, which is done by microcytotoxicity test, which is performed by using antibodies to the HLA class I and class II molecules in microtiter plates. Various classes of HLA class I and class II molecules are used; complement is added following an incubation period and cytotoxicity is measured by using one of the various dyes. If the monoclonal antibody is recognized by the antigen (HLA molecule) on the cell surface, the cell will then by lysed by the complement resulting in the uptake of the dye by the dead cells.

In addition to the major histocompatibility antigens, minor histocompatibility antigens (minor H antigens) are also involved in tissue rejection. However, in this case, graft rejection is slow and is the result of tissue transplantation where the donor and recipients differ at genetic loci other than the HLA. This response is much weaker because of the low number of T lymphocytes involved in this process. It has been reported that most minor H antigens may be bound to self HLA class I molecules since $CD8^+$ T cells respond to these antigens. Proteosomes in cytoplasm digest self proteins and subsequently, digested proteins are transported to rough endoplasmic reticulum where they bind to the HLA class I molecules and are expressed on the cell surface. If these expressed peptides are nonself as is the case in tissue transplantation, they will be recognized by T cells as foreign and an immune response will pursue. The response to minor H antigen resembles immune response produced against viral infections. Expression of minor H antigens on grafted cells causes the destruction of the graft.

Mechanisms of Rejection in Tissue Transplantation

The rejection of a transplant principally emanates from a cell-mediated immune response, which is specific for alloantigens that are primarily HLA molecules expressed on the cells of the transplanted tissue. It involves delayed-type hypersensitivity and cell-mediated cytotoxic reactions and can be divided into various distinct phases which are shown in Fig. 7.1. During the first phase, often defined by some as the afferent phase, the antigenic material shed by the graft is transported through the lymphatics to the draining lymph nodes. The antigens from the graft undergo uptake, processing and presentation in the context of HLA molecules by the APCs. The alloantigens of the graft are presented to both $CD4^+$ and $CD8^+$, cells which after recognition proliferate in response to the antigen. Although both minor and major HLA antigens are recognized, minor histocompatibility antigens cause a weak response. However, a strong response is observed when it is the result of a combination of several minor histocompatibility antigens. The response to the major histocompatibility antigens involves recognizing HLA molecules of the grafted tissue and associated peptide ligand in the cleft of the HLA molecules. The allogeneic cells produce proteins that form the grooves of the HLA class I molecules

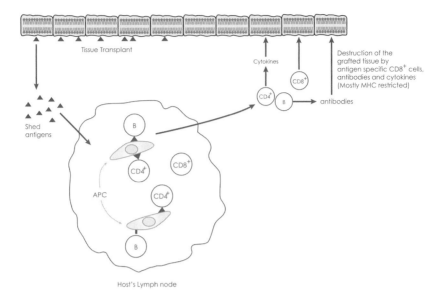

Fig. 7.1 Phases of tissue transplant rejection. The transplanted tissue sheds antigens. These antigens undergo uptake, processing and presentation to the T cells in the secondary lymphoid tissue by APCs, which include macrophages, B cells, Langerhans cells or dendritic cells. This phase results in the production of antibodies and antigen-specific TH and Tc cells. The antibodies and effector cells then migrate to the grafted tissue where TH cells secrete cytokines and which in combination with the antibodies and Tc cells destroy the grafted tissue (*see* Color Insert)

of the donor. In contrast, the allogeneic cells do not synthesize proteins that are present in the grooves of the HLA class II molecule; instead, they are the result of uptake and processing.

The next phase, also referred to as the central phase, involves the activation of the immune effector cells. When lymph nodes are examined following a tissue transplant, there is increased traffic of lymphocytes across the high endothelium of postcapillary venules. This is followed by the activation of TH cells, which results from the recognition of an alloantigen on the APCs in the context of HLA molecules and the required costimulatory signals. The APCs vary according to the transplanted organ but dendritic cells generally serve as APCs for most grafts. This is due to their ubiquitous distribution with high expression of the HLA class II molecules. During this phase, there is also the development of lymphoid follicles and generation of graft-specific cytolytic T cells. The APCs of the host may also enter the graft and then endocytose both of its major and minor histocompatibility alloantigens and present them to the T cells in the context of HLA molecules and the required costimulatory signals. There is a rapid proliferation of T cells in the recipient after recognition of the antigens on the transplanted tissue, and the degree of response varies depending on the organ/tissue transplant. The grafts placed in sites that do not possess lymphatic drainage or vasculature are not rejected. Examples include the brain, testes and anterior chamber of the eye, which are also called immunologically privileged sites. However, the skin, which does not possess major blood vessels at the time of the graft, later acquires the vessels, and lymphocytes of the recipients enter the graft, and its antigens are transported back to the lymph nodes, resulting in the development of an effector response.

The final phase in graft rejection is termed the efferent phase, or the effector phase. After activation in the spleen or lymph nodes, T cells, B cells and monocytes enter the transplanted tissue. The nature of the effector response corresponds with histoincompatibility as well as whether the rejection involves a primary or secondary graft. The most common effector mechanisms involved in tissue rejection include delayed-type hypersensitivity and cytotoxic T cell-mediated lysis. The other mechanisms include ADCC and complement-dependent cytotoxicity, which are less frequent. Histological analysis of rejected tissue has demonstrated the infiltration of $CD4^+$ T cells and macrophages in the graft, and the infiltration is similar to that observed in delayed-type hypersensitivity reaction. T cells cause tissue damage by the elaboration of cytokines, such as IFN-γ, TNF-α, TNF-β, and also by HLA class II-restricted cell-mediated cytotoxicity. Cytokines such as IFN-γ, TNF-α and TNF-β have direct as well as indirect effects in this process. An example of indirect effect is the augmentation of the expression of HLA molecules by these cytokines. IL-2 is required for T-cell proliferation, and the production of $CD8^+$ T cells and the local production of inflammatory cytokines and IFN-γ increase HLA expression in the graft. The recipient's $CD8^+$ T cells play a dominant role in the killing of the graft. During rejection of the kidney, liver, heart and pancreas transplants, the levels of HLA class I and class II molecules increase dramatically. The HLA class I molecule expression is increased in all grafted tissues including the endocrine cells of the pancreas; however, the induction of the HLA class II molecules is more selectively induced as is the case with hepatocytes where there is no induction of HLA class II molecules.

Memory T cells specific for allograft antigens present a serious challenge in organ transplantation. The presence of memory T cells not only results in a robust immune response to a transplanted organ but may also be responsible for a poor response to the immunosuppressive drugs. Lymphoid sequestration of memory $CD4^+$ T cells prolongs the survival of allografts in animal models, which can be achieved by immunosuppressive drugs such as fingolimod, which influences memory CD4 T-cell trafficking.

Adhesion molecules also play a crucial role in allograft rejection. The migration of activated T cells from secondary lymphoid organs to the graft site is regulated by adhesion molecules. Furthermore, these molecules are pivotal in the interaction of T cells with the functional components of tissue allografts that are epithelial in origin.

Role of Innate Immune Response in Allotransplant Rejection

Until recently, most of the attention regarding allograft rejection focused on T cells and acquired immune response. Pattern recognition receptors (PRRs), which sense conserved pathogen-derived molecules that differentiate infectious nonself from self and host-derived molecules liberated from injured tissue, are involved in the sentinel function of innate immune response. It has been suggested that allorecognition and rejection may also be the result of injury during tissue transplantation where PRRs may play a role in this process. The role of NK cells in tissue rejection is also important. They may not be sufficient to reject solid organs directly, but may act as facilitators of solid organ rejection by augmenting early graft inflammation and aiding the function of alloreactive T cells. Lastly, TLRs have also been implicated in playing some role in allograft rejection.

Types of Organ Rejection

The rejection period for a graft varies from tissue to tissue and the type of immune response involved in this process. They are divided into three types: hyperacute rejection, acute rejection and chronic rejection.

Hyperacute Rejection

This is a rare type of rejection that can occur very rapidly within a few days, which is the result of preexisting antibodies to the donor antigens. The events include a big infiltration of neutrophils into the grafted tissue caused by antigen–antibody complexes that activate the complement system and major clotting within capillaries as a result of inflammatory reactions, which does not permit the vascularization of the graft. Most grafts in clinical transplantation are vascularized and are directly linked to the host's circulation. In hyperacute rejection, these preexisting antibodies react with antigens on the vascular endothelial cells and cause immediate death of the graft. The cytotoxic antibodies against HLA antigens are formed due to several factors, which include repeated blood transfusions, several pregnancies where repeated exposure to paternal alloantigens of fetus results in the development of antibodies to these antigens and/or a previous transplant where antibodies are still left against the antigens of the original graft. At the present time, all recipients are screened for the presence of antibodies to the donor antigens, thus minimizing the chances of hyperacute rejection.

Acute Rejection

About 10 days after transplantation, acute rejection of the graft begins as a result of cell-mediated immunity. Acute rejection is a result of infiltration of large numbers of macrophages and lymphocytes into the graft. Helper T-cell activation and proliferation play a major role in this process, and both complement-dependent cell-mediated cytotoxicity and ADCC are involved in the destruction of the graft. Acute rejection could be in the form of acute vascular rejection, acute cellular rejection or both. Acute vascular rejection involves the necrosis of the blood vessel cells of the graft where thrombotic occlusion is not observed, but histologically the pattern is similar to vasculitis. This form of rejection is mediated by IgG antibodies that are directed against the alloantigens of endothelial cells and involves complement activation. T cells and cytokines, which induce endothelial necrosis, also contribute to acute vascular rejection.

Acute cellular rejection involves infiltration of macrophages and lymphocytes into the graft and is evident from the necrosis of the parenchymal cells of the graft. The lysis of the parenchymal cells of the transplanted tissue is achieved by the infiltrating leukocytes. Acute cellular rejection may be the product of several mechanisms including cytolytic T-cell-mediated lysis, NK cell-mediated lysis and activated macrophage-mediated lysis. The acute cellular rejection predominantly involves $CD8^+$ T cytolytic cells that kill the grafted tissue.

Chronic Rejection

In some cases, a slow rejection phase begins many months or even years after transplantation when acute rejection has subsided. The chronic rejection appears to be due to the slow buildup of antibodies against the graft antigens and/or due to cell-mediated immune responses by the recipient. It does not respond well to the immunosuppressive drugs, and a new transplant is needed following chronic rejection reaction.

Fibrosis resulting in the loss of normal organ structures is the hallmark of chronic rejection. The fibrosis may be due to wound healing, which is then followed by the cellular necrosis of acute rejection. However, it must be pointed out that chronic rejection develops many times in the absence of acute rejection. Fibrosis may be a result of several diverse factors such as equation of chronic rejection with chronic delayed-type hypersensitivity reaction, injury to blood vessels and resulting response to chronic ischemia, the proliferation of smooth muscle cells in the intima of arterial walls producing vascular occlusion, or persistent viral infections that will induce cellular immune response.

Transplant Tolerance

The long-term acceptance of an allograft is the basic purpose of tissue transplantation. However, the mechanisms potentially involved in achieving this goal are unknown. A number of hypotheses have been put forward to describe the mechanisms favoring tolerance to an allograft. According to one, in long-term survivors of tissue transplantation, there is clonal deletion of lymphocytes specific for the alloantigens of the graft, and in these recipients, the T cells responsible for attacking and killing the graft have been specifically eliminated. The second hypothesis suggests that T cells specific for the alloantigens of the graft have become anergic and consequently they lose their effector function and are unable to attack the graft in the recipient. The third suggested mechanism is the development of regulatory

T cells in transplant tolerance. The concept of suppressor T cells in achieving transplantation tolerance was introduced many years ago; however, their definite role was elusive since no markers were identified to characterize these cells and to delineate their role in tissue transplantation and immune tolerance. In 1990, Hall and colleagues reported that a subset of T cells (CD4+), which also expressed CD25 marker, transferred tolerance in a specific manner. Later, Sakaguchi and colleagues reported that CD4+CD25+ subset of T cells possessed suppressive properties and consequently the term suppressor T cells was replaced by regulatory T cells. The interaction of tolerogenic dendritic cells with regulatory T cells may downregulate the host's response to an allograft resulting in long-term graft survival. The tolerance induction may be achieved by blocking various cell surface markers including CD4, CD8, CD28, CD40L, CD80 and CD86 and as a result of the proliferation of regulatory T cells. One concept under study is an ex vivo expansion and administration of graft-specific regulatory T cells to the recipients of tissue transplantation. Nonetheless, T-cell regulation to achieve allograft tolerance is a difficult task since allografts are always rejected without immunosuppression.

Procedures for Preventing Graft Rejection

According to the laws of tissue transplantation, an allograft will be rejected. Since a good match between the graft and the recipient aids in prolonged acceptance of the transplanted tissue, one strategy of significance is to minimize the immunogenicity of the graft. The matching of three or four alleles of four HLA-A and -B loci is preferable. Human leukocyte antigen compatibility is also assessed for both HLA class I and class II molecules. During rejection, the immune response results in the destruction of the transplanted tissue. The CD95 (Fas or Apo-1) ligand (FasL) in the allograft is responsible for counteracting rejection by causing the apoptosis of lymphocytes and monocytes carrying the Fas and may induce an immune privileged status in allografts. Although this approach is still under investigation, other approaches have been used in clinical practice to improve the potential of transplant acceptance. Immune suppression before the transplantation of the graft is a major practice to avoid or delay the rejection process. The focus has been to pharmacologically eliminate or reduce T cells, which is achieved by the use of the immunosuppressive drugs or irradiation to suppress the immune response of the recipient. The preexisting antibodies are removed by plasmapheresis, and drugs including rapamycin, 15-deoxyspergualin and brequinar have been used to inhibit antibody synthesis. Other strategies include attempts to induce tolerance in the host. Renal transplant patients who were given multiple blood transfusions accepted kidney transplants better than their counterparts who did not undergo multiple allogeneic blood transfusions. Consequently, in clinical practice, multiple transfusions of blood have been used to induce tolerance prior to transplantation. Other strategies under trials include using peptides obtained from grafts' HLA molecules and administering the recipient with their high doses to induce tolerance. HLA donor molecules, and the antagonism of costimulatory molecules so the transduction signal to T cells is blocked, when they recognize antigens of the graft, thus inducing T-cell anergy have also been employed. The use of soluble CTLA-4 to antagonize the signal transduction between B7 molecules of the graft and CD28$^+$ T cells of the recipient is also of interest since this inhibits the production of IL-2, which induces anergy in graft-activated T cells.

Clinical Issues in Tissue Transplantation

The availability of tissue or organ is the first factor for tissue transplantation. United Network for Organ Sharing (UNOS) is a program designed for the collection and delivery of donor tissues or organs nationwide. The shortage of organs for transplantation, along with the viability of the tissue or organs, continues to be a problem. A graft is either a vascularized or a nonvascularized organ, and in cases of vascular organs, which include the heart, lung, liver and kidney, surgical anastomosis of the vascular system of the recipient is required. In contrast, the nonvascularized organs include islets of Langerhans and they are transplanted in a tissue where the presence of sufficient tissue fluids allows the delivery of nutrients and the removal of waste products. The survival of this tissue is not dependent on its attachment to the vasculature, and its normal function is expected. Successful harvesting of any tissue or organ is important for the desired outcome of tissue transplantation such that there is a minimum time of ischemia for each tissue.

Before a tissue is considered for grafting, it is important that it be matched with the recipient for HLA-A and HLA-B alleles. The graft survival improves with the degree of matching of HLA-A and HLA-B alleles between the donor and the recipient. This matching plays a major role in avoiding a major rejection during the first year after the tissue transplant. Since HLA-C does not play an important role in T-cell recognition, its alleles are not generally matched.

The tissue typing is done to determine the compatibility of the graft and the recipient for HLA and blood groups, and the existence of preexisting antibodies against the grafted donor tissue is also detected by using complement-dependent lymphocytotoxicity assay. A positive crossmatch will suggest a risk for hyperacute rejection and is indicative of a higher risk for vascular rejection immediately after transplantation and consequently the transplantation of this tissue is contraindicated. This problem is not usually observed for liver transplants but is important for heart and kidney transplants. The crossmatch assays are modified for kidney transplantation to increase their sensitivity. The techniques used include ELISA, flow cytometry, B lymphocyte crossmatches and antihuman globulin augmentation. The presence of alloreactive antibodies resulting from pregnancies, blood transfusions and previous transplants is determined in the serum by testing against a random cell panel. The results are expressed as percentage panel reactive antibody, which reflects the degree of sensitization and can vary from nonsensitivity (0%) to high sensitivity (80–100%). The task for finding a crossmatch negative donor decreases as the degree of sensitivity increases. The serum screening procedure is devised to detect HLA-specific antibodies. HLA molecules that have multiple epitopes are divided into two groups, private and public determinants. The examples of private determinants include HLA-A and HLA-B7. The public determinants, also called cross-reactive groups, are antigenic determinants where molecules with different private specificities share an epitope and include A2+A9+A28 and B7+22+27+40. The highly sensitive patients exhibit the same pattern of reactivity against a single or multiple public antigenic determinants. Cellular rejection is mediated by T lymphocytes.

In addition to matching HLA-A and HLA-B, matching of HLA-DR alleles is important. Due to strong linkage disequilibrium between HLA-DR and HLA-DQ, these two loci often match but DP typing is not done. As expected, due to a diverse population, the matching is often difficult in the United States compared to countries where there is more inbreeding. The graft's acceptance is also dependent on the HLA-DR type of the recipient. This is independent of matching at HLA-A, HLA-B alleles. This phenomenon is due to "immune response" (IR) gene effect, because mature T-cell repertoire selection is dependent on the host's HLA-DR molecules. An immune response to a graft antigen will not develop if the host's T cells lack specificity for that alloantigen.

Graft-Versus-Host Disease

Billingham in 1966 described the development of graft-versus-host disease as a result of the presence of immunocompetent cells in the graft, foreignness of the graft to the recipient and the inability of the recipient to mount an effective immune response against the graft. This disease results principally from bone marrow transplantation because the bone marrow contains the pluripotent hematopoietic stem cells, and the presence of immunocompetent cells in the graft allows it to mount an immune response against the recipient, which is called graft-versus-host reaction. This injures the host, and the process is called graft-versus-host disease.

The injury to the immune system of the host results in the graft-versus-host reaction, which could be either acute or chronic, that is, a response as the recipient tries to prevent the rejection of its stem cells. Epithelial cell necrosis characterizes the acute graft-versus-host disease, which mainly affects three organs, the liver, GI tract and skin. Skin rash, diarrhea and jaundice are the clinical manifestations of acute graft-versus-host disease. The later may result in death if the epithelial lining of the GI tract is completely damaged or the necrosis of skin is major. In chronic graft-versus-host disease, there is no acute cell necrosis but atrophy and fibrosis of the liver, GI tract and/or skin take place. The symptoms of chronic graft-versus-host disease also include skin rash, liver abnormalities, inflammation of the skin, mouth lesions, hair loss, indigestion, lung damage and drying of the eyes and mouth. Both acute and chronic graft-versus-host disease increase the risk of infections, and death will result from organ failure. The acute graft-versus-host disease is manifested within 8 weeks of the transplant, whereas the chronic form is seen within 12 weeks. The removal of the mature T cells from the bone marrow not only reduces the development of graft-versus-host disease, but also makes the transplantation less effective. The colony-stimulating factors released by mature T cells assist in the homing of the stem cells.

It has been suggested that single nucleotide polymorphisms in various cytokines may determine the severity of graft-versus-host disease, and these polymorphisms may suppress or enhance the inflammatory responses of the recipient. A role of IL-10 and not of IL-1β, IL-6 and TNF-α is suggested in the outcomes of graft-versus-host disease. It appears that factors other than histocompatibility between the graft and the recipient and the residual number of T cells in the graft may also be responsible for the pathogenesis of graft-versus-host disease. For example, the development of graft-versus-host disease in the recipients of either syngeneic or autologous bone marrow transplants cannot be explained only on the basis of major and minor histocompatibility differences, and the disease develops even when patients are taking cyclosporine. The mechanisms of cyclosporine-induced graft-versus-host disease in autologous bone marrow transplantation may be a result of dysregulation of self-tolerance.

In graft-versus-host disease, immunocompetent cells recognize epithelial lining of the target tissue as foreign, which induces an inflammatory response followed by the death of the target tissue by apoptosis. The apoptosis is not solely dependent on alloreactive T cells derived from the nonidentical donor, as it will take place even if the alloreactive T cells are from the recipient. Although T cells are responsible for the development of the initial inflammatory response, other cell types including NK cells are present on the site of epithelial injury and play a role in graft-versus-host disease.

Clinical Transplantation

Kidney Transplantation

The first kidney transplants were done in the 1950s. In the cases of advanced and permanent kidney failures, kidney transplantation is a potential treatment option where a kidney donor could be any normal healthy adult with normal kidney function. The absolute contraindications include inability to successfully administer the anesthesia or the immunosuppressive therapy to the recipient and the presence of diseases such as cancer, cardiopulmonary problems, infection and/or untreated peptic ulcer. The immunological evaluations performed prior to the grafting include blood grouping, HLA testing, presence of preexisting antibodies and/or T cells to HLA antigens and detection of viral infections. A kidney from blood group A can be transplanted to a recipient with blood group A or AB; for donor blood group B, the recipient should be of blood group B or AB; for donor blood group AB, the recipient must have the same blood type; individuals with blood group O are universal donors. The blood grouping is followed by HLA testing to evaluate the compatibility. Siblings (brothers and sisters) present the optimal chance for an excellent match (six of six HLA). However, kidneys from unrelated donors are also used because of advances in immunosuppressive therapy. Crossmatching is done to determine the presence of preexisting antibodies; a negative crossmatch is accepted by the recipient and a positive crossmatch is rejected by the recipient. The potential donor is also evaluated for cancer, diabetes, renal function, heart disease and/or infection. In addition to the blood tests for renal function, a three-dimensional CAT scan of the kidney and a renal angiogram are performed to exclude any potential abnormalities in the kidney's physical condition and the kidney's blood vessels. Furthermore, preoperative angiogram shows whether the right or the left kidney will be selected for grafting. If a potential donor is not available, cadaveric transplantation is considered. In most cases, the failing kidneys are not removed but the graft is placed generally in the iliac fossa and a different blood supply is used. The renal artery of the donor kidney is generally connected to the external iliac artery of the recipient. The renal vein of the donor kidney is generally linked to the external iliac vein of the recipient. The surgery takes about 3 h. The new kidney is expected to function immediately after surgery but may take up to a few days.

The long-term outcome of the transplant is dependent on the extent of the matching. The success rate decreases as the number of matching antigens decreases from six to five to four and so on. Nonetheless, due to the availability of the immunosuppressive drugs, mismatched grafts, that is, zero antigen-matched organ, also do well, and consequently, good tissue matching is preferred but not absolutely required for the successful outcome of kidney transplantation.

Postoperative immunosuppressive therapy is the most important aspect of the procedure. The most common immunosuppressive regimens include tacrolimus, mycophenolate and prednisone. Other regimens include cyclosporine, sirolimus, azathioprine or monoclonal antibodies. Hyperacute rejection is no longer a problem due to crossmatching before transplantation. Acute rejection will be exhibited in 10–25% of the patients within 60 days of grafting. This may not be an indication of total failure, but rather a reflection of a need for adjusting the immunosuppressive therapy. The symptoms of acute rejection include swelling and tenderness over the graft and a decrease in kidney function. The transplant rejection could also be chronic. Kidney transplantation is a life-extending surgery as it prolongs life from 10 to 15 years on average. As expected, the young benefit more than the elder recipients where life is extended up to 4 years. It is recommended that the kidney transplant be preemptive since extended periods of dialysis negate the period of survival of the grafted kidney.

Liver Transplantation

Liver transplantation was first performed in 1963 in the United States and United Kingdom. However, success was not achieved until 1967 when a recipient survived 1 year after the surgery. It remained an experimental procedure until the 1980s, which also coincided with the introduction of cyclosporine. In the 1970s, patient survival for 1 year was 25%, which has now improved to 85–90%. The indication for liver transplantation is an acute or chronic condition that causes an irreversible dysfunction of the organ with possible death within 2 years. The contraindications for the procedure include the recipient's HIV seropositivity, extrahepatic malignancy, hepatitis B seropositivity, drug and alcohol use, heart and lung disease, advanced age and/or active septic infection. However, exceptions to HIV seropositivity are now considered. ABO blood matching is done to select a suitable donor, and organ sizing is used for the suitability of the graft. The approach of organ sizing has been changed since now livers can be made smaller by surgery, which is particularly useful for transplantation in infants and children when an adult liver is used. Based on the regenerative capacity of the human liver, living donor liver transplantation has become a new option in recent decades. This procedure was originally performed in Brazil in 1986 for a child, but it was eventually realized that adult-to-adult transplantation was also feasible. The transplantation procedure is orthotopic as the damaged liver is removed and the donor liver is placed in the same location. There are three phases of the surgery: (a) hepatectomy, (b) anhepatic phase and (c) postgrafting phase. The hepatectomy involves severing all attachments of ligaments, bile duct, portal vein and hepatic artery; the implanting of the graft involves connecting the inferior vena cava, portal vein and hepatic artery, which is followed by connecting the bile ducts after the blood flow to the grafted liver is restored. On average, the surgery takes 4–7 h.

Human leukocyte antigen typing and leukocyte crossmatching are not done due to relatively short maximum cold ischemic time. ABO-compatible matches are also used under certain conditions since ABO identical grafts have better survival rate than compatible nonidentical grafts. Hemolysis is a major problem in ABO nonidentical liver transplants, which results from graft-versus-host reaction, which is due to the production of antibodies by the donor lymphocytes against ABO antigens of the recipient. Hemolysis is observed between 4 and 7 days after the transplantation. The symptoms include fever, increases in serum bilirubin levels and reticulocyte count and a decrease in serum hemoglobin levels. Hemolysis can be addressed by hydration, diuretics, blood transfusion and plasmapheresis. Human leukocyte antigen typing is also not performed because the liver is not very antigenic. Although expression of HLA class I molecules is minimal on hepatocytes, the presence of Kupffer cells, vascular endothelial cells and other inflammatory cells may cause rejection of the graft, therefore rejection is carefully monitored.

There are only a few reports of hyperacute rejection after liver transplantation. This may be due to the ability of the Kupffer cells to remove cytotoxic antibodies formed against the graft because of their reticuloendothelial function. Acute rejection is the more common form of rejection, which is manifested within 7–10 days after liver transplantation and exhibits symptoms of fever, malaise, pain, tachycardia and hepatomegaly. Mental disorientation in patients has also been reported during acute rejection. Liver biopsy is performed to confirm acute rejection that is generally mild in nature, and lymphocytic infiltration is observed in the portal tracts under the endothelium of the sinusoids.

Chronic rejection after liver transplant is rare (<3%), which is attributed to the recognition of acute rejection and the ability of the liver to regenerate itself. During chronic rejection, there is progressive failure of liver function, and humoral immune response is involved in this process as the liver exhibits fibrosis, arteriolar thickening and loss of bile ductules resulting in the loss of liver function, portal hypertension and jaundice. Chronic rejection can now be treated with immunosuppressive therapy and a new transplant is not needed. The immunosuppressive drugs for liver transplantation include corticosteroids in combination with tacrolimus, cyclosporine or sirolimus. The risk of chronic rejection decreases over time, but the immunosuppressants are taken on a permanent basis. This may be attributed to genotypic chimerism in the bone marrow of patients undergoing liver transplantation. The most critical period is within 3 months after the transplantation, and the 5-year survival rate is 76%.

Pancreas Transplantation

The pancreas was first transplanted in 1966. This was a transplant of multiple organs where kidney and duodenum were also transplanted in a 28-year-old woman; she exhibited a decrease in sugar levels immediately after transplantation, but died 3 months later due to pulmonary embolism. The first partial pancreatic transplant, in which the donor was a living relative, was performed in 1979 but until 1990, it was considered an experimental procedure.

The transplantation of the pancreas is mostly performed in individuals with Type I diabetes and is a life-enhancing and not a life-saving procedure. The indications for the transplant include diabetes, neuropathy, nephropathy or retinopathy. In addition to transplantation of the pancreas by itself, the procedure could also be simultaneously performed as pancreas–kidney transplantation or pancreas transplantation after kidney transplantation. The simultaneous pancreas–kidney transplantation is performed when both organs are from the same deceased person; the majority of pancreas transplantation (>90%) are simultaneous pancreas–kidney transplantation. The pancreas transplantations could be performed with either a whole organ, a segment or islets of Langerhans. Ideally, the transplant should be performed before the development of any diabetic complications. The pancreas transplant donors and recipients are first matched for blood group compatibility. The details of the requirements have already been described in reference to the kidney transplants. Rh factor is not considered for blood typing. The HLA testing is done for six markers including class I HLA-A, HLA-B and HLA-C and class II HLA-DP, HLA-DQ and HLA-DR as the number of proteins that these six markers encode is between 10,000 and 13,800. The panel-reactive antibody test is also performed to rule out the presence of preexisting antibodies.

The candidates for pancreas transplantation are evaluated for renal disease, diabetic retinopathy, coronary artery disease, gastroparesis, coronary artery disease, stroke, autonomic neuropathy and peripheral vascular disease. Gastroparesis (deficient gastric emptying) affects the use of immunosuppressive agents after transplantation, and other drugs including cisapride or metoclopramide are administered to treat the problem. Each candidate is thoroughly tested clinically before determining the suitability of the procedure. The tests include blood chemistry, CBC, liver function test, kidney function, chest radiography, exercise scintigraphy and, in some cases, stress cardiac ultrasound and/or cardiac arteriography. One of the most important tests is assessing the potential recipient for an existing infection, which may include hepatitis B and C, HIV, cytomegalovirus, tuberculosis and Epstein–Barr virus.

There are many different procedures used for pancreas transplantation, and there is no one standard protocol used in all transplant centers. The important considerations, however, are that the arterial blood flow supply to the pancreas and duodenal segment, and venous outflow from the pancreas via the portal vein should be adequate. The recipient's right common or external iliac artery is used to restore vascularization of the artery in the pancreas. The Y graft of the tissue is anastomosed end-to-side and the venous vascularization is performed either systemically or portally, but mostly it is done with systemic venous drainage.

Desired immunosuppression is more challenging for pancreas transplantation than for kidney or liver transplantation. T-cell alloimmune rejection response is a major problem, and lifelong immunosuppressive therapy must be provided. A short course of immunosuppressive therapy using intravenous treatment with antilymphocytic polyclonal antibodies or monoclonal antibodies, including Orthoclone OKT3, daclizumab, basiliximab, or alemtuzumab, is helpful, which provides protection from early acute rejection. Specifically, the antibody treatment is important when either other immunosuppressive agents such as cyclosporine, tacrolimus or sirolimus are administered or subtherapeutic doses are used until the function of the grafted pancreas improves. The maintenance immunosuppressive agents include prednisone, azathioprine, cyclosporine, tacrolimus or sirolimus. The current survival rate for pancreas transplantation is 97.6% after 1 year, the functional survival rate of pancreas transplant after 1 year is 83% and the success of a transplant is dependent on the age of the donor and HLA match.

Heart Transplantation

Ancient mythology and biblical references mention heart transplantation, but after the initial work of Alexis Carrel in the early 1900s, dog heart transplants were performed at Mayo Clinic in 1933. The work of Norman Shumway at Stanford University, reported in 1959, led to the first human transplant by Christian Bernard in 1967 in South Africa. By the 1970s, the enthusiasm diminished due to poor survival rate after heart transplantation until the discovery of cyclosporine.

Today, the indication for cardiac transplantation is the presence of end-stage heart failure or severe coronary artery disease. In general, normal functioning heart is removed from a recently deceased donor and is placed in a recipient patient. The recipient's original heart is taken out mostly by an orthotopic procedure. Although xenografts and artificial hearts have also been used for transplantation, they have exhibited limited success. Cardiac donors are heart-beating, brain-dead cadavers without any history of myocardial disease. The recipients have advanced irreversible heart disease, which may include congenital heart disease, heart valve disease, cardiomyopathy, life-threatening arrhythmias and/or coronary artery disease. The recipient could be less suitable if other circulatory disease is present. If cardiac function is normal, advanced donor age is no longer a contraindication. Donors and recipients are tested for ABO blood group compatibility and are matched for heart size. The recipients are also screened for the presence of preexisting HLA antibodies. For heart transplants, HLA matching is not routinely done despite concern from some cardiologists. One limiting factor is the period of viability of the heart, which does not allow enough time for HLA matching. Increased compatibility improves the outcome despite suggestions that the availability of an excellent immunosuppressive regimen will result only in a slight improvement in the outcome if HLA matching is done.

A heart is transplanted in a heterotopic or orthotopic position. Heterotopic transplant is preferred in patients with severe pulmonary hypertension. Orthotopic transplantation is done according to the technique developed by Shumway and Lower, or as a bicaval anastomosis. Immunosuppressive therapy is initiated right after the transplant surgery. There are several immunosuppressive regimens, which include both pretransplant induction therapy and posttransplant maintenance therapy. Each transplant center has its own preference for the choice of immunosuppressive drugs, which include cyclosporine, tacrolimus, sirolimus and/or corticosteroids. Hyperacute rejection has been observed only rarely in allograft cardiac transplants, but could occur immediately after restoring the blood flow for about a week. During the first month after transplantation, endomyocardial biopsies are performed weekly to detect rejection, and the frequency diminishes with time. Acute rejection is expected, and the incidence diminishes after 6–12 months of transplantation but could take place at any time after that as well. Treatment of acute rejection includes corticosteroid and polyclonal or monoclonal antibodies depending on the severity of the rejection. Lymphocytes, lymphoblasts and monocytes are the predominant cells in acute rejection. Late deaths, 1 year after transplant, are attributed to chronic rejection, which may also involve humoral mechanisms. Currently, the average patient and graft survival rates are 81.8 and 69.8% at 1 and 5 years, respectively, and 10-year survival rates have been reported for a significant number of patients.

Lung Transplantation

The first human lung transplant was performed in 1963. Today, it is considered an excellent alternative for children with end-stage lung disease. Single-lung, double-lung and living donor lobar lung transplantation have been performed. The lung transplantation is recommended in patients with untreatable end-stage disease attributed to multiple etiologies. Congenital heart disease, pulmonary vascular disease and/or idiopathic pulmonary hypertension result in end-stage lung disease in children less than 1 year old. Cystic fibrosis results in end-stage lung disease in children who are 1–10 years old and is responsible for 36% of the total lung transplants. Pulmonary fibrosis, congenital cardiac disease and chronic lung disease of infancy also commonly lead to lung transplants. Donors and recipients are matched for size, thoracic dimensions and ABO blood group. Crossmatching is also done to exclude the possibility of the presence of preexisting antibodies. Absolute contraindications include the presence of malignancy, tuberculosis, infection, neuromuscular disease, renal malfunction and/or immunodeficiency disorders. The potential donors are screened for infections, smoking habits and should be less than 60 years old. The pediatric surgical technique most commonly used in the United States is bilateral sequential procedure with telescoping anastomoses. Lower lobe of the lung is used if a living lobar family donor is used, and this is no longer an emergency procedure, but rather an elective one. Immunosuppression after the lung transplantation is achieved by using cyclosporine, tacrolimus, sirolimus, everolimus, corticosteroids, IL-2 receptor antagonists and/or azathioprine. Acute pulmonary rejection is treated with short-time increased corticosteroid therapy or by using polyclonal or monoclonal antibodies including antilymphocytic immunoglobulins or OKT3. The symptoms of rejection include flu-like symptoms, fever, breathing difficulties, nausea, chest pain and decreased pulmonary function. Chronic rejection is manifested in about 50% of the patients and is presented as bronchiolitis obliterans. Survival rates are 70.7, 54.8, and 42.6% at 1, 3, and 5 years, respectively.

The first combined heart–lung transplant was performed in 1981 at Stanford University. Combined heart–lung donors need to satisfy both the requirements already described separately. Combined heart–lung transplant is recommended in patients with congenital problems affecting these organs, pulmonary hypertension and/or cystic fibrosis. The recipients for the combined transplant are recommended to be less than 55 years old. Survival rates are 79, 66, and 54% at 1 month, 1 year and 3 years, respectively, after transplantation.

Bone Marrow Transplantation

Bone marrow and stem cell transplantation is used to treat both malignant and nonmalignant diseases. The foundation for the bone marrow transplantation was set by E. Donnell Thomas between the 1950s and 1970s at the Fred Hutchinson Cancer Research Center, and the first successful allogeneic bone marrow transplant was performed in 1968. The stem cell transplant could be from an autologous or an allogeneic source. Autologous stem cells are obtained from the patient and are stored in the freezer. The patient's bone marrow is either damaged by high doses of chemotherapy or subject to irradiation due to malignancy. Transplanting autologous bone marrow comes with a lower risk of graft-versus-host disease (GVHD), infection and rejection. Allogeneic stem cell transplantation requires HLA matching as well as an immunosuppressive regimen. The indications for stem cell transplant include nonmalignant diseases such as inherited immune disorders, inherited metabolic disorders, marrow failure and autoimmune disease, and various malignant leukemias and lymphomas.

The most compatible matches come from fully matched family members. Routine testing involves matching of six antigens, HLA-A, HLA-B, HLA-C, HLA-DR, HLA-DQ and HLA-DP, each of which have two alleles, and the recipient and the donor are considered a mismatch if all six antigens are different. The haplotype donors match only three of the six antigens. It is important to match all six antigens when the donor is not related to the recipient. The transplantation procedure is divided into five phases, which include conditioning, stem cell processing and infusion, neutropenic phase, engraftment and postengraftment period. The patient undergoes chemotherapy or radiation during the conditioning phase to create space or destroy the malignant cells. This could be achieved by the administration of cyclophosphamide, busulfan, intravenous immunoglobulin (ALG) or by total lymphoid irradiation.

A new treatment approach is non-myeloablative allogeneic human stem cell transplant. Low doses of chemotherapy and radiation are used for this procedure, which does not eliminate all bone marrow cells. In this case, the recipient benefits from the graft-versus-tumor effects of the non-myeloablative transplant, and the recipients are treated with high doses of immunosuppressive agents during the early stages of the procedure. The result of this treatment is the presence of a state of mixed chimerism right after the procedure, since stem cells of both the recipient and the donor coexist. The T cells from the donor marrow eradicate the stem cells of the recipient as the dose of the immunosuppressive therapeutic regimen is decreased, which induces GVHD and the graft-versus-tumor effect. This procedure also results in a low mortality rate and is recommended for high-risk patients but is not widely available at the present time.

The stem cells must home themselves in the bone marrow since they cannot be placed at a particular location in the bone marrow. There are a limited number of "niches" within marrow cavities and they are occupied, and this concern is addressed by using the conditioning phase to vacate sites where the stem cells can be placed. T cells are depleted to lessen the severity of the GVHD. The neutropenic phase represents the period with an increased risk of infection due to a totally deficient immune response and a period of poor healing. This is followed by the engraftment phase where healing begins, and fever and infections are not a major problem with the exception of some viral infections. This period could last up to several weeks. This is the period where graft cells may mount a rejection response against the host (GVHD). Allogeneic stem cells are readily rejected, and this could happen even if the host is minimally immunocompetent. The last phase is characterized as the postengraftment period, which could last from months to years. This phase manifests the development of tolerance, development of the immune system and control of the chronic host-versus-graft disease. The rejection could result from acute or chronic GVHD, which requires intensive immunosuppressive therapy. Acute GVHD is treated with high doses of corticosteroids, while chronic GVHD can be treated by various immunosuppressive therapeutic regimens.

Bibliography

Abt P, Shaked A. 2003. The allograft immune response. Graft. 6:71–79.
Alegre ML, Najafian N. 2006. Costimulatory molecules as targets for the induction of transplantation tolerance. Curr Mol Med. 6:843–857.

Antin JH, Ferrara JL. 1992. Cytokine dysregulation and acute graft vs host disease. Blood. 80:2964–2968.

Arkelov A, Lakkis FG. 2000. The alloimmune response and effector mechanisms of allograft rejection. Semin Nephrol. 20:95–102.

Azuma H, Tilney NL. 1995. Immune and nonimmune mechanisms of chronic rejection of kidney allografts. J Heat Lung Transplant. 14:S136–S142.

Bellgrau D, Gold D, Selawry H, Moore J, et al. 1995. A role for CD95 ligand in preventing graft rejection. Nature. 377:630–632.

Benichou G, Fedoseyeva EV. 1996. The contribution of peptides to T cell allorecognition and allograft rejection. Int Rev Immunol. 13:231–243.

Benjamin LC, Allan JS, Madsen JC. 2002. Cytokines in immunity and allograft rejection. Crit Rev Immunol. 22:269–279.

Betkowski AS, Graff R, Chen JJ, Hauptman PJ. 2002. Panel- reactive antibody screening practices prior to heart transplantation. J Heart Lung Transplant. 21:644–650.

Billingham RE. 1966. The biology of graft-versus-host reactions. Harvey Lect. 67:21–78.

Bishop DK, Shelby J, Eichwald EJ. 1992. Mobilization of T lymphocytes following cardiac transplantation. Evidence that CD4- positive cells are required for cytotoxic T lymphocyte activation, inflammatory endothelial development, graft infiltration and acute allograft rejection. Transplantation. 53:849–857.

Bishop DK, Li W, Chan SY, Ensley RD, et al. 1994. Helper T lymphocytes unresponsiveness to cardiac allografts following transient depletion of CD4-positive cells. Implications for cellular and humoral responses. Transplantation. 58:576–584.

Brent L. 1997. History of Transplantation Immunology. Academic Press, San Diego, CA.

Broelsch CE, Burdelski M, Rogiers X, Gundlach M, et al. 1994. Living donor for liver transplantation. Hepatology. 20:49S–55S.

Bromberg JS, Murphy B. 2001. Routes to allograft survival. J Clin Inv. 107:797–798.

Brook NR, Nicholson ML. 2003. Kidney transplantation from non heart-beating donors. Surgeon. 1:311–322.

Bucin D. 1988. Blood transfusion in renal transplantation – The induction of tolerance by incompatibility for class I antigen. Med Hypothesis. 27:19–27.

Chalasani G, Li Q, Konieczny BT, Smith-Diggs L, et al. 2004. The allograft defines the type of rejection (acute vs chronic) in the face of an established effector immune response. J Immunol. 172:7813–7820.

Codarri L, Vallotton L, Ciuffreda D, Venetz JP, et al. 2007. Expansion and tissue infiltration of an allospecific CD4+CD25+CD45RO+IL-7R (alpha) high cell population in solid organ transplant recipients. J Exp Med. 204:1533–1541.

Cooper DK. 1969. Transplantation of the heart and both lungs. I. Historical review. Thorax. 24:383–390.

Doyle AM, Lechler RI, Turka LA. 2004. Organ transplantation: Half-way through the first century. J Am Soc Nephrol. 15:2965–2971.

Dreger P, Kloss M, Petersen B, Haferlach T, et al. 1995. -Autologous progenitor cell transplantation: Prior exposure to stem cell-toxic drugs determines yield and engraftment of peripheral blood progenitor cell but not of bone marrow grafts. Blood. 86: 3970–3978.

El-Asady RS, Rongwen Y, Hadley GA. 2003. The role of CD103 ([alpha]E[beta] 7 integrin) and other adhesion molecules in lymphocyte migration to organ allografts: Mechanisms of rejection. Curr Opin Organ Transplant. 8:1–6.

Ferrara JL, Deeg HJ. 1991. Graft-versus-host disease. NEJM. 324:667–674.

Ferrara JLM, Cooke KR, Pan L, Krenger W. 1996. The immunopathophysiology of acute graft-versus-host-disease. Stem cells. 14:473–489.

Field EH, Matesic D, Rigby S, Fehr T, et al. 2001. CD4+CD25+ regulatory cells in acquired MHC tolerance. Immunol Rev. 182:99–112.

Flye MW (Ed.). 1989. Principles of Organ Transplantation. W. B. Saunders, Philadelphia, PA.

Galili U. 1993. Interaction of the natural anti-Gal antibody with α-galactosyl epitopes: A major obstacle of xenotransplantation in humans. Immunol Today. 14:480–482.

Game DS, Lechler RI. 2002. Pathways of allorecognition: Implications for transplantation tolerance. Transpl Immunol. 10:101–108.

Gebel HM, Bray RA. 2000. Sensitization and sensitivity: Defining the unsensitized patients. Transplantation. 69:1370–1374.

Goulmy E, Schipper R, Pool J, Blokland E, et al. 1996. Mismatches of minor histocompatibility antigens between HLA-identical donors and recipients and the development of graft-versus-host-disease after bone marrow transplantation. NEJM. 334:281–285.

Griepp RB, Ergin MA. 1984. The history of experimental heart transplantation. J Heart Transplant. 3:145–145.

Griffith BP, Hardesty RL, Trento A, Bahnson HT. 1985. Asynchronous rejection of heart and lungs following cardiopulmonary transplantation. Ann Thorac Surg. 40:488–493.

Griffith TS, Brunner T, Fletcher SM, Green DR, et al. 1995. Fas ligand-induced apoptosis as a mechanism of immune privilege. Science. 270:11889–1192.

Hakim NS. 2003. Recent developments and future prospects in pancreatic transplantation. Exp Clin Transplant. 1:26–34.

Hall B, Jelbart ME, Gurley KE, Dorsch SE. 1990. Specific unresponsiveness in rats with prolonged cardiac allograft survival after treatment with cyclosporine. III. Further characterization of CD4$^+$ suppressor cell and its mechanisms of action. J Exp Med. 171:141–157.

Hall BM, Chen J, Robinson C, Xy H, et al. 2002. Therapy with mab to CD25 blocks function of CD4+CD25+ T regulatory cells which maintain transplantation tolerance. Nephrology. 7:A111–A111.

Hayry P, Isoniemi H, Yilmaz A, Mennander K, et al. 1993. Chronic allograft rejection. Immunol Rev. 134:33–81.

Hayry P. 1996. Pathophysiology of chronic rejection. Trans Proc. 28:7–10.

Kahan BD. 1991. Transplantation timeline. Mankind's three millennia-one maverick's three decades in the struggle against biochemical individuality. Transplantation. 51:1–21.

Kang SM, Tang Q, Bluestone JA. 2007. CD4$^+$CD25$^+$ regulatory T cells in transplantation: Progress, challenges and prospects. Am J Transplant. 7:1457–1463.

Kaufman DB, A. 2006. Pancreas transplantation. Koffron www.emedicine.com/med/topic 2605.htm.

Kerman RH, Orosz CG, Lorber MI. 1997. Clinical relevance of anti-HLA antibodies pre and post transplant. Am J Med Sci. 313:275–278.

Kitchens WH, Uehara S, Chase CM, Coluin RB, et al. 2006. The changing role of natural killer cells in solid organ rejection and tolerance. Transplantation. 81:811–817.

Klein J. 1986. Natural History of the Major Histocompatibility Complex. Wiley, New York.

Krensky AM. 2004. Immunologic tolerance. Ped Nephrol. 16:675–679.

Krensky AM, Clayberger C. 2004. Prospects for induction of tolerance in renal transplantation. Ped Nephrol. 8:772–779.

Krieger NR, Yin DP, Fathman CG. 1996. CD4$^+$ but not CD8$^+$ cells are essential for allorejection. J Exp Med. 184:2013–2018.

LaRosa DF, Rahman AH, Turka LA. 2007. The innate immune system in allograft rejection and tolerance. J Immunol. 178:7503–7509.

Lattmann T, Hein M, HOrber S, Ortmann J, et al. 2005. Activation of pro-inflammatory and anti-inflammatory cytokines in host organs during chronic allograft rejection: Role of endothelin receptor signaling. Am J Transplant. 5:1042–1049.

Lau CL, Palmer SM, Posther KE, Howell DN, et al. 2000. Influence of panel-reactive antibodies on post transplant outcomes in lung transplant recipients. Ann Thorac Surg. 69:1520–1524.

Lau HT, Yu M, Fontana A, Stoeckert CJ. 1996. Prevention of islet allograft rejection with engineered myoblasts expressing FasL in mice. Science. 273:109–112.

Majno G. 1975. The Healing Hand. Man and Wound in the Ancient World. Harvard University Press, Cambridge, MA.

Mancini MC. 2006. Heart-lung transplantation. www.emedicine.com/med/topic2063.htm.
Mancini MC, Gangahar DM. 2006. Heart transplantation. www.emedicine.com/med/topic 3187.htm.
Mandanas RA. 2006. Graft vs. host disease. www.emedicine.com/medtopic926.htm.
Manzarbeitia C. 2006. Liver transplantation. www.emedicine.com/med/topic3510.htm.
Matthews LG. 1968. S.S. Cosmas and Damian – Patron Saints of medicine and pharmacy. Their cult in England. Med Hist. 12:281–288.
Medawar PB. 1986. Memoir of a Thinking Radish. Oxford University Press, Oxford.
Merrill JP, Murray JE, Harrison JH, Guild WR. 1984. Landmark article Jan. 28, 1956. Successful homotransplantation of the human kidney between identical twins. JAMA. 251:2566–2571.
Morris PJ. 2004. Transplantation – A medical miracle of the 20th century. NEJM. 351:2678–2680.
Murphy B, Krensky AM. 1999. HLA-derived peptides as novel immunomodulatory therapeutics. J Am Soc Nephrol. 10:1346–1355.
Norman DJ. 1998. Expected clinical outcomes/risk factors. In: Norman DJ, Suke WN, Thorofare NJ, Eds. Primer on Transplantation, Am. Soc. Transplant Physicians, Thorofare, NJ, pp. 245–250.
Paul LC. 1995. Immunobiology of chronic renal transplant rejection. Blood Purif. 13:206–218.
Petersdorf EW, HansenJA, Martin PJ, Woolfrey A, et al. 2001. Major-histocompatibility-complex class I alleles and antigens in hematopoietic cell transplantation. NEJM. 345:1794–1800.
Philip T, Gugliemli C, Hagenbeek A, Somers R, et al. 1995. Autologous bone marrow transplantation as compared with salvage chemotherapy in relapses of chemotherapy-sensitive non-Hodgkin's lymphoma. NEJM. 333:1540–1545.
Pinderski LJ, Kirklin JK, McGiffin D, Brown R, et al. 2005. Multi-organ transplantation: Is there a protective effect against acute and chronic rejection? J Heart Lung Transplant. 24:1828–1833.
Prescilla RP, Mattoo TK. 2006. Immunology of transplant rejection. www.emedicine.com/ped/ topic2841.htm.
Rana RE, Arora BS. 2002. History of plastic surgery in India. J Postgrad Med. 48:76–78.
Reitz BA, Wallwork JL, Hunt SA, Pennock JL, et al. 1982. Heart-lunch transplantation: Successful therapy for patients with pulmonary vascular disease. NEJM. 306:557–564.
Rocha PN, Plumb TJ, Crowley SD, Coffman TM. 2003. Effector mechanisms in transplant rejection. Immunol Rev. 196:51–64.
Roopenian D, Choi EY, Brown A. 2002. The immunogenomics of minor histocompatibility antigens. Immunol Rev. 190:86–94.
Rosenberg AS, Singer A. 1992. Cellular basis of skin allograft rejection: An in vivo model of immune-mediated tissue destruction. Ann Rev Immunol. 10:333–358.
Rossini AA, Greiner DL, Mordes JP. 1999. Induction of immunologic tolerance for transplantation. Physiol Rev. 79:99–41.
Roush W. 1995. Xenotransplantation. New ways to avoid organ rejection and buoy hopes. Science. 270:234–235.
Rubinstein P. 2001. HLA matching for bone marrow transplantation – How much is enough? NEJM. 345:1842–1844.
Ruiz P, Sarwar S, Suterwala MS. 2006. Graft versus host disease. www.emedicinecom/ped/topic 893.htm.
Sablinski T, Sayegh MH, Hancock WW, Kut JP, et al. 1991. Differential role of CD4+ cells in the sensitization and effector phases of accelerated graft rejection. Transplantation. 51:226–231.
Sade RM. 2005. Transplantation at 100 years: Alexis Carrel, pioneer surgeon. Ann Thorac Surg. 80:2415–2418.
Sakaguchi S. 2006. Regulatory T cells: Meden again. Immunol Rev. 212:1–5.
Sakaguchi S, Ono M, Setoguchi R, Yagi H, et al. 2006. Foxp3+CD25+Cd4+ natural regulatory T cells in dominant self tolerance and autoimmune disease. Immunol Rev. 212:8–27.
Salvalaggio PRO, Camirand G, Ariyan CD, Deng S, et al. 2006. Antigen exposure during enhanced CTLA-4 expression promotes allograft tolerance in vivo. J Immunol. 176:2292–2298.
Sanfilippo F. 1998. Transplantation tolerance – The search continues. NEJM. 339:1700–1702.

Sayegh MH, Carpenter CB. 1996. Role of indirect allorecognition in allograft rejection. Int Rev Immunol. 13:221–229.

Sayegh MH, Turka LA. 1998. The role of T cell costimulatory activation pathways in transplant rejection. NEJM. 338:1813–1821.

Sayegh MH, Carpenter CB. 2004. Transplantation 50 years later – Progress, challenges and promises. NEJM. 351:2761–2766.

Scheinfeld NS, Kuechle MK. 2007. Graft versus host disease. www.emedicine.com/derm/toic478.htm.

Shapiro AMJ, Lakey JRT, Ryan EA. 2000. Islet transplantation in seven patients with type 1 diabetes mellitus using a glucocorticoid-free immunosuppressive regimen. NEJM. 343: 230–238.

Sho M, Kishimoto K, Harada H, Livak M, et al. 2005. Requirements for induction and maintenance of peripheral tolerance in stringent allograft models. Proc Nat Acad Sci. 102:13230–13235.

Silverstein AM. 1988. A History of Immunology. Academic Press, San Diego, CA.

Socie G, Mary JY, Lemann M. 2004. Prognostic value of apoptotic cells and infiltrating neutrophils in graft-versus-host disease of the gastrointestinal tract in humans: TNF and Fas expression. Blood. 103:50–57.

Spierings E, Vermeulen CJ, Vogt MH, Doerner LE, et al. 2003. Identification of HLA class II restricted H-Y-specific T-helper epitope evoking CD4$^+$ T-helper cells in H-Y-mismatched transplantation. Lancet. 362:610–615.

Strom TB. 2004. Is transplantation tolerable? J Clin Inv. 113:1681–1683.

Sullivan KM. 1999. Graft-versus-host disease. In: Thomas ED, Ed. Hematopoietic Cell Transplantation. 2 nd Ed. Blackwell Science, Boston, MA.

Sykes M. 1996. Immunobiology of transplantation. FASEB J. 10:721–730.

Tambur AR, Pamboukian SV, Costanzo MR, Herrera ND, et al. 2005. The presence of HLA-directed antibodies after heart transplantation is associated with poor allograft outcome. Transplantation. 80:1019–1025.

Terasaki PE. (Ed.). 1991. A History of Transplantation: Thirty-Five Recollections. UCLA Tissue Typing Laboratory Press, Los Angeles, CA.

Thomas ED, Lochte HL Jr., Lu WC, Ferrebee JW. 1957. Intravenous infusion of bone marrow in patients receiving radiation and chemotherapy. NEJM. 257:491–496.

Tilney NL. 2000. Transplantation and its biology: From fantasy to routine. J Appl Physiol. 89:1681–1689.

Valujskikh A, Lakkis FG. 2003. In remembrance of things past: Memory T cells and transplant rejection. Immunol Rev. 196:65–74.

Valujskikh A. 2006. The challenge of inhibiting alloreactive T-cell memory. Am J Transplant. 6:647–651.

Van Buskirk AM, Pidwell DJ, Adam PW, Orosz CG. 1997. Transplantation immunology. JAMA. 278:1993–1999.

Van Twuyver E, Mooijaart RJ, tenBerge IJ, Van der Horst AR, et al. 1991. Pretransplantation blood transfusion revisited. NEJM. 325:1210–1213.

Zhang Q, Chen Y, Fairchild RL, Heeger PS, et al. 2006. Lymphoid sequestration of alloreactive memory CD4 T cells promotes cardiac allograft survival. J Immunol. 176:770–777.

Zheng XX, Markees TG, Hancock WW, Li Y, et al. 1999. CTLA4 signals are required to optimally induce allograft tolerance with combined donor specific transfusion and anti-CD154 monoclonal antibody treatment. J Immunol. 162:4983–4990.

Chapter 8
Acquired Immune Deficiency Syndrome

Introduction

In 1959, a Bantu man died of an unidentified illness in Belgian Congo, and later analysis of his blood samples confirmed him to be the first case of HIV infection. Human immunodeficiency virus-1 (HIV-1) and -2 (HIV-2), which cause AIDS, evolved from strains of simian immunodeficiency virus SIVepz and SIVsm, respectively. SIVepz infects a subspecies of chimpanzees and SIVsm infects sooty mangabeys but AIDS is not observed in these hosts. Humans may have been exposed to SIV, but these strains are not able to adapt, infect and transmit between humans. It is suggested that HIV-1 originated from a subspecies of chimpanzees as HIV-1 groups M and N have their origins in a distinct population of chimpanzees in Cameroon. The spread may have started after World War II. It is not known how many people developed this disease in the 1970s or before that, but in 1978 gay men in the United States and Sweden and heterosexuals in Haiti and Tanzania exhibited symptoms that were later identified as the acquired immune deficiency syndrome.

In 1981, cases of AIDS were diagnosed in young gay men in the United States, and the disease was called gay-related immune deficiency (GRID). Physicians in New York and Los Angeles observed cases of *Pneumocystis carinii* pneumonia and Kaposi's sarcoma, which led the Centers for Disease Control and Prevention (CDC) to monitor young men, women and babies with severe immune deficiency. It was soon realized that this was not a local phenomenon, but rather a global epidemic. Fourteen nations reported cases of AIDS in 1982 and at the same time reports emerged of a hemophilic patient developing AIDS due to blood transfusion and infected infants born to mothers with AIDS. It was not until late 1982 when this condition began to be diagnosed as AIDS, and in 1983, reports surfaced that the disease may be passed to women through heterosexual sex. In May 1983, Dr. Luc Montagnier of the Pasteur Institute in France reported that they had isolated a new virus, which may be responsible for AIDS that received little attention, and the virus was later named lymphadenopathy-associated virus (LAV). In 1984, Robert Gallo of the National Cancer Institute isolated human T-cell lymphotropic virus-III (HTLV-III), and this virus was considered to be responsible for AIDS. However, in 1985, LAV and HTLV-III were identified to be the same, and in 1986, it was renamed

M.M. Khan, *Immunopharmacology*, DOI: 10.1007/978-0-387-77976-8_8,
© Springer Science+Business Media, LLC 2008

human immunodeficiency virus (HIV). Attempts were underway to develop a test that could detect the presence of the virus in the blood since its transmission was not accompanied by immediate symptoms for the diagnosis.

AIDS is now a pandemic, and according to the Joint United Nations Program on HIV/AIDS, 58 million people have been infected with HIV and 25 million have died worldwide. In the twenty-first century, it remains an ultimately fatal disease that is difficult to treat, but the development of new drugs has improved the outlook of patients infected with the virus in the United States and Western Europe.

Human Immunodeficiency Virus

AIDS is a result of immunodeficiency attributed to the HIV lentiviruses which, in contrast to the herpes viruses, replicate constantly. There is no period of viral latency after infection unless some infected cells contain nonreplicating virus. HIV infects only humans and chimpanzees and has two major families, HIV-1 and HIV-2. HIV-1 is responsible for the spread of the infection worldwide, while HIV-2 is endemic in Western Africa. HIV-1 and HIV-2 share about 40% of the genome. HIV-1 has at least five subfamilies or clades. An HIV particle is about 100–150 billionths of a meter in diameter, and it is a retrovirus surrounded by a viral envelope (coat of lipid bilayer) that projects spikes of gp120 and gp41 proteins. A layer called the matrix, which is made up of the protein p17, is present under the viral envelope. The viral core (capsid) is composed of protein p24 in which are contained three enzymes, reverse transcriptase, integrase and protease, which are required for HIV replication. Two copies of the genome (RNA) are contained in the nucleocapsid core. HIV is composed of nine genes, three of which are the structural genes gag, pol and env. Gag encodes major structural proteins of the virus, pol encodes the reverse transcriptase, the proteases and the viral integrase, and env encodes the proteins that are responsible for the attachment of the virus and entry to the cell. The other six genes, tat, rev, nef, vif, vpr and vpu, are responsible for the translation of regulatory proteins required for infection. The regulatory proteins enhance virion production and counter host defense. A sequence called the long-term repeat is present on each of the RNA strands and serves as a control mechanism for HIV replication. (Fig. 8.1)

Human Immunodeficiency Virus Replication

The replication of HIV takes place only in the human cell. The process begins after the recognition of CD4 molecule on human cells by the virus and the glycoprotein gp120, along with gp41, infects any target cell that expresses CD4 receptors. The gp120 portion binds to CD4 with high affinity and gp41 mediates fusion of the virus with the membrane of the target cell. The interactions involve the trimeric envelope complex (gp160), CD4 and a chemokine receptor. The chemokine receptors are generally either CCR5 or CXCR4 but others may also interact. The

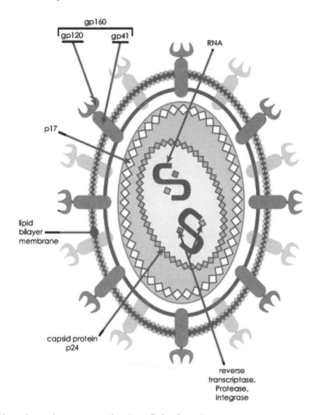

Fig. 8.1 HIV, a schematic representation (*see* Color Insert)

gp120 and gp41 portions collectively form gp160 spike with three transmembrane glycoproteins (gp41) and three extracellular glycoproteins (gp120). The fusion initiates with the high-affinity binding of CD4-specific domains of gp120 to the N-terminal membrane-distal domain of CD4. After the binding of gp120 to CD4, there is a structural change in the envelope resulting in the exposure of the chemokine-binding domain of gp120. This allows the gp120–CD4 complex to interact with chemokine receptors, which results in the penetration of gp41 into the cell membrane. A hairpin structure is formed as a result of the collapse of the extracellular portion of gp41, which assists the virus in entering the host cell.

After the entry of the virus into the cell, single-stranded RNA is separated from the viral proteins, and cDNA is then transcribed from RNA by reverse transcriptase. A second strand of cDNA is synthesized to make double-stranded viral DNA, and this viral double-stranded DNA may remain in the cytoplasm as unintegrated circular DNA or migrate to the nucleus. In the nucleus, a complete genomic copy of viral DNA is integrated into the host cell genome by viral integrase. After integration, infection can be latent without active viral replication, which requires upregulation of T cells. The upregulation of T cells results in the activation of transcription factor NF-κB, which is needed to actively produce the HIV. The mRNA is made from the integrated provirus and is spliced into smaller chunks, which produce Tat and Rev, two regulatory proteins. Tat stimulates new virus production and Rev inhibits mRNA splicing. Full-length mRNA now produces structural proteins including gag and env, which surround the genomic RNA, and nucleocapsids are formed. At the cell surface, the transmembrane envelope and structural proteins assemble, which are concentrated in cholesterol-rich lipid rafts, resulting in the packaging of new virus particles. The final step in this process is assembly and release as the plasma membrane of the host cell is the site of the assembly of the new HIV virions. After migrating through the endoplasmic reticulum, the env polyprotein (gp160) enters the Golgi complex, where protease processes gp160 into two glycoproteins gp41 and gp120. These molecules are transported into the cell membrane of the virus and gag (P55) and gag–pol (P160) proteins associate on the inside of the membrane. The polyproteins of HIV are cleaved into individual functional HIV proteins and enzymes by HIV proteases during maturation, and viral particles are now completely assembled and are released by budding from the host cells.

Human Immunodeficiency Virus and Disease

HIV spreads by contact with body fluids, and the common modes of infection include sexual intercourse and contaminated needles used for intravenous drug delivery. Previously, therapeutic use of infected blood and infected breast milk for the baby were also modes of transmission but both have declined due to better screening for blood and AIDS education of expectant mothers. The virus is carried in infected CD4+ T cells, macrophages or as free viruses in blood, semen, vaginal fluids and milk.

   In general, viral infections are acute and limited, resulting in the development of lasting protective immunity. This is not the case with HIV, and an immune response that could completely eliminate the virus is rarely developed. Nonetheless, during the first few months after HIV infection, most people mount an immune response that includes both the humoral and cellular response. During this phase, the number of CD8+ cells increases up to 20-fold above the normal levels, but the levels of CD4+ cells fall sharply in the first few weeks after infection, which results in the suppression of immune response induced by CD4+ cells. Individuals who can produce HIV-specific CD4+ cells for a longer duration have lower viral load than individuals who cannot produce a healthy HIV-specific CD4+ cell response. Due to their ability to produce IFN-γ and IL-2, individuals producing HIV-specific CD4+ cell response can effectively control HIV loads. HIV can alter the function of CD4+ cells even without infecting them.

   During the early phase of infection, HIV-specific CD8+ cells are produced and control the viral load, but within weeks they die and only memory CD8+ cells are left. Only in some individuals does a strong HIV-specific CD8+ cell response continue to exist, resulting in a good control of the viral load. The gradual loss of

HIV-specific CD8$^+$ function may result from continued mutation and high levels of viral turnover, which gradually diminishes the capability of the cell to recognize the genetic sequences of the virus due to killing of some of the CD8 T cells' repertoire by HIV itself.

Four to eight weeks after HIV infection, antibody responses begin to develop; they are predominantly directed against the circulating virus and some antibodies may destroy the virus-infected cells as well. However, the antibody response is unable to continue to neutralize HIV because of the rapid mutation of the virus. The development of the initial cellular and humoral response leads to a clearance of much of the viremia and a rebound of CD4$^+$ cells.

After a period of apparent quiescence of the disease, known as clinical latency or the asymptomatic period, most of the patients who are infected with HIV will eventually develop AIDS. But this is not a silent period because there is a gradual decline in the number and function of CD4$^+$ cells. If the number of peripheral CD4$^+$ cells falls below 200 cells/mm^3, the risk of opportunistic infection and ultimate death increases. The opportunistic infections appear anywhere between 3 and 15 years or more after the primary infection, but some patients who are characterized as long-term nonprogressors (LTNP) have demonstrated the ability to be infected with the virus for more than 20 years without significant decline in CD4 count or function or any other symptoms. These patients are not a homogeneous group based on viral load and specific immune response against HIV as some in this group are infected with an inefficiently replicating virus while others are infected with normal-replicating HIV. The patients in the LTNP group who are infected with normal-replicating virus produce robust humoral and cell-mediated immunity. A subgroup of LTNP exhibits signs of progression without any changes in clinical and laboratory parameters over an extended period and are called long-term survivors (LTS). The α-defensin protein may play a role in the inactivation of HIV, and CD38 subset of CD8 T cells may play a role in the progression of the disease as nonprogressors have lower levels of CD38 subset of CD8$^+$ cells. The regulatory T cells accumulate in the lymphoid tissues of patients infected with HIV. The interaction between regulatory T cells and HIV promotes the survival of regulatory T cells resulting in the inhibition of cell-mediated immune response. The CD25$^+$ regulatory T cells obtained from the peripheral blood and the lymph nodes of HIV patients who have high viral load and/or low CD4$^+$ T-cell count are potent suppressors of HIV-specific CD8$^+$ T cells.

Human Immunodeficiency Virus Transmission

Most commonly HIV is spread among adults through sexual intercourse. The entry points of the virus include rectum, penis, vagina, vulva and mouth (after oral sex). Damage to the lining of these organs may increase the chances of acquiring the infection. Other sexually transmitted infection that may result in ulcers or inflammation also increases the likelihood of transmission. The dendritic cells may initiate the transmission process, carrying the virus from the point of its origin to the lymph nodes where HIV infects other lymphoid cells. DC-SIGN is a type II transmembrane protein with an external C-type lectin domain expressed on the surface of several subsets of immature and some mature dendritic cells and is implicated in this process.

After reaching the lymph nodes, CD4$^+$ cells are rapidly infected and replication of the virus continues. During the initial phase of infection, the virus is spread throughout the body via blood that contains many viral particles. The flu-like symptoms are first observed in about 70% of the patients 2–4 weeks after HIV infection. At this stage, HIV titer is reduced due to the development of virus-specific CD8$^+$ cells and due to humoral immune response, which generally causes a return to the normal numbers of CD4$^+$ cells. As the HIV continues to replicate, a person may stay free of HIV-related symptoms for years. The high rate of mutation makes it impossible for the body to completely eliminate the HIV. Independent of mutation, certain subsets of HIV-recognizing killer T cells are not present or lack optimal function, and there is an inhibition of IFN secretion and cytotoxic T-cell activity due to impairment in the function of CD4$^+$ cells. The HIV is also protected from immune surveillance when it hides within the chromosomes of the infected cells.

Several billion virus particles are produced every day as a result of rapid replication of HIV, and a wide range of variants are also produced as a result of recombination among various strains of HIV. It has been reported that the strains of virus in the same person during early and late stages of infection differ in their virulence. They infect and kill different cells. As the disease progresses, the virulence increases, and the spectrum of the target cells widens, which may be due to the ability of virus to utilize additional coreceptors for infection as it gets to the later stages.

A number of mechanisms have been proposed regarding the ability of HIV to destroy or disable CD4$^+$ cells. A very large number (billions) of CD4$^+$ cells are destroyed by HIV every day and finally the body is unable to regenerate these cells. The proposed mechanisms include direct T-cell killing, apoptosis, innocent bystanders, anergy and damage to precursor cells. The direct T-cell killing is a result of HIV, which infects CD4$^+$ cells, replicates in them and interferes with their normal function. The presence of HIV proteins in CD4$^+$ cells and on the surface of uninfected cells leads to apoptosis. Uninfected CD4$^+$ cells are also killed by antibody-dependent cell-mediated cytotoxicity since they exhibit HIV particles on their cell surface. Cytotoxic T cells (CD8$^+$) are also involved in the killing of HIV-infected CD4$^+$ cells. It is interesting to note that while HIV also infects monocytes and macrophages, it does not kill these cells, and infected monocytes and macrophages carry HIV infection to the brain.

Although the net effect of HIV infection is immunodeficiency, the infection generally results in the increased activation of the immune response but the overall impact of this immune hyperactivity is negative. The activated CD4$^+$ cells play a pivotal role in the replication of the virus; this activation of CD4$^+$ cells enhances the secretion of various cytokines, some of which contribute to muscle wasting. A superactive humoral response against HIV impairs the body's ability to mount antibody response against other pathogens, and the activation of immune response by other pathogens further induces HIV replication in the infected individuals. All these events induce the progression of the HIV disease.

Criteria for Diagnosis

According to the CDC, the diagnosis of AIDS constitutes certain opportunistic infections, neoplasms, encephalopathy or wasting syndrome in the presence of HIV infection. In 1993, the CDC expanded the criteria to also include CD4$^+$ T-cell count below 200 cells/μl in the presence of HIV infection. The most common opportunistic infections include pneumocystis carinii pneumonia, pneumonitis, toxoplasmosis, mycobacterial disease, recurrent herpes simplex virus infection and/or cytomegalovirus infection. Kaposi's sarcoma is the most common form of cancer. HIV-related nervous system diseases include acute septic meningitis, AIDS dementia complex, subacute encephalitis, HIV encephalopathy and CNS opportunistic infections and neoplasm.

Clinical Strategies for the Treatment of AIDS

Reverse Transcriptase Inhibitors

Zidovudine (AZT)

Zidovudine (3′-azido-3′-deoxythymidine) is a thymidine analog that inhibits the activity of the reverse transcriptase in HIV-1, HIV-2 and in a number of other retroviruses. It needs to be phosphorylated by cellular thymidine kinase after it diffuses into the cell before it can exert its inhibitory effect on the reverse transcriptase. Its selectivity is attributed to its greater affinity to HIV reverse transcriptase than to human DNA polymerase. Thymidine kinase phosphorylates zidovudine to zidovudine 5-monophosphate, which is phosphorylated by thymidylate kinase to zidovudine 5-diphosphate followed by its phosphorylation by nucleoside diphosphate kinase to the pharmacologically active zidovudine 5′-triphosphate. Thymidine kinase is S-phase-specific, which causes zidovudine to be more potent in activated lymphocytes as opposed to the resting cells. It has higher activity against lymphoblasts and monocytes than the cells previously infected. Zidovudine's principal target are the lymphocytes, in which it is more active than the macrophages. It has no effect on cells that are already infected with the virus. Zidovudine also becomes incorporated into the transcribed DNA strand, and as a consequence prevents further HIV DNA synthesis. The resistance to the drug results from site-directed mutagenesis at codons 41, 44, 67, 70, 118, 210, 215 and 219 of viral reverse transcriptase.

Zidovudine is rapidly absorbed with an oral bioavailability of 60–70%. The peak plasma concentrations are reached within 1 h. The absorption varies widely, and food intake has a retarding effect. The elimination half-life of the triphosphate metabolite is 3–4 h as opposed to 1 h for the parent compound. It is metabolized in the liver and is converted to glucuronide form, 5-glucuronyl zidovudine. Zidovudine is used for the treatment of HIV infection or exposure in adults, children and pregnant mothers (to prevent mother-to-child transmission). For effective treatment, the drug is always administered in combination with other anti-HIV drugs. As a monotherapeutic agent, it is 67% effective in controlling the risk of transmission from infected pregnant mother to fetus.

Side Effects and Drug Interactions

During initiation of the zidovudine therapy, the common side effects include severe headache, nausea, emesis, fatigue, malaise and myalgia, but these symptoms diminish with time. Other side effects include nail pigmentation, esophageal ulceration, hepatitis, neurotoxicity and bone marrow suppression.

Fluconazole, probenecid and atovaquone increase the risk of myelotoxicity by zidovudine. This may be attributed to an increased plasma concentration of zidovudine in the presence of these drugs, perhaps through their inhibitory effects on glucuronose transferase. Rifabutin and rifampin decrease plasma concentrations, and clarithromycin decreases the absorption of zidovudine. Zidovudine and stavudine should not be used in combination because they compete for intracellular phosphorylation.

Didanosine (Dideoxyinosine, DDI)

Didanosine is a synthetic purine nucleoside analog that inhibits the activity of reverse transcriptase in HIV-1, HIV-2, other retroviruses and zidovudine-resistant strains. A nucleobase carrier helps transport it into the cell where it needs to be phosphorylated by 5′-nucleotidase and inosine 5′-monophosphate phosphotransferase to didanosine 5′-monophosphate. Adenylosuccinate synthetase and adenylosuccinate lyase then convert didanosine 5′-monophosphate to dideoxyadenosine 5′-monophosphate, followed by its conversion to diphosphate by adenylate kinase and phosphoribosyl pyrophosphate synthetase, which is then phosphorylated by creatine kinase and phosphoribosyl pyrophosphate synthetase to dideoxyadenosine 5′-triphosphate, the active reverse transcriptase inhibitor. Dideoxyadenosine triphosphate inhibits the activity of HIV reverse transcriptase by competing with the natural substrate, deoxyadenosine triphosphate, and its incorporation into viral DNA causes termination of viral DNA chain elongation. It is 10–100-fold less potent than zidovudine in its antiviral activity, but is more active than zidovudine in nondividing and quiescent cells. At clinically relevant doses, it is not toxic to hematopoietic precursor cells or lymphocytes, and the resistance to the drug results from site-directed mutagenesis at codons 65 and 74 of viral reverse transcriptase.

The oral bioavailability of didanosine is about 35–45%. Food may decrease its absorption by 50% or more. As a consequence, it should be administered a minimum of 30 min before or 2 h after eating. Peak plasma concentrations are reached within 1 or 2 h of administration after a chewable tablet and delayed release capsule, respectively. The elimination half-life of the triphosphate metabolite is 35–40 h as opposed to about $1\frac{1}{2}$ h for the parent compound. The drug is not metabolized significantly and is excreted by glomerular filtration and tubular secretion. Didanosine is used in combination with other antiretroviral agents for HIV-infected adults and children. The drug is used in combination with other classes of anti-HIV drugs.

Side Effects and Drug Interactions

The major dose-limiting toxicities of didanosine include peripheral neuropathy and pancreatitis. The neuropathy is typically symmetrical distal sensory neuropathy, which is reversible, and typically causes paresthesias, numbness and pain in lower extremities. Didanosine also causes retinal changes and optic neuritis. Other adverse effects include diarrhea, skin rash, headache, insomnia, seizures, hepatic toxicity, elevated hepatic transaminases and asymptomatic hyperuricemia.

Allopurinol increases didanosine plasma concentrations and their coadministration is not recommended. Ganciclovir, tenofovir and disoproxil also increase didanosine plasma concentrations, and dose reduction is recommended. Conversely, methadone decreases didanosine plasma concentrations, and appropriate doses for the combination have not been established. Didanosine should not be administered with drugs that cause pancreatic or neurotoxicity. Ribavirin increases its risk of toxicity and should not be coadministered.

Zalcitabine

Zalcitabine ($2',3'$-dideoxycytidine), a reverse transcriptase inhibitor, is a synthetic pyrimidine nucleoside analog that is active against HIV. Its entry into the cell is by carrier-mediated transport and passive diffusion. Inside the cell, deoxycytidine kinase converts it to dideoxycytidine $5'$-monophosphate, which is then converted to diphosphate by deoxycytidine monophosphate kinase, followed by its conversion by nucleoside diphosphate kinase to dideoxycytidine $5'$-triphosphate, which is the active metabolite. Zalcitabine is more potent than other nucleoside analogs in resting cells because it is more efficiently phosphorylated in nondividing cells. It inhibits reverse transcriptase by competing with deoxycytidine $5'$-triphosphate, and it also incorporates into viral DNA, resulting in the termination of viral DNA growth. Dideoxycytidine $5'$-triphosphate also inhibits the activity of cellular DNA polymerase-β and mitochondrial DNA. The resistance to the drug results from site-directed mutagenesis at codons 65, 69, 74 and 184 of viral reverse transcriptase.

After oral administration, the bioavailability of zalcitabine is more than 80%. Food slightly interferes with its absorption. Sixty to eighty percent of the compound is excreted unchanged in the urine. Its (dideoxycytidine $5'$-triphosphate) peak concentrations are at 2–3 h. The primary metabolite is dideoxyuridine, which is <15% of the administered dose. Zalcitabine is indicated in combination with other antiretroviral agents for the treatment of HIV infection.

Side Effects and Drug Interactions

The adverse effects of zalcitabine include peripheral neuropathy, oral ulceration and stomatitis. Additional side effects may include elevated hepatic transaminases, arthralgias, myalgias, fatigue, headache, fever and cardiomyopathy.

Zalcitabine does not interact with zidovudine, and lamivudine inhibits its phosphorylation. It should not be administered with other drugs that cause neuropathy or pancreatitis including didanosine and stavudine.

Stavudine (d4T)

Stavudine (2′,3′-didehydro-2′,3′-dideoxythymidine), a reverse transcriptase inhibitor, is a synthetic nucleoside analog that is active against HIV and needs to be phosphorylated intracellularly before it can exert its antiretroviral effects. Thymidine kinase phosphorylates stavudine to stavudine 5′-monophosphate, which is then phosphorylated by thymidylate kinase to stavudine 5′-diphosphate followed by its phosphorylation by nucleoside diphosphate kinase into the pharmacologically active stavudine 5′-triphosphate. Stavudine triphosphate inhibits the activity of HIV reverse transcriptase by competing with the natural substrate, thymidine triphosphate. Furthermore, its incorporation into viral DNA causes termination of viral DNA chain elongation. It also inhibits cellular DNA polymerase β and γ and suppresses mitochondrial DNA synthesis and is most potent in replicating cells. Stavudine and zidovudine compete for thymidine kinase with zidovudine being more potent than stavudine. The resistance to the drug results from site-directed mutagenesis at codons 41, 44, 67, 70, 118, 210, 215 and 219 of viral reverse transcriptase.

The oral bioavailability of stavudine in adults is 86%. Food does not interfere with its absorption. Peak plasma concentrations are reached within 1h. The renal elimination accounts for 40% of its clearance. In addition to glomerular filtration, it undergoes active tubular secretion. Sixty percent of the remaining drug is eliminated by endogenous pathways. Its binding to plasma proteins is less than 5%. Stavudine in combination with other antiretroviral agents is indicated for the treatment of HIV-1 infection. It causes suppression of HIV and sustained increase in CD4$^+$ cells.

Side Effects and Drug Interactions

In combination with other antiretroviral agents, stavudine has caused fatal lactic acidosis in some patients. It is also associated with motor weakness in which case it should be discontinued. Peripheral neuropathy is the most common toxicity associated with stavudine, which is more prevalent at high doses (4 mg/kg per day). Neuropathy in these patients generally is associated with numbness, tingling or pain in feet or hands. Patients treated with the combination of stavudine and didanosine may also exhibit liver function abnormalities (hepatic steatosis) and pancreatitis. It may also be associated with the etiology of HIV lipodystrophy syndrome.

The use of stavudine in combination with isoniazid, vincristine, phenytoin and ethambutol may increase the risk of peripheral neuropathy and pancreatitis. It competes with zidovudine for phosphorylation and should not be used in combination.

Emtricitabine

Emtricitabine, a reverse transcriptase inhibitor, is a synthetic nucleoside analog with activity against HIV-1, HIV-2 and HBV. Following its entrance into the cell by passive diffusion, deoxycytidine kinase and cellular kinases phosphorylate it into emtricitabine 5'-triphosphate, which is the active form of the drug, and it competes with the natural substrate of the reverse transcriptase. It also incorporates into viral DNA and terminates the elongation of viral DNA. Emtricitabine has low affinity for human DNA polymerases. Since it is chemically related to lamivudine, both drugs share a number of properties including the development of site-directed mutagenesis of viral reverse transcriptase at codon 84 where the single amino acid substitution changes the methionine residue to valine.

After oral administration, emtricitabine is rapidly absorbed with a bioavailability of 93%, and it could be administered with or without food. The peak plasma concentration occurs 1–2 h after the oral dose. It does not significantly bind to plasma proteins, and its elimination half-life is 8–10 h. Following glomerular filtration and active tubular secretion, it is primarily excreted unmetabolized in urine. In combination with other antiretroviral agents, emtricitabine is recommended for the treatment of HIV infection.

Side Effects and Drug Interactions

The toxicity of emtricitabine is rather minimal. The side effects include hepatitis, pancreatitis, hyperpigmentation of the skin, elevated hepatic transaminases, headache, diarrhea and nausea. Skin discoloration has been reported in high frequency in emtricitabine-treated groups versus control groups. The mechanism and clinical significance of skin discoloration, described as hyperpigmentation of the palms and soles, are not known. There are no reports regarding adverse drug interactions with emtricitabine.

Lamivudine

Lamivudine (-2',3'-dideoxy-3'-thiacytodine), a reverse transcriptase inhibitor, is a synthetic nucleoside analog with activity against HIV-1, HIV-2 and HBV. It enters the cell via passive diffusion where deoxycytidine kinase converts it to its monophosphate form, which is then converted to the diphosphate by deoxycytidine monophosphate kinase, followed by its conversion by nucleoside diphosphate kinase to lamivudine 5'-triphosphtate, which is the active form of the drug. It inhibits reverse transcription via DNA chain termination of the nucleotide analog into viral DNA. The resistance to the drug results from site-directed mutagenesis of viral reverse transcriptase with single amino acid substitutions, M184 V and M184 I, changing the methionine residue to either valine or isoleucine.

After oral administration, lamivudine is rapidly absorbed, and its bioavailability is about 86%. Food slows down its absorption, and it is excreted unchanged in the urine. The drug crosses the placenta, and the transfer to placenta does not appear to be altered by zidovudine. Its concentrations are higher in the male genital tract in comparison with the circulation. In combination with other antiretroviral therapy, lamivudine is recommended for the treatment of HIV infection. It inhibits plasma HIV-1 RNA concentrations but resistance develops rapidly when used as monotherapy.

Side Effects and Drug Interactions

The toxicity of lamivudine is rather minimal. The side effects include headache, nausea, neutropenia, malaise and fatigue, nasal signs and symptoms and cough. Pancreatitis has been reported in children, which has been fatal in some cases.

Lamivudine and emtricitabine are not used in combination since they are similar in their pharmacological effects and in the development of resistance. Since lamivudine is predominantly excreted in urine, its concurrent administration with drugs that utilize the same pathway for clearances should be carefully monitored. Lamivudine is also an inhibitor of the phosphorylation of zalcitabine.

Abacavir

Abacavir [(1S, *cis*)-4-[2-amino-6-(cyclopropylamino)-9H-purin-9-yl]-2-cyclo-pentene-1-methanol], a reverse transcriptase inhibitor, is a synthetic carbocyclic nucleoside analog with activity against HIV. After it enters the cells, adenosine phosphotransferase converts it to a monophosphate, which is then converted to (–) carbovir-3'-monophosphate followed by its conversion to di- and triphosphate by cellular kinases. The triphosphate is the active form of the drug that inhibits the activity of HIV reverse transcriptase by competing with the natural substrate GTP, and it also incorporates into viral DNA and terminates the elongation of viral DNA. It is a weak inhibitor of human DNA polymerase α, β and γ. The resistance to the drug results from site-directed mutagenesis of viral reverse transcriptase. The specific codon substitutions include K65R, L74 V, Y115F and M184 V. Furthermore, mutations at codons 41, 210 and 215 have also been reported.

After oral administration, abacavir is rapidly absorbed, and its bioavailability is about 83%. Food does not interfere with its absorption, and it is metabolized by alcohol dehydrogenase to 5'-carboxylic acid derivative and to 5'-glucuronide by glucuronidation. Abacavir does not affect the cytochrome P-450 system. In combination with other antiretroviral drugs, abacavir is indicated for the treatment of HIV-1 infection. It is more potent than other nucleoside reverse transcriptase inhibitors in reducing HIV plasma concentration and increasing CD4+ count.

Side Effects and Drug Interactions

The most serious and sometimes fatal side effect associated with abacavir is hypersensitivity reaction, which is a multiorgan clinical syndrome involving one or more of the following symptoms: fever, rash, GI problems (nausea, vomiting, diarrhea and abdominal pain), constitutional component (malaise, fatigue, aches and pain) and/or respiratory component (cough, dyspnea, pharyngitis). The onset of symptoms has a median time of about 11 days; in most cases the hypersensitivity reaction takes place within 6 weeks. If the symptoms appear, the drugs need to be immediately discontinued and can never be resumed. The hypersensitivity reaction to abacavir is associated with the HLA-B5701 gene, and the reaction takes place in 2–9% of the patients, which results in death in 4% of the patients. The other serious side effects associated with abacavir alone or in combination with other nucleoside analogs are lactic acidosis and severe hepatomegaly with steatosis; some of these cases may be severe and fatal.

There are no serious drug interactions associated with abacavir with the exception that high ethanol consumption may increase its plasma levels and affect its elimination.

Tenofovir

Tenofovir disoproxil fumarate is a prodrug, which is a fumaric acid salt of bis-isopropoxycarbonyloxymethyl ester derivative of the active compound, and requires initial hydrolysis and is converted to tenofovir. It is a reverse transcriptase inhibitor, which is a nucleotide analog, active against HIV-1, HIV-2 and HBV. Cellular kinases then phosphorylate it to tenofovir diphosphate, which inhibits the activity of HIV reverse transcriptase by competing with the natural substrate of the reverse transcriptase deoxyadenosine 5′-triphosphate, and it also incorporates into viral DNA and terminates the elongation of viral DNA. The affinity of tenofovir is low for human DNA polymerases α, β and γ. In the presence of suboptimal doses of tenofovir during replication, the resistance to the drug results from site-directed mutagenesis of viral reverse transcriptase; only one codon, K65R, is involved in resistance.

The oral availability in the absence of food is 25%, which is increased after a meal that is high in fat content (700–1000 calories containing 40–50% fat). Its plasma half-life is 14–16 h. Tenofovir is not a substrate of the cytochrome P-450 system. After an IV administration, 70–80% is found unchanged in urine. Its elimination is by glomerular filtration and active tubular secretion. Therefore, its concurrent administration with drugs that utilize the same pathway for clearance should be monitored. In combination with other antiretroviral agents, tenofovir is indicated for the treatment of HIV infection.

Side Effects and Drug Interactions

Adverse effects are not a major concern with the use of tenofovir. The occurrence of acute renal failure and Fanconi syndrome is rare. Lactic acidosis and severe hepatomegaly with steatosis, including fatal cases, have also been observed.

A combination of tenofovir and didanosine is not recommended since tenofovir increases the AUC of didanosine. It also reduces the AUC of atazanavir.

Efavirenz

Efavirenz is a nonnucleoside reverse transcriptase inhibitor specific for HIV-1. After binding to a site distant from the active site on the HIV-1 reverse transcriptase, it disrupts catalytic activity of the enzyme by causing a conformational change and does not compete with deoxynucleoside triphosphates. Efavirenz does not inhibit HIV-2 reverse transcriptase and human DNA polymerases α, β, γ and δ. The resistance to the drug develops rapidly from site-directed mutagenesis specifically at codon 103, and also at codons 100, 106, 108, 181, 190 and 225 of viral reverse transcriptase. This resistance will be applicable for all nonnucleoside transcriptase inhibitors.

After oral administration, efavirenz is rapidly absorbed from the GI tract, and the peak plasma levels are attained within 5 h, but at higher doses (>1600 mg), the absorption is diminished and high-fat meals increase the bioavailability up to 22%. It is highly bound (99.5%) to plasma proteins predominantly albumin and is metabolized by the cytochrome P-450 system (CYP3A4, CYP2B6). Efavirenz is hydroxylated, and hydroxylated metabolites undergo glucuronidation and it induces its own metabolism by the activation of the cytochrome P-450 system. The half-life after a single dose is 52–76 h and after multiple doses it is 40–55 h. It is excreted both in the urine and the feces. In combination with other antiretroviral agents, efavirenz is recommended for the treatment of HIV infection. Efavirenz-containing regimen has proved to be superior over other drug combinations. Furthermore, its use is convenient and effective. Efavirenz is well tolerated and is effective in patients in whom other therapeutic regimens have not been successful.

Side Effects and Drug Interactions

The most serious side effects of efavirenz are psychiatric symptoms, rash and nervous symptoms. The psychiatric symptoms include suicide thoughts, depression, paranoia, manic disorders and aggressive behavior; the rashes include maculopapular skin eruptions and life-threatening Stevens-Johnson syndrome has also been reported. The neurological symptoms are difficulty in concentration, insomnia, dizziness, confusion, agitation, hallucinations and amnesia. Additional side effect may include an increase in cholesterol and hepatic transaminase levels.

Efavirenz inhibits the plasma levels of indinavir, saquinavir and amprenavir and increases the concentrations of ritonavir and nelfinavir. It also lowers the plasma levels of methadone, phenytoin, carbamazepine and phenobarbital. Drugs that stimulate the cytochrome P-450 system will increase its clearance and should not be coadministered.

Nevirapine

Nevirapine is a member of the dipyridodiazepinone class of chemicals and is a nonnucleoside reverse transcriptase inhibitor that induces a conformational change in HIV-1 reverse transcriptase. Although the conformational change is at a distance from its active site, it disrupts its catalytic activity. It blocks both RNA-dependent and DNA-dependent DNA polymerase activity but does not affect the activity of the template or nucleoside triphosphate. Nevirapine does not inhibit HIV-2 reverse transcriptase or human DNA polymerases α, β or γ. The resistance to the drug results from site-directed mutagenesis at codons 103 or 181, and also at 100, 106, 108, 188 and 190 of viral reverse transcriptase. The development of resistance to one nonnucleoside reverse transcriptase implies that HIV will also be resistant to the rest of the drugs in this class.

After oral administration, nevirapine is rapidly absorbed with a bioavailability of 93%, and peak plasma concentrations are achieved in 4 h. Food or antacids do not interfere with its absorption. It is very lipophilic, crosses the placenta and its presence has been reported in breast milk. Nevirapine is mainly metabolized by the cytochrome P-450 system (CYP3A4 and CYP2B6) to hydroxylated metabolites, and after metabolism, the primary route of excretion is through urine. It has an elimination half-life of 25–30 h. Nevirapine can induce its own metabolism by stimulating the cytochrome P-450 system, which results in the reduction of the half-life of subsequent doses. In combination with other antiretroviral agents, nevirapine is recommended for the treatment of HIV infection in adults and children. It should not be administered alone since resistance develops rapidly.

Side Effects and Drug Interactions

The adverse effects associated with the use of nevirapine include rash, mild macular or papular eruptions, pruritus, elevated hepatic transaminases and hypersensitivity reaction. There is also a risk of hepatitis/hepatic failure, which may be associated with muscle ache, fatigue, malaise and/or renal dysfunction. The use of nevirapine is rarely associated with Stevens–Johnson syndrome but is potentially fatal.

Since nevirapine is metabolized by the cytochrome P-450 system and induces 3A4 and 2B6, other drugs that are also metabolized by these isoenzymes will have low plasma levels when given in combination. It also increases the clearance of methadone and results in methadone withdrawal. Nevirapine decreases the plasma concentrations of norethindrone, ethinyl estradiol and protease inhibitors (HIV).

Delavirdine

Delavirdine is a synthetic nonnucleoside reverse transcriptase inhibitor which after directly binding to HIV-1 reverse transcriptase blocks RNA-dependent and DNA-dependent DNA polymerase activities. It disrupts the catalytic activity of the enzyme after causing a conformational change in HIV-1 reverse transcriptase and does not compete with deoxynucleoside triphosphates. The resistance to delavirdine results from site-directed mutagenesis at codons 103 or 181, and also at 106, 188 and 236 of viral reverse transcriptase. As with nevirapine, the developed resistance will be applicable to the entire class of these drugs (nonnucleoside reverse transcriptase inhibitors).

After oral administration, delavirdine is rapidly absorbed with a bioavailability of 85%; peak plasma concentrations are achieved in 1 h. It may be administered with or without food, and its absorption may be inhibited by proton pump inhibitors, H2 receptor antagonists and achlorhydria. Delavirdine is metabolized by the cytochrome P-450 system (CYP3A4), and the major metabolic pathways are N-dealkylation and pyridine hydroxylation. Ninety-eight percent of the drug is bound to plasma albumin, and its elimination half-life is about 6 h, although considerable variability in different patients has been noted. Delavirdine inhibits its own metabolism by inhibiting CYP3A4 activity. In combination with at least two other antiretroviral agents, delavirdine is recommended for the treatment of HIV infection. However, it is not widely used because of its short half-life.

Side Effects and Drug Interactions

The most common side effects associated with the administration of delavirdine are macular, papular, erythematous pruritic rashes involving trunk and extremities and severe dermatitis. Fatal hepatitis is not associated with its use, but elevated hepatic transaminases have been reported. Other rare side effects of delavirdine include Stevens–Johnson syndrome and neutropenia.

Delavirdine should not be used in combination with drugs that are CYP3A4 substrates such as pimozide, midazolam, triazolam, amiodarone, propafenone and ergot derivatives. Inducers of the hepatic P-450 system, rifampin, rifabutin, phenobarbital, phenytoin or carbamazepine, should not be used in combination with delaviridine. It also increases the plasma levels of HIV protease inhibitors.

Human Immunodeficiency Virus Protease Inhibitors

In the life cycle of HIV, its RNA is translated into a polypeptide chain that is composed of several individual proteins including protease, integrase and reverse transcriptase, but in this form these enzymes are not functional. They must be cleaved by viral proteases from the assembled sequence in order for them to become functional. These posttranslational modifications allow the enzymes to facilitate the production of new viruses. The protease itself is made up of two 99-amino-acid monomers, and an aspartic acid residue in the monomer is required for the cleavage. The protease inhibitors inhibit the enzyme protease and consequently interfere with viral replication and maturation by preventing proteases from cleaving proteins into peptides. In humans, these drugs inhibit cleavage of HIV gag and pol polyproteins, which are part of the essential viral structural components, P7, P9, P17 and P24, and protease as well as other enzymes. As a result, the protease inhibitors interfere with the maturation of HIV virus particles.

Saquinavir

Saquinavir is a peptide-like substrate analog that inhibits HIV protease after binding to its active site and is active against both HIV-1 and HIV-2 maturation. It blocks splicing of the viral polyproteins, which results in the production of immature viral particles that lack the ability to infect other cells. The resistance to saquinavir is associated with mutations in protease genes G48 V and L90 M, whereas secondary mutations are associated with codons 36, 46, 82, 84, 101, 154 and 184; multiple mutations are necessary to render strong resistance to the drug.

Its bioavailability is about 4% after a single dose (600 mg) of its hard gelatin form (Invirase) in the presence of a high-fat meal, and this low bioavailability is attributed to incomplete absorption and first-pass metabolism. Ninety-eight percent of saquinavir is bound to plasma proteins, and its penetration into the brain is limited by P-glycoprotein transporter in the capillary endothelial cells of the blood–brain barrier. Saquinavir has a short half-life and is metabolized by the intestinal and hepatic cytochrome P-450 system where more than 90% of the hepatic metabolism is mediated by the isoenzyme CYP3A4, resulting in an inactive hydroxylated compound. The metabolites are excreted primarily through feces and biliary system, whereas the urinary excretion is minimal.

Saquinavir is available as Invirase, which is its hard gelatin form, and Fortovase, which is its soft gelatin form. In combination with other antiretroviral agents and another protease inhibitor ritonavir, Invirase is recommended for the treatment of HIV infection, and due to its low bioavailability, considerably higher doses are recommended. The two preparations, Invirase and Fortovase, are not bioequivalent and could not be used interchangeably. Although Fortovase can be used as a sole protease inhibitor in a combination therapy, Invirase could be used only in combination with ritonavir.

Side Effects and Drug Interactions

The adverse effects of saquinavir are mild but predominantly they are related to GI discomfort including abdominal discomfort, nausea, diarrhea and vomiting. Lipodystrophy may result from its long-term use. Rarely, saquinavir causes confusion, weakness, ataxia, seizures, headache and liver abnormalities.

The pharmacokinetics of saquinavir is modified by agents that alter isoenzyme CYP3A4 of the cytochrome P-450 system and P-glycoprotein transporter. It should not be administered with midazolam, triazolam and ergot derivatives. The plasma concentrations of saquinavir are lower when coadministered with efavirenz, nevirapine or rifampin. Ritonavir reverses the effects of nevirapine on saquinavir. The coadministration of astemizole, terfenadine, amiodarone, bepridil, quinidine, propafenone or flecainide with saquinavir is also not recommended due to its potential for serious and/or life-threatening reactions.

Ritonavir

Ritonavir is a peptidomimetic protease inhibitor with activity against the HIV-1 and HIV-2 proteases. Following the binding of ritonavir to the viral protease, the enzyme is no longer able to process the gag–pol polyprotein precursors, resulting in the production of immature HIV particles that lack the capability to infect other cells. The resistance to the drug results from site-directed mutagenesis of viral protease. The mutations take place at codon 82, 84, 71 or 46, but codons 20, 32, 54, 63, 84 and 90 may also be involved; mutations at multiple codons are necessary to render strong resistance to the drug.

The peak levels of ritonavir are reached between 2 and 4 h after oral administration, and its peak levels decrease by 23% and absorption decreases by 7% compared to the fasting conditions. Ritonavir is 98–99% bound to plasma proteins (α_1-acid glycoproteins), and its penetration into the brain is limited by P-glycoprotein transporter in the capillary endothelial cells of the blood–brain barrier. Ritonavir is metabolized by the cytochrome P-450 system isoenzyme cytochrome P450, family 3, subfamily A (CYP3A) where its major metabolite is isopropyl thiazole oxidation metabolite (M-2), which is equivalent to the parent drug in its pharmacological activity but has lower plasma levels, and the parent drug and its metabolites are mainly excreted in feces. In combination with other antiretroviral agents, ritonavir is recommended for the treatment of HIV infection. Monotherapy with ritonavir is shown to have the greatest decrease in viral RNA levels (60%). It is combined with other HIV protease inhibitors since it inhibits their metabolism via its effects on CYP3A4.

Side Effects and Drug Interactions

The adverse effects of ritonavir comprise GI discomfort including abdominal pain, nausea, diarrhea, vomiting, asthenia and neurological disturbances. Taste perversion, peripheral paresthesias and lipodystrophy including elevated levels of triglycerides and cholesterol have also been reported with its administration.

Ritonavir increases the plasma levels of triazolam, pimozide, midazolam, ergot derivatives, propafenone and amiodarone by delaying their elimination since it is a very potent inhibitor of CYP3A4. Rifampin, due to its ability to induce CYP3A4, will reduce the plasma levels of ritonavir, and their coadministration is not recommended. Since ritonavir is also an inhibitor of CYP2D6, its coadministration with most antidepressants, certain antiarrhythmics and narcotic analgesics should be carefully monitored.

Indinavir

Indinavir is a peptidomimetic hydroxy ethylene protease inhibitor that is ten times more potent against HIV-1 enzyme than HIV-2. Following the binding of indinavir to the viral protease, the protease is no longer able to process the gag–pol polyprotein precursors, resulting in the production of immature HIV particles that lack the capability to infect other cells. The resistance to the drug develops from site-directed mutagenesis at codons 46, 82 and 84 of viral protease. Furthermore, codons 10, 20, 24, 46, 54, 63, 71, 82, 84 and 90 may also be involved; resistance results from multiple and variable substitutions at these codons.

Indinavir is absorbed rapidly during fasting, and peak plasma concentrations are reached in 0.8 h. A meal high in fat and protein calories causes a 77% decrease in plasma concentrations of indinavir but a light meal does not affect its plasma concentration. Sixty percent of indinavir is bound to plasma proteins and its half-life is 1.8 h. Indinavir is metabolized by the cytochrome P-450 system (CYP3A4) to a glucuronide conjugate and other oxidative metabolites, and most of the drug and its metabolites are excreted in feces. The problems of its short half-life and interference caused by food could be addressed when it is coadministered with ritonavir. In combination with other antiretroviral agents, indinavir is indicated for the treatment of HIV infection.

Side Effects and Drug Interactions

Some patients receiving indinavir exhibit nephrolithiasis/urolithiasis including flank pain that may be accompanied by hematuria. The frequency of nephrolithiasis is dependent on the period of treatment with indinavir. Other side effects associated with indinavir include insulin resistance, hyperglycemia, asymptomatic hyperbilirubinemia, HIV lipodystrophy syndrome and skin abnormalities. Indinavir should not be coadministered with drugs that affect the cytochrome P-450 system (CYP3A4). Antacids are not recommended within 2 h of its administration, specifically didanosine containing an antacid buffer.

Nelfinavir

Nelfinavir is a nonpeptide protease inhibitor that is active against both HIV-1 and HIV-2. Following the binding of nelfinavir to the HIV protease, the protease is no longer able to process the gag–pol polyprotein precursors, resulting in the production of immature HIV particles that lack the capability to infect other cells. The resistance to the drug results from site-directed mutagenesis of viral protease. The high level of resistance is associated with codon 30; other viral protease mutations have been reported at codons 35, 46, 71, 77 and 88.

The peak levels of nelfinavir are reached between 2 and 4 h after oral administration, and the content of fat in food affects its absorption. The drug is more than 98% protein-bound, and its half-life in plasma is between 3.5 and 5 h. Its penetration into the brain is limited by P-glycoprotein transporter in the capillary endothelial cells of the blood–brain barrier. After a single oral dose of 750 mg, 82–86% of the drug remains unchanged while the rest is metabolized by the cytochrome P-450 system, primarily by isoenzyme CYP2C19 and some by isoenzymes CYP3A4 and CYP2D6, resulting in one major and several minor oxidative metabolites, all of which along with the parent drug are excreted in feces. In combination with other antiretroviral agents, nelfinavir is indicated for the treatment of HIV infection. It is used in children and pregnant women due to its relatively mild side effects.

Side Effects and Drug Interactions

The most common side effect associated with the use of nelfinavir is diarrhea; other less frequent adverse effects include elevated triglycerides and cholesterol plasma levels, nausea, rash and hyperglycemia.

The drugs that inhibit CYP3A will increase the plasma concentrations of nelfinavir and the drugs that induce CYP3A or CYP2C19 will reduce its plasma concentration and therapeutic effect. Its coadministration with drugs that are metabolized by CYP3A will increase the plasma concentration of other drugs that could augment its adverse effects. Nelfinavir also reduces the levels of ethinyl estradiol and norethindrone by inducing hepatic drug-metabolizing enzymes and of zidovudine by induction of glucuronyl S-transferase. No clinically significant drug interactions have been reported with most of the reverse transcriptase inhibitors.

Lopinavir

Lopinavir is an inhibitor of HIV protease, with a structure similar to ritonavir, and is active against both HIV-1 and HIV-2. Following the binding of lopinavir to the HIV protease, the protease is no longer able to process the gag–pol polyprotein precursors, resulting in the formation of immature HIV particles that lack the capability to infect other cells. The resistance to the drug ultimately results from site-directed mutagenesis at codons 10, 20, 24, 36, 46, 53 and 73 of viral protease. This drug is available only in a coformulation with ritonavir, which helps increase the plasma levels of lopinavir.

Due to first-pass metabolism, if administered alone lopinavir has varied plasma levels. This problem is overcome by adding ritonavir to the formulation. Its oral absorption is rapid, and its bioavailability is increased by food rich in fat content. Lopinavir is metabolized by cytochrome P-450 isoenzyme CYP3A4. Most of the drug in the plasma is bound to α_1-acid glycoprotein. In combination with other antiretroviral agents, lopinavir is indicated for the treatment of HIV infection.

Side Effects and Drug Interactions

The adverse effects of the administration of this coformulation include GI problems such as nausea, diarrhea, vomiting and loose stool and increased plasma cholesterol and triglyceride levels.

St. John's wart, rifampin, efavirenz, nevirapine and amprenavir will lower plasma concentrations of lopinavir due to their effect on cytochrome P-450 enzyme CYP3A4. Lopinavir increases plasma concentrations of ergot derivatives, triazolam, midazolam, and propafenone and should not be given together.

Amprenavir and Fosamprenavir

Amprenavir is a nonpeptide protease inhibitor that is active against both HIV-1 and HIV-2; fosamprenavir is the prodrug for amprenavir and has better bioavailability. After binding to the active site of the viral protease, it inhibits the processing of viral gag and gag–pol polyprotein precursors, resulting in the production of immature HIV particles that lack the capability to infect other cells. The resistance to the drug results from site-directed mutagenesis primarily at codons 50 and 84, and also at codons 10, 32, 46, 54 and 90.

The oral absorption of amprenavir is rapid, and peak concentrations are reached between 1 and 2 h; administration of both amprenavir and fosamprenavir with food is not a concern. Fosamprenavir is dephosphorylated to amprenavir in intestinal mucosa. Amprenavir is 90% bound to plasma protein with most to α_1-acid glycoprotein. It is metabolized by the cytochrome P-450 system, CYP3A4, in the liver, and more than 90% of the drug is excreted after its metabolism in feces. In combination with other antiretroviral agents, amprenavir/fosamprenavir are indicated for the treatment of HIV infection.

Side Effects and Drug Interactions

The adverse effects associated with amprenavir comprise GI symptoms including nausea, diarrhea and vomiting; other side effects include headache, fatigue and skin eruptions.

The drugs that induce cytochrome P-450 system isoenzyme CYP3A4 will inhibit the blood levels of amprenavir. It decreases the plasma levels of methadone and delavirdine and increases the plasma levels of rifabutin, ketoconazole and atorvastatin. This may be the result of the ability of amprenavir to inhibit as well as induce CYP3A4.

Tipranavir

Tipranavir is a nonpeptide protease inhibitor that is active against HIV-1. Following the binding of tipranavir to the HIV protease, the protease is no longer able to process the gag–pol polyprotein precursors, resulting in the production of immature HIV particles that lack the ability to infect other cells. The resistance to the drug results from site-directed mutagenesis at codons 10, 13, 33, 36, 45, 71, 82 and 84 of viral protease.

The oral absorption of tipranavir is limited, and the bioavailability is increased with a high-fat meal. Its binding to plasma proteins is more than 99.9% where it binds to both α_1-glycoprotein and albumin. Tipranavir is metabolized by cytochrome P-450 isoenzyme CYP3A4. Ritonavir decreases its first-pass clearance, and most of the drug is excreted in feces. In combination with other antiretroviral agents, tipranavir coadministered with ritonavir is indicated for HIV infection in patients who have received prior HIV treatment or have highly resistant HIV strains.

Side Effects and Drug Interactions

The adverse effects after its coadministration with ritonavir include diarrhea, nausea, vomiting and bronchitis. Tipranavir induces P-glycoprotein, resulting in a number of drug–drug interactions. When administered in combination with ritonavir, it inhibits CYP3A and consequently will increase the plasma levels of the drugs metabolized by CYP3A. Its coadministration with amiodarone, quinidine, propafenone, flecainide, bepridil, terfenadine and astemizole is not recommended as this combination could cause potential life-threatening cardiac arrhythmias. Other drugs including ergot derivatives, rifampin and cisapride also are contraindicated in combination with tipranavir.

Atazanavir

Atazanavir is an azapeptide protease inhibitor. It is active against HIV-1. Following the binding of atazanavir to the HIV protease, the protease is no longer able to process the gag–pol polyprotein precursors. This results in the production of immature HIV particles that lack the capability to infect other cells. The resistance to the drug results from site-directed mutagenesis of viral protease. The resistance is associated with codon 50. In combination with other antiretroviral agents, atazanavir is indicated for the treatment of HIV infection.

Its oral absorption is rapid, and the peak levels are reached in about $2\frac{1}{2}$ h. Food affects its absorption with a 70% increase after a light meal and a 30% increase after a meal rich in fat content. Atazanavir is 86% bound to plasma proteins including both α_1-glycoproteins and albumin. Food also helps overcome the interindividual pharmacokinetic variability associated with this drug. Atazanavir is extensively metabolized through monooxygenation and deoxygenation. The cytochrome P-450 isoenzyme CYP3A4 is involved in the metabolism. Most of the metabolites are excreted in feces with some in urine.

Side Effects and Drug Interactions

The adverse effects of atazanavir include fever, jaundice/scleral icterus, myalgia and diarrhea. Its coadministration is not recommended with the drugs that induce cytochrome P-450 isoenzyme CYP3A4. Ritonavir increases plasma concentrations of atazanavir. It is an inhibitor of isoenzymes CYP3A4, CYP2C8 and UGT1A1. The coadministration of atazanavir with calcium channel blockers, HMG-CoA reductase inhibitors, immunosuppressants and phosphodiesterase 5 inhibitors should be carefully monitored.

Darunavir

Darunavir is a nonpeptide second-generation protease inhibitor of HIV-1. In combination with ritonavir, it is superior to lopinavir plus ritonavir. Darunavir is a selective inhibitor of the cleavage of HIV-encoded gag–pol polyproteins in infected cells, resulting in the formation of immature virions. It is not known to inhibit the activity of other protease, nucleoside and nonnucleoside reverse transcriptase inhibitors. Darunavir has a half-life of 15 h and is metabolized by cytochrome P-450 3A (CYP3A), and therefore its levels are greatly increased since it must be coadministered with ritonavir. It does not induce P-glycoprotein, which is the case with tipranavir, and as a consequence, it has fewer drug–drug interactions. Darunavir is indicated for the treatment of patients who are infected with strains that are resistant to more than one protease inhibitor. The side effects associated with the use of darunavir include nausea, diarrhea, moderate increase in the levels of lipids and transaminases, mild to moderate rash, headache, abdominal pain and vomiting. Severe rashes are rare but have been reported.

Drugs Inhibiting Viral Binding

Enfuvirtide

Enfuvirtide, a biomimetic peptide with a sequence matching the gp41 receptor on the HIV membrane, blocks the entry of HIV into the host cell by inhibiting membrane fusion that allows the entry of the virus into $CD4^+$ cells. It is active only against HIV-1 and the resistance develops as a result of mutation in the drug-binding region of gp41.

Enfuvirtide requires parenteral administration, and the peak levels are reached in about 4 h. Its bioavailability is 84%, and the drug is 92% protein-bound in plasma. Its half-life is 3.8 h, and enfuvirtide is metabolized in liver, which is not at significant levels. Enfuvirtide is approved for the treatment of HIV in patients whose response to antiretroviral therapeutic regimen is not satisfactory, and there is indication that HIV is still replicating.

Side Effects and Drug Interactions

Irritation at the site of injection is a common problem that may include pain, swelling, redness and induration; nodules or cysts at the site of injection have also been reported. No drug interactions are known.

Drugs Stimulating the Immune Response

The use of cytokines in AIDS has already been described in chapter 2.

Future Anti-Human Immunodeficiency Virus Drugs Under Development

Table 8.1 shows other classes of potential anti-HIV drugs that are under development.

Table 8.1 Future Potential Target Mechanisms of Anti-HIV Therapy

Inhibitors of membrane fusion
Inhibitors of integration of provirus
Inhibitors of transcription provirus
Inhibitors of translation of viral mRNA (antisense oligonucleotides)
Inhibitors of posttranslational glycosylation
Antagonists of chemokine receptors

Combination Therapy for AIDS

The Department of Health and Human Services published guidelines for the use of antiretroviral agents in HIV-infected adults and adolescents in October 2006. The panel confirmed that the regimens that demonstrated the best results included one nonnucleoside reverse transcriptase inhibitor (in the presence or absence of ritonavir boosting) plus two nucleoside reverse transcriptase inhibitors. The panel suggested that the choice of an antiretroviral regimen should be individualized depending on the patient and the specific pharmacological effects of the drug. The latest recommendations of the panel for naïve patients are shown in Table 8.2, and physicians are suggested to choose a regimen by selecting one component each from columns A and B. In selected setting, other options are also available. The choice of the regimen is modified as the resistance to the drugs develops.

Table 8.2 Suggested Combination Therapy for HIV

		A	B
Preferred regimen	Efivarenz	Atazanavir + ritonavir	Tenofovir/ emtricitabine
		Fosamprenavir + ritonavir	Zidovudine/ lamivudine
		Lopinavir/ ritonavir	
Alternative regimen	Nevirapine	Atazanavir	Abacavir/ lamivudine
		Fosamprenavir	Didanosine/ lamivudine
		Fosamprenavir + ritonavir	
		Lopinavir/ ritonavir	

Source: Guidelines for the use of antiretroviral agents in HIV-1-infected adults and adolescents developed by the DHHS Panel, October 10, 2006.

Vaccines for HIV Infection

AIDS does not fit the paradigm for classical vaccines for a number of reasons. The classical preventive vaccines enhance natural immunity against microbes that change a little or none at all, whereas HIV mutates. There is some level of initial replication and dispersal at the point of entry before the virus reaches its target tissue, which allows the immune system to mount a response and possibly eliminate it. This response leaves some memory T cells and an individual recovers from the infection and reinfection or prior immunization produces a robust response to prevent the disease, but there are no recovered HIV patients. Most vaccines are whole-killed or live-attenuated organisms but killed HIV does not retain antigenicity. In response to HIV initially, there is an increase in the numbers of HIV-specific $CD8^+$ and $CD4^+$ T cells that reduce the virus levels. Furthermore, the binding antibodies are produced about 6–12 weeks after infection, and the neutralizing antibodies are not secreted until after a reduction in virus levels due to the production of $CD8^+$ cells; however, at this stage there are rapid genetic mutations in the virus envelope, which greatly reduces the neutralizing effects of the antibodies. Another challenge that HIV possesses is its ability to establish pools of latently infected resting $CD4^+$ T cells during very early stages of infection, resulting in an indefinite infection that takes the ability of the vaccine to eradicate the virus away and as a consequence, virtually no person clears HIV infection. Other viral infections do not follow this path where initial viral replication does not result in the establishment of permanent viral reservoirs. Lastly, a lack of suitable animal models to test HIV vaccines continues to be a challenge.

Nonetheless, scores of strategies had been employed since the identification of HIV, despite its enormous genetic diversity and unique features. A critical problem is the fact that the structures of the HIV's outer envelope and monomeric gp120 are different, and the outer membranes of circulating HIV hide its epitope so that genetically engineered gp120 is unable to mimic the response. Some of the immunogens used in an attempt to produce broadly neutralizing antibodies included whole-killed HIV, pseudovirions, live vector viruses (non-HIV viruses engineered to carry genes encoding HIV proteins), naked DNA containing one or more HIV genes, HIV peptides and most notably viral surface proteins such as gp120, but with limited success. Since the identification of gp120 on the envelope of the virus as the binding site for CD4, which allows the attachment and entry of the virus into human cells, major effort has been devoted to develop a vaccine that contains genetically engineered gp120 and the larger glycoprotein gp160 to inhibit the infection, but this approach failed to provide protection against HIV infection. More recently, emphasis has been placed on using novel immunogen developed by modifying viral envelope that may induce broadly neutralizing antibodies. The goals of some of these strategies are to mimic native trimer on the virion surface, redirect immune responses to conserved confirmational epitopes and redirect responses away from variable epitopes.

As opposed to the development of vaccines that produce antibodies, the development of T-cell vaccines is a novel concept. These vaccines are designed to induce primarily T-cell responses that will control the viral proliferation and viral levels during the early stages of infection and will delay the disease progression. These vaccines will not prevent infection but will inhibit HIV levels and protect uninfected memory $CD4^+$ T cells. It is expected that HIV-infected patients receiving this vaccine may remain disease-free for a prolonged period of time. A number of current HIV vaccine trials have focused on $CD8^+$ cell-mediated products that

Table 8.3 Some AIDS Vaccines Under Development

Vaccine	Composition	Trial Phase
Canarypox + envelope	Gag, pro, env (E+gp120) (B,E)	Phase III
Adenovirus type 5	Gag, pol, nef (B)	Phase IIb
Naked DNA followed by adenovirus type 5	Gag, pol, nef (B) Env (A,B,C) + gag, pol (B), env (A,B,C)	Phase II
Canarypox + lipopeptides	Gag, pol, nef, env (B) + CD8$^+$ cells epitopes (B)	Phase II
DNA-polylactide coglycolide + envelope	Gag, env + gp140(B)	Phase I
DNA + modified vaccinia Ankara	Gag, pol, nef, tat, env (C)	Phase I
DNA + peptides	Gag (B) many T cell epitopes $\pm$ IL-12 or IL-15 or GM-CSF	Phase I
Fowl pox + modified vaccinia Ankara	Env, gag, nef, pol, rev, tat (B)	Phase I

employ either viral vectors alone or in combination with DNA plasmids that contain viral genes. For example, a canarypox vector vaccine in combination with a gp120, which boosts for both internal and HIV envelope proteins, is in Phase III trials. Another vaccine that contains altered adenovirus type 5, which cannot replicate but transmits HIV's gag, pol and nef genes, is in Phase IIB trials. A different approach is a two-step design where a vaccine made up of naked DNA is first administered to prime an immune response against both the internal and external HIV proteins followed by a booster shot of inactivated adenovirus vector, which may induce specific responses to HIV envelope proteins and internal proteins. A list of some AIDS vaccines under development is presented in Table 8.3.

Although none of the vaccines will be able to eradicate the disease, it is expected that they will be able to prolong the disease-free period by reducing the viral levels and as a consequence will also reduce transmission.

Bibliography

Arrighi JF, Poin M, Garcia E, Escola JM, et al. 2004. DC-SIGN-mediated infectious synapse formation enhances X4 HIV-1 transmission from dendritic cells to T cells. JEM. 200:1279–1288. http:www.aidsinfo.nih.gov/guidelines.

Bang LM, Scott LJ. 2003. Emtricitabine: An antiretroviral agent for HIV infection. Drugs. 63:2413–2424.

Barreiro P, Rendon A, Rodriguqz-Novoa S, Soriano V. 2005. Atazanavir: the advent of a new generation of more convenient protease inhibitors. HIV Clin Trails. 6:51–60.

Barre-Sinoussi F, Chermann JC, Rey F, Nugeyre MT. et al. 1983. Isolation of a T-lymphotropic retrovirus from a patient at risk for acquired immune deficiency syndrome (AIDS). Science. 220:868–871.

Berger EA, Murphy PM, Farber JM. 1999. Chemokine receptors as HIV coreceptors: Role in viral entry, tropism and disease. Annu Rev Immunol. 17:657–700.

Boshoff C, Weiss R. 2002. AIDS – related malignancies. Nat Rev Cancer 2:373–382.

Bradsley-Elliott A, Plosker GL. 2000. Nelfinavir: An update on its use in HIV infection. Drugs. 59:581–620.

Cao Y, Qin L, Zhang L, Safrit J, et al. 1995. Virologic and immunologic characterization of long term survivors of human immunodeficiency virus type 1 infection. NEJM. 332:201–208.

Chan D, Kim P. 1998. HIV entry and its inhibition. Cell. 93:681–684.

Chun TW, Engel D, Berrey MM, Shea T. et al. 1998. Early establishment of a pool of latently infected, resting CD4(+) T cells during primary HIV infection. Proc Natl. Acad Sci USA. 95:8869–8873

Clotet B, Bellos N, Molina JM, Cooper D, et al. 2007. Efficacy and safety of darunavir-ritonavir at week 48 in treatment-experienced patients with HIV-1 infection in POWER 1 and 2: A pooled subgroup analysis of data from two randomized trials. Lancet. 369:1169–1178.

Coffin J, Haase A, Levy A, Montagnier L. et al. 1986. What to call the AIDS virus? Nature 321: 10–10.

Coffin JM. 1995. HIV population dynamics in vivo. Implications for genetic variation, pathogenesis and therapy. Science. 267:483–489.

Collier AC, Coombs RW, Schoenfeld DA, Bassett RL. et al. 1996. Treatment of human immunodeficiency virus infection with saquinavir, zidovudine, and zalcitabine. AIDS clinical trials group. N Engl J Med 334:1011–1017.

Connor EM, Sperling RS, Gelber R, Kiselev P, et al. 1994. Reduction of maternal-infant transmission of human immunodeficiency virus type 1 with zidovudine treatment. Pediatric AIDS clinical trials group protocol, 076 study group. NEJM. 331:1173–1180.

Coovadia H. 2004. Antiretroviral agents – how best to protect infants from HIV and save their mothers from AIDS. N Engl J Med 351:289–292.

Cvetkovic RS, Goa KL. 2003. Lopinavir/ritonavir: a review of its use in the management of HIV infection. Drugs. 63:769–802.

DeClercq E. 2003. Clinical potential of the acyclic nucleoside phosphonates cidofovir, adefovir and tenofovir in treatment of DNA virus and retrovirus infections. Clin Microbiol Rev. 16: 569–596.

Deeks SJ. 2006. Antiretroviral treatment of HIV infected adults. BMJ. 332:1489–1489.

DeLuca A, DiGiambenedetto S, Maida I, Nunez M, et al. 2004. Triple nucleoside regimens versus efavirenz. NEJM. 351:717–719.

Dolin R, Amato DA, Fischl MA, Pettinelli C, et al. 1995. Zidovudine compared with didanosine in patients with advanced HIV type 1 infection and little or no previous experience with zidovudine. AIDS Clinical Trial Group. Arch Int Med. 155:961–974.

Doyon L, Tremblay S, Bourgon L, Wardrop E, et al. 2005. Selection and characterization of HIV-1 showing reduced susceptibility to the non-peptide protease inhibitor tipranavir. Antiviral Res. 68:27–35.

Dudley MN. 1995. Clinical pharmacokinetics of nucleoside antiretroviral agents. J Infect Dis. 171:S99–S112.

Ellis JM, Ross JW, Coleman CI. 2004. Fosamprenavir: A novel protease inhibitor and prodrug of amprenavir. Formulary. 19:151–160.

Embretson J, Zapancic M, Ribas JL, Burke A. et al. 1993. Massive convert infection of helper T lymphocytes and macrophages by HIV during the incubation period of AIDS. Nature. 362:359–362.

Fauci AS. 1996. Host factors and the pathogenesis of HIV-induced disease. Nature. 384:529–534.

Feldman C. 2005. Pneumonia associated with HIV infection. Curr Opin Infect Dis. 18:165–170.

Fischl MA, Stanley K, Collier AC, Arduino JM, et al. 1995. Combination and monotherapy with zidovudine and zalcitabine in patients with advanced HIV disease. The NIAID AIDS Clinical Trials. Ann Int Med. 122:24–32.

Flexner C. 1998. HIV-protease inhibitors. N Engl J Med. 338:1281–1292.

Flexner C. 2000. Dual protease inhibitor therapy in HIV-infected patient: Pharmacologic rationale and clinical benefits. Annu Rev Pharmacol Toxicol. 40:649–674.

Flexner C. 2005. Antiretroviral agents and treatment of HIV infection. In: Brunton LL, Ed. Goodman and Gilman's, the Pharmacological Basis of Therapeutics. 11th Ed., The McGraw-Hill Companies.

Flexner C, Bate G, Kirkpatrick P. 2005. Tipranavir. Nat Rev Drug Discov. 4:955–956.

Furman PA, Fyfe JA, St. Clair MH, Weinhold K, et al. 1986. Phosphorylation of 3-azido-3 deoxythymidine and selective interaction of the 5' triphosphate with human immunodeficiency virus reverse transcriptase. Proc Natl Acad Sci USA. 83:8333–8337.

Gallant JE, Gerondelis PZ, Wainberg MA, Shulman NS, et al. 2003. Nucleoside and nucleotide analogue reverse transcriptase inhibitors: A clinical review of antiretroviral resistance. Antivir Ther. 8:489–506.

Gallant JE, DeJesus E, Arribas JR, Pozniak AL, et al. 2006. Tenofovir DF, Emtricitabine and Efavireuz vs. Zidovudine, Lamivudine and Efavirenz for HIV. NEJM. 354:251–260.

Gallo RC. 2005. The end or the beginning of the drive to an HIV preventive vaccine: a view from over 20 years. Lancet. 366:1894–1898.

Gao F, Bailes E, Robertson DL, Chen Y. et al. 1999. Origin of HIV-1 in the chimpanzee Pan troglodytes. Nature. 397:436–441.

Greene WC. 1991. The molecular biology of human immunodeficiency virus type 1 infection. N Engl J Med. 324:308–317.

Greene WC. 1993. AIDS and the immune system. Sci Am. 269:98–105.

Greene WC, Peterlin BM. 2002. Charting HIV's remarkable voyage through the cell: Basic science as a passport for future therapy. Nat Med. 8:673–680.

Gulick RM, Mellors JW, Havilir D, Eron JJ, et al. 1997. Treatment with indinavir, zidovudine, and lamivudine in adults with human immunodeficiency virus infection and prior antiretroviral therapy. NEJM. 337:734–739.

Gulick RM, Ribaudo HJ, Shikuma CM, Lustarten S, et al. 2004. Triple-nucleoside regimens versus efavirenz-containing regimens for the initial treatment of HIV-infection. NEJM. 350: 1850–1861.

Guss DA. 1994. The acquired immune deficiency syndrome: an overview for the emergency physician, Part I. J Emerg Med. 12:375–384.

Guss DA. 1994. The acquired immune deficiency syndrome: an overview for the emergency physician, Part II. J Emerg Med 12:491–497.

Hammer SM, Katzenstein DA, Hughes MD, Gundacker H, et al. 1996. A trial comparing nucleoside monotherapy with combination therapy in HIV infected adults with CD4 cell counts from 200 to 500 per cubic millimeter. AIDS clinical trials group study, 175 study team. NEJM. 335:1081–10990.

Hammer SM, Saag, MS, Schechter M, Montaner JSG, et al. 2006. Treatment for adult infection: 2006. Recommendations of the International AIDS Society-USA Panel. JAMA. 296:827–843.

Heeney JL, Dalgleish AG, Weiss RA. 2006. Origins of HIV and the evolution of resistance to AIDS. Science. 313: 462–466.

Hervey PS, Perry CM. 2000. Abacavir: A review of its clinical potential in patients with HIV infection. Drugs. 60:447–479.

Ho DD, Pomerantz RJ, Kaplan JC. 1987. Pathogenesis of infection with human immunodeficiency virus. N Engl J Med. 317:278–286.

Ho DD, Neumann AU, Perelson AS, Chen W. et al. 1995. Rapid turnover of plasma virions and CD4 lymphocytes in HIV infection. Nature. 373:123–126.

Hoxie JA, Alpers JD, Rackowski JL, Huebner K. et al. 1986. Alterations in T4 (CD4) protein and mRNA synthesis in cells infected with HIV. Science. 234:1123–1127.

Johnston MI, Fauci AS. 2007. An HIV vaccine – evolving concepts. N Engl J Med. 356: 2073–2081.

Kahn JO, Walker BD. 1998. Acute human immunodeficiency virus type 1 infection. N Engl J Med. 331:33–39.

Kaplan SS, Hicks CB. 2005. Safety and antiviral activity of lopinavir/ritonavir-based therapy in human immunodeficiency virus type 1 (HIV-1) infection. J Antimicrob Chemother. 56: 273–276.

Keet IP, Krol A, Klein MR, Veugelers P. et al. 1994. Characteristics of long term asymptomatic infection with human immunodeficiency virus type 1 in men with normal and low CD4+ cell counts. J Infect Dis. 169:1236–1243.

Kestler H, Kodama T, Ringler D, Marthas M. et al. 1990. Induction of AIDS in rhesus monkeys by molecularly cloned simian immunodeficiency virus. Science. 248:1109–1112.

King JR, Acosta EP. 2006. Tipranavir: A novel nonpeptide protease inhibitor of HIV. Clin Pharmacokinet. 45:665–682.

Kinter A, McNally J, Riggin L, Jackson R, et al. 2007. Suppression of HIV-specific T cell activity by lymph node CD25⁺ regulatory T cells from HIV-infected individuals. Proc Natl Acad Sci USA. 104:3390–3395.

Kuritzkes DR. 2004. Preventing and managing antiretroviral drug resistance. AIDS Patient Care STDs. 18:259–273.

Laessig KA, Lewis LL, Hammerstrom TS, Jenny-Avital ER, et al. 2006. Tenofir DF and emtricitabine vs. zidovudine and lamivudine. NEJM. 354:2506–2508.

Lalezari JP, Henry K, O'Hearn M, Montaner JSG, et al. 2003. Enfuvirtide on HIV fusion inhibitor, for drug resistant HIV infection in North and South America. NEJM. 348:2175–2185.

Lawn SD. 2004. AIDS in Africa. The impact of coinfections on the pathogenesis of HIV-1 infection. J Infect Dis. 48:1–12.

Lazzarin A, Clotet B, Cooper D, Reynes J, et al. 2003. Efficacy of enfuvirtide in patients infected with drug resistant HIV-1 in Europe and Australia. NEJM. 348:2186–2195.

Lee LM, Karon JM, Selik R, Neal JJ, et al. 2001. Survival after AIDS diagnosis in adolescents and adults during the treatment era, United States 1984–1997. JAMA, 285:1308–1315.

Letvin NL, Daniel MD, Sehgal PK, Desrosiers RC. et al. 1985. Induction of AIDS like disease in macaque monkeys with T-Cell tropic retrovirus STLV-III. Science. 230:71–73.

Levy JA. 1993. HIV pathogenesis and long term survival. AIDS. 7:1401–1410.

Levy JA. 1993. Pathogenesis of human immunodeficiency virus infection. Microbial Rev. 57: 183–289.

Lohse N, Hansen ABE, Gerstoft J, Obel N. 2007. Improved survival in HIV-infected persons: Consequences and perspectives. J Antimicrob Chemother. 60:461–463.

Luft BJ, Chua A. 2000. Central nervous system Toxoplasmosis in HIV pathogenesis, diagnosis and therapy. Curr Infect Dis Rep. 2:358–362.

MacArthur RD. 2007. Darunavir: Promising initial results. Lancet. 369:1143–1144.

Manfredi R, Sabbatani S. 2006. A novel antiretroviral class (fusion inhibitors) in the management of HIV infection. Present features and future perspectives of enfuvirtide (T-20). Curr Med Chem. 13:2369–2384.

Markel H. 2005. The search for effective HIV vaccines. N Engl J Med. 353:753–757.

Martin AM, Nolan D, Gaudieri S, Almeida CA, et al. 2004. Predisposition to abacavir hypersensitivity conferred by HLA-B*57021 and a haplotypic Hsp70-Hom variant. Proc Natl Acad Sci USA. 101:4180–4185.

Martinez E, Arnaiz JA, Podzamczer D, Dalmau D, et al. 2003. Substitution of nevirapine, efavirenz or abacavir for protease inhibitors in patients with human immunodeficiency virus infection. NEJM. 349:1036–1046.

Martinez-Picado J, DePasquale MP, Kartsonis N, Hanna GJ, et al. 2000. Antiretroviral resistance during successful therapy of human immunodeficiency virus type 1 infection. Proc Natl Acad Sci USA. 97:10948–10953.

Marx JL. 1982. New disease baffles medical community. Science. 217:618–621.

McLead GX, Hammer SM. 1992. Zidovudine: Five years later. Ann Intern Med. 117:487–501.

Mellors JW, Rinaldo CR, Gupta P, White RM, et al. 1996. Prognosis in HIV-1 infection predicted by the quantity of virus in plasma. Science. 272:1167–1170.

Mellors JW, Munoz A, Giorgi JV, Margolick JB, et al. 1997. Plasma viral load and CD4+ lymphocytes as a prognostic markers of HIV infection. Ann Intern Med. 126:946–954.

Nilsson J, Boasso A, Velilla PA, Zhang R, et al. 2006. HIV-1 driven regulatory T cell accumulation in lymphoid tissues is associated with disease progression in HIV/AIDS. Blood. 108: 3808–3817.

Oldfield V, Keating GM, Plosker G. 2005. Enfuvirtide: A review of its use in the management of HIV infection. Drugs. 65:1139–1160.

Palella FJ Jr., Delaney KM, Moorman AC, Loveless MO, et al. 1998. Declining morbidity and
mortality among patients with advanced human immunodeficiency virus. HIV outpatient study
investigators. N Engl J Med. 338:853–860.
Pantaleo G, Demarest JF, Soudeyns H, Graziosi C, et al. 1994. Major expansion of CD8+ T
Cells with a predominant V beta usage during the primary immune response to HIV. Nature.
370:463–467.
Perelson AS, Neumann AU, Markowitz M, Leonard JM, et al. 1996. HIV-1 dynamics in
vivo: Virion clearance rate, infected cell-life span, and viral generation time. Science. 271:
1582–1586.
Perry CM, Faulds D. 1997. Lamivudine: A review of its antiviral activity, pharmacokinetic prop-
erties and therapeutic efficacy in the management of HIV infection. Drugs. 53:657–680.
Perry CM, Noble S. 1998. Saquinavir soft-gel formulation: A Review of its use in patients with
HIV infection. Drugs. 55:461–486.
Perry CM, Noble S. 1999. Didanosine: An updated review of its use in HIV infection. Drugs.
58:1099–1135.
Piscitelli SC, Gallicano KD. 2001. Interactions among drugs for HIV and opportunistic infection.
NEJM. 344:984–996.
Plosker GL, Noble S. 1999. Indinavir: A review of its use in the management of HIV infection.
Drugs. 58:1165–1203.
Pollard VW, Mallim MH. 1998. The HIV-1 Rev protein. Annu Rev Microbial. 52:491–532
Popovic M, Sarngadharan MG, Read E, Gallo RC. 1984. Detection, isolation, and continuous
production of cytopathic retroviruses (HTLV-111) from patients with AIDS and pre AIDS.
Science. 224:497–500.
Rivero A, Mira JA, Pineda JA. 2007. Liver toxicity induced by nono-nucleoside reverse transcrip-
tase inhibitors. J Antimicrob Chemother. 59:342–346.
Sataszewski S, Morales-Ramirez J, Tashima KT, Rachlis A, et al. 1999. Efavirenz plus zidovudine
and lamivudine, efavirenz plus indinavir, and indinavir plus zidovudine and lamivudine in the
treatment of HIV-1 infection in adults. Study 006 team. NEJM. 341:1865–1873.
Shafer RW, Smeaton LM, Robbins GK, Gruttola V, et al. 2003. Comparison of four drug regimens
and pairs of sequential three-drug regimens as initial therapy for HIV-1 infection. NEJM.
349:2304–2315.
Shankar P, Russo M, Harnisch B, Patterson M, et al. 2000. Impaired function of circulating
HIV-specific CD8+ T cells in chronic human immunodeficiency virus infection. Blood. 96:
3094–3101.
Spence RA, Kati WM, Anderson KS, Johnson KA. 1995. Mechanism of inhibition of HIV-1
reverse transcriptase by nonnucleoside inhibitors. Science. 267:988–993.
Sperling RS, Shapirom DE, Coombsm RW, Todd JA, et al. 1996. Maternal viral load, zidovudine
treatment, and the risk of transmission of human immunodeficiency virus type 1 from mother
to infant. N Engl J Med. 335:1621–1629.
Steinbrook R. 2004. The AIDS epidemic in 2004. N Engl J Med. 351:115–117.
Swainston-Harrison T, Scott LJ. 2005. Atazanavir: A review of its use in the management of HIV
infection. Drugs. 65:2309–2336.
Tonks A. 2007. Quest for the AIDS vaccine. BMJ. 334:1346–1348.
Wang JH. Janas AM, Oson WJ, Kewal Ramani VN, et al. 2007. CD4 coexpression regulates
DC-SIGN-mediated transmission of human immunodeficiency virus type 1. J Virol. 81:
2497–2507.
Watts DH, Brown ZA, Tartaglione T, Burchett SK, et al. 1991. Pharmacokinetic disposition of
zidovudine during pregnancy. J Infect Dis. 163:226–232.
Wei X, Ghosh SK, Taylor ME, Johnson VA, et al. 1995. Viral dynamics in human immunodefi-
ciency virus type 1 infection. Nature. 373:117–122.
Wire MB, Shelton MJ, Studenberg S. 2006. Forsamprenavir: Clinical pharmacokinetics and drug
interactions of the amprenavir prodrug. Clin Pharmacokinet. 45:137–168.

Wood E, Hogg RS, Yip B, Harrigan PR, et al. 2003. Is there a baseline CD4 cell count that precludes a survival response to modern antiretroviral therapy? AIDS. 17:711–720.

Wyatt R, Sodroski J. 1998. The HIV-1 envelope glycoproteins: Fusogens, antigens, and immunogens. Science. 280:1884–1888.

Young SD, Britcher SF, Tran LO, Payne LS, et al. 1995. Erfavirenz: A novel highly potent nonnucleoside inhibitor of the human immunodeficiency virus type 1 reverse transcriptase. Antimicrog Agents Chemother. 39:2602–2605.

Chapter 9
Regulatory T Cells and Disease State

Introduction

The immune response is designed to protect human and other organisms from disease-causing agents; it also protects against detrimental responses to self. The immune system needs to be strictly regulated because of its ability to produce inflammatory mediators, killer cells and antibodies, which are synthesized to eliminate the invading organisms, but can also harm other normal cells. Consequently, an immune response not only can produce autoimmunity but also is capable of producing other diseases due to its ability to attack normal cells that are damaged mainly by the inflammatory cytokines in a collateral damage. A number of regulatory mechanisms keep these harmful effects of the immune response in check.

Immune response is triggered by antigen presentation to the TCR in the context of MHC molecules, resulting in the activation and proliferation of CD4$^+$ cells and secretion of cytokines. This activation also causes CD4$^+$ T cells to express a number of cell surface receptors including CD25 (IL-2 receptor), CTLA-4 and CD40 ligand (CD40L). This growth of antigen-activated T cells was suggested to be controlled by suppressor T cells. Niels Jerne proposed that immune response may be inhibited by special lymphocytes, which pointed to the existence of suppressor T cells. The presence of such cells was initially suggested by the observation that injection of polyclonal T lymphoblasts from a parent to F_1 hybrid blocked allograft rejection in rats. Similarly, injection with myelin basic protein (MBP)-reactive cloned T-cell line abrogated autoimmune encephalomyelitis specific for this antigen. The studies of the regulation of the anti-MBP response in autoimmune encephalomyelitis provided further foundational work in identifying the regulatory T (Treg) cells. This model system allowed the understanding of the recognition of TCR peptides by Treg cells.

Regulatory T cells have been defined as having either suppressor or regulatory functions. The term *regulatory* has been preferred over *suppressor* because of questions about I–J-regulated suppressor T cells. Gershon initially suggested that T cells could also have a regulatory function in addition to its role as helper cells for antibody synthesis. The initial focus was on soluble factors secreted by suppressor T cells as immune regulatory agents, and some of these soluble suppressor factors were characterized as MHC-restricted. However, due to a lack of a definite cell surface marker on suppressor T cells, their existence remained controversial. The

M.M. Khan, *Immunopharmacology*, DOI: 10.1007/978-0-387-77976-8_9,
© Springer Science+Business Media, LLC 2008

concept of the presence of suppressor T cells suffered further with the discovery of TH_1 and TH_2 cell subsets.

Studies by Nishizuka and Sakakura led to the conclusions that the suppression of the disease was due to the actions of thymus-derived lymphocytes and the splenocytes providing protection against the disease also originated in the thymus. However, they did not attempt to isolate the suppressor factor from the thymus. Later studies by other investigators demonstrated that only a small number of Treg cells were required to stop the development of autoimmune disease in mouse models, but the number of cells required to inhibit antibody response was much higher. The splenic T cells from mice with disease were able to transfer autoimmune disease to the newborn or adult nu/nu mice. $CD4^+CD8^-$ cells were later identified as the effector and suppressor T-cell population. It was found that the removal of suppressor T cells from lymphoid cells will result in the disease and the readministration of these cells will induce self-tolerance and suppression of autoimmunity. The suppressor T cells are a very small number of cells among $CD4^+$ T lymphocytes that also express CD25 antigen, and selective depletion of $CD4^+CD25^+$ T cells results in multiple manifestations of autoimmune diseases. This is a result of depletion of Treg cells from either thymus or the peripheral lymphoid tissue.

$CD4^+$ T cells mediate suppressor function independent of $CD8^+$ T cells, and these cells classified as Treg cells maintain self-tolerance and suppress responses to foreign antigens. Although various types of Treg cells have been found, the most attention has been paid to $CD4^+CD25^+$ T cells that predominantly originate in the thymus, and their centralized production is referred to as "the third function of thymus." They are mature T cells with a distinct function, and humans lacking $CD4^+CD25^+$ T cells exhibit severe defects in controlling autoimmune response and have abnormal immune regulation, resulting in autoimmune and allergic diseases.

Based on the immunosuppressive activity, the many different types of Treg cells include natural $CD4^+CD25^+$ T cells, peripheral Treg cells, IL-10-secreting Tr1 cells, TGF-β-secreting TH_3 cells, $CD8^+CD28^-$ T cells, $CD8^+CD122^+$ T cells, Qa-1-restricted $CD8^+$ T cells, γ/δ T cells and NKT cells. The production of different types of Treg cells is distinct; some are produced as a part of the innate immune response while others are produced in response to an antigen as the acquired immune response develops with a selective participation of cytokines. Their function varies and includes regulation of autoimmunity, allergic responses, infection, inflammation and transplant tolerance. The mechanisms of action of Treg cells are also diverse and range from cell–cell interaction involving CTLA-4, perforin, TGF-β, Lag3, granzyme B-dependent killing, regulation of dendritic cells, IL-10-mediated suppression to Treg cell-mediated IL-12 consumption.

Types of Regulatory T Cells

Naturally Occurring Treg Cells

The Treg cells have been classified on the basis of their site of origin or mechanism of action (Table 9.1). The site of origin is for the naturally occurring Treg cells in thymus, and the main mechanism of action is via cell–cell interaction.

Table 9.1 Types of Treg Cells

Cell type	Phenotype
Natural Treg cells	CD4$^+$ CD25$^+$ FoxP3$^+$
Peripheral Treg cells	CD4$^+$, CD25$^{+/-}$ FoxP3$^-$/FoxP3$^+$
Tr1 cells	CD4$^+$ CD25$^{+/-}$, FoxP3$^-$ ROG
Th3 cells	CD4$^+$, CD25$^+$, FoxP3$^+$/?
CD8 regulatory cells	CD8$^+$, CD25$^+$, FoxP3$^+$
Qa-1-restricted CD8$^+$	CD8$^+$, CD25$^+$, CD28$^+$
CD8$^+$ CD28$^-$	CD8$^+$, CD25$^+$, FoxP3?
NKT	Vα24$^+$Vβ11$^+$, CD4$^+$/CD4$^-$/CD8$^+$

These Treg cells constitutively express CD4, CD25 (which is the α chain of IL-2 receptors), Foxp3 (forkhead–winged-helix) transcription factor and surface CD152. A defect in the Foxp3 gene results in the hyperactivation of CD4$^+$ T cells. Foxp3 is also expressed on CD4$^+$CD25$^+$ peripheral T cells and CD4$^+$CD8$^-$CD25$^+$ thymocytes but is not expressed on other thymocytes, T cells and B cells. Naturally occurring Treg cells do not require antigen exposure for their suppressive effector function; however, their generation and some of their activity may require TGF-β. A number of costimulatory signals and cytokines are also involved in the generation of Treg cells, which may include B7, TNF family molecules, CD40L, PD-1, IL-2, TGF-β or TNF-α, and these mechanisms are independent of the avidity of TCR. This results in the induction of Treg cells through a genetic program with concomitant expression of CD25 and Foxp3$^+$ or negative selection of thymocytes. Natural Treg cells possess a broad TCR repertoire that has higher affinity than other T cells for the MHC class II self-peptide ligands that have selected them positively in the thymus. They do not produce inflammatory cytokines and inhibit the activation, proliferation and differentiation of a number of cell types including CD4$^+$ cells, CD8$^+$ cells, B cells, NK cells, NKT cells and dendritic cells.

If natural Treg cells cannot be generated in thymus or have a deficient function, this results in a number of autoimmune diseases. Impaired Treg cell generation is observed in children with thymic hypoplasia resulting from 22q-2 deletion syndrome. Autoimmune polyendocrinopathy candidiasis ectodermal dystrophy (APECED) is due to mutation in a gene called transcription factor autoimmune regulator (AIRE), and important self-antigens on thymic medullary epithelial cells are regulated by AIRE. In the absence of AIRE, T cells that recognize self-antigens do not undergo negative selection and consequently are not deleted. Patients suffering from rheumatoid arthritis and MS exhibiting a reduced number of TCR excision circles (Trec) have suppressed activity of the thymus and its output. In juvenile rheumatoid arthritis, there are reduced numbers of Trec, suggesting premature aging of the thymus. Patients with autoimmune disease have early aged thymus, which results in the poor development of Treg cells and an escape of non-Treg cells with an autoreactive TCR.

Regulatory T cells exhibit different developmental stages; one group (CD4$^+$ CD25$^+$) expresses high CD62L and CCR7 levels and inhibits inflammation after binding to antigen-draining lymph nodes. Another subgroup (CD4$^+$CD25$^+$ or CD4$^+$ CD25$^-$) expresses $\alpha\epsilon$ β7 integrin and suppresses local immune reactions after homing to nonlymphogenic tissues at sites of inflammation. The expression of CD25 on natural Treg cells varies from none, low, intermediate to high, suggesting a shift of expression based on the degree of injury or inflammation.

Various other cell surface markers are also expressed on CD4$^+$ Treg cells including CD45RB^{Low}, CD62L, CD103, CD152 (cytolytic T lymphocyte antigen-4 or CTLA-4) and glucocorticoid-induced TNF receptor (GITR family-related gene). The known phenotypic markers expressed on naturally occurring Treg cells are shown in Table 9.2. Many of these markers are associated with activated/memory cells, and it appears that naturally occurring Treg cells may be similar to memory T cells and are usually in an antigen-primed state. The Treg cells exhibit broad antigen specificity and have enhanced ability to recognize self-antigen rather than other T-cell subsets, and their production, maintenance and function are IL-2-dependent. The development of Treg cells in thymus is blocked if there is a defect in Foxp3/FOXP3 gene that controls the production of these cells. They suppress various immune cells involved in both the innate and acquired immune responses; IL-2 and TCR stimulation is required to express their suppressive effects on T helper cell proliferation and IL-2 production but the subsequent immune suppression is not antigen-specific. Interestingly, the normal T-cell inducers do not cause the proliferation of naturally arising CD4$^+$CD25$^+$ Treg cells or IL-2 secretion. However, they respond to very high doses of IL-2, mature dendritic cells as APCs or anti-CD28 and as a result proliferate and secrete IL-2.

The cell–cell interaction is the critical mechanism of suppression by the natural Treg cells where CTLA-4, GITR and PD-110 play a role in the contact-dependent suppression. A number of molecules expressed on natural Treg cells including CTLA-4, CD80, CD86 and CD223 are inhibitory molecules. For cell–cell interaction, a competition for APCs and specific MHC–peptide antigenic complexes is required.

Treg cells express all three chains (α, β, γ) of the high-affinity IL-2 receptor. Signaling for IL-2 receptor is mediated through induction of JAK1 and JAK3, which results in the phosphorylation and activation of STAT3 and STAT5. The translocation of these activated transcriptional factors to the nucleus results in the functional effects mediated by IL-2. Stimulation of IL-2 receptors also results in the activa-

Table 9.2 Phenotypic Markers of Naturally Occurring Treg Cells

CD4
CD25
CD38
CD122
CD45^{low}
CD45RO
CTLA-4
GITR
CD62L
CD95
CD103
TLRs 4–8

tion of other pathways including MAPK and phosphatidylinositol 3-kinase (PI3 K). IL-2 regulates self-tolerance through its involvement in the development and homeostasis of CD4$^+$CD25$^+$ Treg cells. Treg cells have a distinct IL-2R-mediated signal transduction pathway where, while the JAK–STAT-dependent transduction pathways are not altered, downstream signaling of PI3 K is not observed. This difference in transduction pathways is associated with the expression of PTEN (phosphatase and tensin homolog) and is correlated with the hypoproliferative response of Treg cells. PTEN is a lipid phosphatase, which is a catalyst for the reverse reaction of PI3 K. Consequently, PTEN negatively activates the induction of downstream signal transduction pathway. In Treg cells, the expression of PTEN is unaltered as opposed to activated T cells where it undergoes downregulation. IL-2-induced T-cell proliferation is dependent on PI3 K-mediated signal transduction. PTEN deletion allows the expansion of Treg cells in the presence of IL-2 without compromising their regulatory activity in maintaining homeostasis and self-tolerance.

IL-2 signaling directly targets the Foxp3 gene in Treg cells. This is accomplished as a result of binding between a specific site present in the first intron of the Foxp3 gene and STAT3 and STAT5 proteins. This signaling pathway is specific for CD4$^+$CD25$^+$ Treg cells since an IL-2-induced expression of the Foxp3 gene is not seen in CD4$^+$CD25$^-$ cells. This lack of effect is not due to the absence of IL-2Rα because conditions that do not require the presence of IL-2Rα such as stimulation with CD3 and treatment with high concentrations of IL-2 do not alter the effects of IL-2 on Foxp3 gene expression in CD4$^+$CD25$^-$ cells. Consequently, IL-2 affects CD4$^+$CD25$^+$ Treg cells in a unique manner, which is mediated via expression of the Foxp3 gene.

Peripheral (Adaptive) Treg Cells

The peripheral (adaptive) Treg cells develop in the periphery and the stimulus for their generation is either an ongoing immune response or exposure to tolerogenic dendritic cells. Adaptive Treg cells develop from naïve precursors or mature T cells, and their specificities lie in antigens other than the ones that come in contact in the thymus, such as food antigens, pathogens, parasites and bacterial flora. Their mechanism of action is mediated via suppressive effects of cytokines (IL-10 or TGF-β). The peripheral (adaptive) Treg cells are not generated in thymus and are specific for both foreign and self-antigens. This is in contrast to the thymus-derived Treg cells, which are specific for antigens seen in thymus. Two models have been suggested for the generation of peripheral Treg cells. According to the linear model, after antigen recognition, naïve T cells are activated and differentiated into effector and Treg cells. Alternatively, a parallel model suggests that after activation, naïve T cells remain uncommitted, and their development into effector and Treg cells is in parallel. As a result, the development of effector cells is faster than peripheral Treg cells. IL-2 is required for the development and differentiation of both types of cells. Non-Treg cells can differentiate into CD4$^+$CD25$^+$ Treg cells in the periphery and can function like natural Treg cells, that is, suppression of

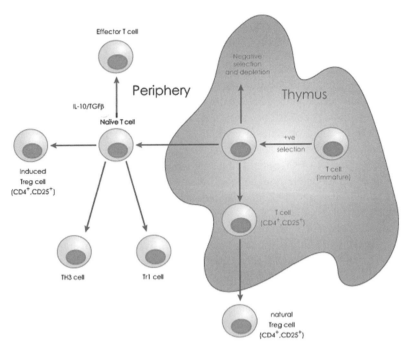

Fig. 9.1 Development of CD4⁺ regulatory T cells. Natural CD4⁺CD25⁺ Treg cells develop in the thymus as a result of positive selection between TCR and host antigens. The thymus-derived Treg cells are specific for antigens seen in the thymus. The autoreactive T cells undergo negative selection and are depleted by apoptosis. The acquired Treg cells develop in the periphery from naïve precursors, and their specificity lies in antigens other than the ones that come in contact with the thymus. Tr1 cells are induced in the periphery when naïve T cells are exposed to an antigen in the presence of IL-10. They are not identified by a particular cell surface marker. TH3 cells are generated from naïve CD4⁺ T cells as a result of low doses of antigen via the oral route, and they secrete TGF-β. (This diagram is based on one published by Lohret et al, 2006) (*see* Color Insert)

T-lymphocyte proliferation by cell–cell interaction independent of cytokines. The only main requirement for the generation of these natural Treg-like cells in the periphery is activation of CD4⁺ T cells, and their production can be achieved by exposure to oral, intravenous or subcutaneous antigens or continued exposure to superantigen. The development of natural Treg-like cells is possible from peripheral CD4⁺CD25⁺ T cells under conditions requiring either TGF-β or both TGF-β and TCR activation, which results in CD4⁺CD25⁻ cells expressing Treg cell function and Foxp3. The treatment of human peripheral blood lymphocytes with allogeneic dendritic cells in the presence of IL-10 also results in the development of natural Treg-like cells in the periphery, as is the case with the treatment of CD4⁺CD25⁻ T cells with anti-CD3 and anti-CD28. The induction of Foxp3 is independent of TGF-β, and peripheral Treg cells look identical to natural Treg cells in phenotype, function and gene expression. They differ from Treg cells only in their require-ment for TCR and CD28 for induction. The induction of peripheral Treg cells requires immunogenic antigen exposure and the combination of an antigen directed at dendritic cells and anti-CD40 antibody. Immune response is limited against foreign antigens, and collateral damage to healthy tissue is avoided as a result of early development of Foxp3 Treg cells. The reduction in the number of effector and Treg cells may result from apoptosis and peripheral migration. However, in addition to these short-term Treg cells, long-persisting Treg cells can also be induced. This induction is a result of exposure to low levels of antigen, which could be achieved without inflammation. The generation of these Treg cells may be less efficient in the beginning but they may have prolonged presence and consequently play a role in tolerance to autoantigens.

Lymphopenia promotes expansion of Treg cells. Lymphocyte activation is regulated by competition with general lymphocyte population. In the absence of such competition, both regulatory and effector cells develop sequentially as a result of weak signals. In lymphopenia, the appearance of Treg cells parallels the recovery, and as soon as the accumulation of cells seizes, the Treg cells are expressed. Foxp3$^+$ cells may contribute to the development of homeostasis, and after the cell number reaches equilibrium, the generation of Treg cells begins. In the periphery, generation of antigen-specific Treg cells could be rapid when other T cells are present, resulting in IL-2 secretion and the production of Treg cells. Peripheral Treg cells play an important role in tolerance, tumor immunity and microbial defense. The suppressive effects of these cells are mediated via production of immunosuppressive cytokines, IL-10 and TGF-β. However, some peripherally induced Treg cells that express Foxp3 also act by cell–cell interaction. Treg cells suppress immune response during infection to avoid tissue damage but this may prolong the infection. Tumor-infiltrating CD4$^+$CD25$^+$ Treg cells are suppressive in nature and are found in increased numbers in human cancers.

The generation of peripheral Treg cells may result in the establishment of homeostasis after its disruption. Examples include infection, autoimmune diseases, certain forms of cancers and immunodeficiency syndrome. The mechanisms by which peripheral Treg cells induce self-tolerance and homeostasis may involve cytokines or cell–cell interaction. The cytokines IL-10 and TGF-β and molecules such as CTLA-4 are involved in the effector mechanisms of Treg cells. The indirect effects of Treg cells may be mediated via APCs or NK cells. More specifically, the assembly of immunological synapse between APCs and effector cells is modulated by Treg cells, which may be mediated via direct or indirect mechanisms.

Bone marrow is also a significant reservoir for Treg cells. Treg cells enter in the bone marrow and are retained through CSCR4/CSCL12 signals. Functional stromal-derived factor (CXCL12) is strongly expressed in the bone marrow and is the ligand for CXCR4. Human bone marrow CXCL12 expression is suppressed by G-CSF, and this causes the migration of Treg cells from bone marrow to the peripheral blood. This also explains improvement in autoimmune disease and graft-versus-host response after treatment with G-CSF.

Tr1 Cells

CD4$^+$ type 1 Treg (Tr1) cells were initially characterized after isolation of the peripheral blood lymphocytes from patients who suffered from combined immunodeficiency and have received successful HLA-mismatched bone marrow transplant, resulting in their generation from naïve CD4$^+$ cells following the development of antigen-specific immune response. Tr1 cells can exist naturally or may be induced. These cells are induced in the periphery when naïve T cells are exposed to an antigen in the presence of IL-10. Tr1 cells are generated from naïve CD4$^+$ T cells after they are activated by TCR and CD28. IL-10-producing Tr1 cells are also produced by treatment with a combination of anti-IL-4, anti-IL-12 antibodies, dexamethasone and active vitamin D3 (Table 9.3). Furthermore, IL-10-secreting Tr1 cells are generated when naïve CD4$^+$ T cells are treated with immature dendritic cells, IFN-α or immunosuppressive agents. Tr1 cells are not identified on the basis of any particular surface marker and also do not constitutively express Foxp3; however, they may express markers associated with TH$_2$ cells and repressor of GATA (ROG). They do express high levels of surface CD152 as is the case with natural Treg cells. Tr1 cells could be either CD4$^+$ or CD8$^+$, proliferate poorly and migrate to the inflamed tissue, secrete high amounts of IL-10, TGF-β and IL-5, low concentrations of IL-2 and IFN-γ, but do not secrete IL-4. In response to signaling through TCR, IL-2R and CD28, they exhibit anergy and suppress antigen-specific proliferation of naïve CD4$^+$ cells. The immune suppression is mediated via cytokines and not via cell–cell interaction. In animal models, Tr1 cells regulate mucosal tolerance, diabetes and responses to transplant antigens, infectious agents and allergens after they migrate to the inflamed or injured site. Their role is pivotal in the maintenance of peripheral tolerance. A member of the TNF superfamily, OX40 ligand, inhibits the generation of Tr1 cells, and as a consequence, OX40L augments immunity and abrogates tolerance.

TH$_3$ Cells

Antigen-specific Treg cells, TH$_3$, are generated from naïve CD4$^+$ T cells as a result of low doses of antigen via the oral route. This phenomenon is just opposite to hyporesponsiveness resulting from anergy or deletion, caused by exposure to high antigen doses. Oral tolerance results from such exposure due to interaction of dietary antigens with GI immune apparatus. TH$_3$ can also be induced in the periphery. They are TGF-β-producing cells, which may also express Foxp3. TH$_3$ cells suppress antigen-specific responses and can transfer tolerance. Their mechanism of action

Table 9.3 Inducers for the Generation of Tr1 Cells

TCR
CD28
α IL-4 + α IL-12 antibodies
Dexamethasone
Active vitamin D3
Immature dendritic cells
IL-10
IFN-α

is mediated via TGF-β, and defects in TH$_3$ cells may be associated with the development of autoimmune disease.

Natural Killer T Cells

Since NKT cells express a TCR, they are defined as T lymphocytes and are distinct from NK cells, although they both share CD161 or NKR-P1. They also differ from T lymphocytes and other Treg cells because they lack the ability to interact with peptide antigens in a classical MHC class I- or class II-restricted manner. Instead, they recognize antigens in the context of a glycolipid, which is a nonclassical antigen-presenting molecule called CD1d. NKT cells have two main subsets CD4$^+$ or CD4$^-$ but some of the NKT cells are also CD8$^+$. The subsets of NKT cells are present along with other lymphocytes, but their numbers are tissue-dependent. In mice, their numbers are most prevalent in liver and show lower frequencies in bone marrow, spleen, thymus, blood and lymph nodes.

They recognize a class of antigens that is not recognized by T lymphocytes. After recognition of the antigen, they are activated within 1–2 h of TCR ligation and produce both TH$_1$ (IFN-γ, TNF-α) and TH$_2$ (IL-4, IL-13) cytokines. Many glycolipids including glycophosphatidylinositol, gangliosides, and phosphoethanolamine can activate NKT cells. Although their cytokine production pattern is TH$_0$-like, they can produce either a TH$_1$ or a TH$_2$ immune response.

NKT cells express receptors for both NK lineage and T cells (TCRαβ) and thus are a unique population of lymphocytes. Tumor cells expressing lipid antigens are recognized and killed by NKT cells. These lipid antigens related to the glycolipid α-galactosylceramide are presented to NKT cells in the context of MHC (Ib, CD1d) molecules. In addition to their ability to kill tumor cells, they also regulate autoimmune diseases. Their effector function as NK cells and the ability to secrete a number of cytokines including IFN-α, TGF-β, IL-4 and IL-10 are enhanced following antigen recognition by TCR in the context of MHC class Ib molecules. On the basis of the secretion of cytokines, it appears that NKT cells may be involved in modulating both innate- and TH$_2$-dependent acquired immune response. In animal models, NKT cells have been shown to inhibit the onset of type 1 diabetes mellitus and MS, and their depletion accelerates the development of the disease. In humans, there is also an association of NKT cells and autoimmune disease. A decrease in the frequency of Vα24-JαQ NKT cells is associated with relapse in patients with MS, and patients with diabetes also have lower expression of Vα24-JαQ NKT cells.

NKT cells also play a role in allograft tolerance, which involves NKT cell-dependent allospecific Treg cell generation. They induce cardiac allograft tolerance and inhibit graft-versus-host disease. NKT cells may also play a role in tumor rejection and are also required for IL-12-mediated cancer therapy. α-GalCer, a glycolipid recognized by NKT cells, causes rejection of a number of tumor cell lines via activating NKT cells.

NKT cells are unique regulatory cells although they also act as effector cells. Their regulatory role is reflected by the pattern of cytokines they secrete, interaction with dendritic cells and in their small numbers. Interaction with dendritic cells suggests their role in acquired immune response. It seems that the phenotype of these cells may be distinct at different locations such as thymus versus liver, which may be a deciding factor in their ability to promote or suppress the immune response. Human $CD4^+$ NKT cells produce a high TH_2/TH_1 cytokine ratio with distinct expression of cytotoxic and chemokine receptors as opposed to $CD4^-$ NKT cells. The type of signal that NKT cells receive may also determine whether they will produce pro- or anti-inflammatory response. Examples include the production of IFN-γ when cross-linked with IL-12 or anti-NK1.1 and IL-4 production after cross-linking with IL-7; therefore the cytokines secreted by NKT cells may be dependent on the type of TCR stimulation. The OCH analog of α-GalCer induces TH_2 responses and C-glycoside induces TH_1 responses. Alternatively, the products produced by NKT cells may be the same but different physiological or pathological circumstances may interpret them differently and thus NKT cells are not responsible for the final impact. The example includes the immediate production of both IFN-γ and IL-4 following α-GalCer stimulation, which are the products of preformed mRNA, but IL-4 secretion stops within a few hours while NKT cells continue to secrete IFN-γ for another 2–3 days. This suggests that this differential response may be attributed to the temporal nature of the interactions of NKT cells with other immune cells.

$CD4^+CD25^+$ T cells and NKT cells exist as natural suppressor cells from early fetal life before antigen exposure and play an important role in immune regulation during innate and/or primary immune responses. The NKT cells recognize glycoproteins via Vα chain of TCR$\alpha\beta$, which is expressed by tumor cells, pathogens, injured apoptotic cells and blast cells. The primary immune response balance of TH_1 and TH_2 cells is modulated by NKT cells via their secretion of IL-4 and IL-10. Naturally occurring $CD4^+CD25^+$ T-cell subset is also present in the peripheral lymphoid system and without antigen stimulation can affect the primary immune response. Their effect is mediated via cell–cell interaction as well as the secretion of TGF-β. Antigen recognition activates NKT cells, which respond by secreting IL-13. IL-13 receptors are expressed on certain myeloid cells, and these cells as a result of IL-13 binding and signal transduction produce TGF-β and suppress the $CD8^+$ CTLs. This results in the suppression of tumor immunity since $CD8^+$ CTLs kill tumor cells.

Regulatory $CD8^+$ T Cells

Most of the information about Treg cells involves $CD4^+$ sublineage with regulatory activity. Although $CD8^+$ T cells were initially suggested to be suppressor T cells, identification of $CD8^+$ Treg cells has not received equal attention. Following the identification of $CD4^+$ Treg cells, a subpopulation of $CD8^+$ T cells was identified, which suppressed T helper cell and B cell responses in an MHC-dependent manner, requiring the expression of HLA Class Ib MHC molecule Qa-1 on target cells. $CD8^+$

Treg cells in mice are divided into at least two groups: Qa-1 restricted and Qa-1 nonrestricted; Qa-1 is an equivalent of human HLA-E.

CD8$^+$ Treg cells are activated by autologous CD4$^+$ T cells after their induction during the primary immune response, differentiating into functional suppressor T cells. Their effector function is prominent during the secondary immune response as well as memory-based immune responses. CD8$^+$ Treg cells may be responsible for producing suppression of autoimmunity after the patient recovers from the first episode of the disease and consequently will resist to a relapse and may decrease the severity of the symptoms in future episodes of the same disease. Qa-1-restricted CD8$^+$ Treg cells recognize their target through their TCRαβ in an MHC-restricted manner. Some Qa-1 self-peptide-expressing activated T cells are downregulated by these cells, but this is not the case for all activated T cells. Both types of Qa-1 receptors, TCR and CD94/NKG2, can be expressed on CD8$^+$ Treg cells. CD94/NKG2 is a C-type lectin receptor present on NK and CD8$^+$ cells. The TCR can recognize Qa-1 complex on induced CD4$^+$ cells, resulting in suppressor activity, and CD94/NKG2 recognizes Qa-1/Qdm ligands. NKG2 receptors play a dual role as they can either enhance suppression in response to NKG2C, E or H or inhibit suppression in response to NKG2A or NKG2B. A non-Qa-1-restricted CD8$^+$CD28$^-$ Treg cell subset has been identified, which mediates suppression via APCs.

Human CD8$^+$ Treg cells express CD25, CD69, CTLA-4 and Foxp3. They secrete IL-4, IL-5, IL-13 and TGF-β, but do not secrete IFN-γ and contribute to immunoregulation. Naïve CD8$^+$CD25$^-$ cells are considered to differentiate into CD8$^+$ Treg cells when presented with an antigen. CD8$^+$CD28$^-$ Treg cells are induced in the presence of IL-10. IL-10 may be involved in the downregulation of dendritic cell costimulation as well as in the upregulation of ILT-3 and ILT-4. An additional human Treg subset has been identified, which includes CD8$^+$, LAG-3$^+$ (lymphocyte activation gene-3, an MHC class II-binding CD4 homologue), CD25$^+$, Foxp3$^+$ and CCL4$^+$, and it suppresses T-cell responses via secretion of chemokine CC chemokine ligand-4.

CD8$^+$ Treg cells are generated in neonatal life when T lymphocytes enter into nonlymphoid tissue and maintain tolerance during adulthood. The thymus does not contribute to this antigen-specific tolerance, and the continued presence of the antigen is necessary to maintain tolerance. Antibody-mediated inhibition of T-cell migration abrogates this tolerance. TGF-β1 may play a role in the upregulation of this TCR$^+$CD8$^+$ subset-mediated tolerance. Furthermore, granzyme B is activated in this TCR$^+$CD8$^+$ Treg cell subset that has been implicated in the induction of cell death of effector T cells by CD4$^+$CD25$^+$ Treg cells. This naturally occurring TCR$^+$CD8$^+$ Treg cell subset is induced by self-antigens that are expressed in neonatal mice on parenchymal cells. They maintain tolerance during adult life as a result of downregulating effector function of T cells. This mechanism is independent of CD4$^+$ T cells.

The generation and function of CD8$^+$ Treg cells is less defined than that of CD4$^+$ Treg cells. It is well established that Treg cells induce tolerance by controlling autoreactive T cells that were not deleted in the thymus. Although CD4$^+$CD25$^+$

Treg cells are generated in the thymus, the site of origin of CD8$^+$ Treg cells has not been determined.

Regulatory T Cells in the Mucosal System

Tolerance is an important goal of the immune response in the GI tract. Harmful pathogens are recognized by the mucosa-associated lymphoid tissue to protect the epithelial layer from their deleterious effects. Moreover, the mucosa-associated lymphoid tissue develops tolerance against dietary and bacterial antigens. In normal individuals, a number of regulatory cell types control inflammatory response when pathogenic bacteria and viruses attack intestinal mucosa. A lack of appropriate regulatory responses that limit inflammation in the gut results in the development of inflammatory bowl disease. A number of Treg cell subsets including CD45Rbl0, CD4$^+$CD25$^+$, CD4$^+$, CD103$^+$, CD4$^+$Tr1, CD4$^+$CD8$^+$IELS, TCRαβ$^+$CD8$^+$IELS and CD8$^+$CD28$^-$ cells may be involved in mucosal immunity. CD4$^+$Tr1 are present in intestinal mucosa, and their immunosuppressive effects are mediated via IL-10, which they produce in large amounts. They are induced by IL-10, secreted by intestinal epithelial cells and other Treg cells in intestinal mucosa but their antigenic specificity and TCR repertoire have not been identified. The generation of mucosal CD4$^+$ Tr1 cells is negatively modulated by a subset of dendritic cells. The role of these cells in the prevention of human inflammatory bowel disease has not yet been established. Another subset of Treg cells associated with mucosal immunity is CD4$^+$TH3. These cells are present in human intestinal mucosa and play a role in controlling inflammation in the gut, but their mechanisms of induction are unknown. In addition to CD4$^+$CD25$^+$ Treg cells, several other subpopulations of CD8$^+$ Treg cells also contribute to mucosal immunity.

The intestinal epithelial cells play an important role in the generation of these Treg cells that maintain tolerance in the mucosal immune system because of their ability to serve as APCs. The intestinal epithelial cells process and present antigenic fragments in the context of MHC molecules to the TCR. Several regulatory subsets, which include both CD4$^+$ and CD8$^+$ T cells, are involved in oral tolerance, and most of the immune suppression that they cause is mediated via IL-4, IL-10 and TGF-β. The mechanism of tolerance resulting from antigen exposure in lamina propria is different from oral tolerance. This tolerance is not dependent on perforin, and these cells act like CD4$^+$CD25$^+$ or CD8$^+$CD28$^-$ Treg cells and express CD8, CD101 and CD103. These CD8$^+$ Treg cells are not present in patients with inflammatory bowel disease, which may be due to the epithelial cell glycoprotein gp180, a molecule expressed on all normal intestinal epithelial cells. The interaction of the gp180/CD1d complex on intestinal epithelial cells with a subset of CD8$^+$ Treg cells results in oligoclonal expansion of CD8$^+$ Treg cells in intestinal mucosa.

T-Cell Vaccination and Regulatory T Cells

Anti-idiotypic and antiergotypic Treg cells are activated after T-cell vaccination and are regulators of the immune response. The Treg cells induced by activated

T-cell vaccines that are not anti-idiotypic are called antiergotypic. They proliferate in response to autologous T cells after their activation, and their presence does not require T-cell vaccination or prevalence of an autoimmune disease. They are widely distributed in thymus, spleen and lymph nodes in naïve rats, and they do not need antigen exposure. Antierg T cells include both TCRα/β$^+$ and TCRγ/δ$^+$ T cells, and CD8$^+$ markers are present on naïve antierg T cells. IFN-γ and TNF-α are secreted by TCRγ/δ$^+$ antierg T cells following activation of T cells. No detectable levels of cytokines are produced by TCRα/β$^+$ antierg T cells as they proliferate in response to T-cell activation. Cell–cell interaction is involved for interaction between antierg T cells and activated stimulator T cells. Antierg T cells can also recognize ergotype on the cell surface of macrophages and other APCs, but this results in a much weaker response as opposed to the activated stimulator T cells. The naïve TCRα/β$^+$CD8$^+$ and TCRγ/δ$^+$ antiergotypic T cells respond in a classical MHC class I-restricted manner, and the B7 and CD28 molecules are involved in this recognition process. CD4$^+$CD25$^+$ Treg cells do not play a role in antiergotypic response.

A regulatory cell needs to meet two conditions in order for it to be called an ergotype. It must be expressed and presented by the activated and not the resting cells, resulting in the activation of antierg T cells; TCR, CD25 and HSP60 epitopes are some examples of ergotypes. Only activated T cells can present ergotypic TCR peptides to antierg T cells, despite the expression of TCR on resting T cells.

Anti-idiotypic Treg cells utilize unique TCR CDR3 peptides on the cell surface of effector cells. Anti-idiotypic T cells generated after T-cell vaccination are cytolytic T cells, which are CD8$^+$ that kill after interaction with target TCRs in the context of MHC class I molecules. This CD8$^+$ anti-idiotypic T-cell response is responsible for the depletion of circulating autoreactive T cells. Furthermore, CD4$^+$ Treg cell responses also occur in response to T-cell vaccination in addition to the generation of CD8$^+$ anti-idiotypic T cells. The production of CD4$^+$ Treg cells in anti-idiotypic response may be responsible for the production of T-cell vaccination-induced clinical effects. The precise mechanisms relating to the involvement of CD4$^+$ Treg cells in these processes have not yet been established. However, interaction with ergotypes including IL-2 receptors and heat-shock protein 60 may result in the induction of CD4$^+$ Treg cell responses.

In preliminary clinical trials of MS, the therapeutic effects of T-cell vaccination involve CD4$^+$ Treg cells that are produced as a result of repeated immunization with irradiated autologous T cells selected for autoantigens. The MS patients receiving T-cell vaccination produce two different populations of Treg cells that differ in their expression of Foxp3 gene and cytokine production and may have different mechanisms of action. Most of the cells have an abundant expression of Foxp3 gene and produce IL-10 and IFN-γ, while a small number of cells produce only IL-10 and have very low levels of Foxp3 gene. After T-cell vaccination, Treg cells expressing CD4CD25Foxp3 may be derived from the naturally occurring CD4$^+$CD25$^+$ Treg cells. The T-cell vaccination results in the upregulation and proliferation of CD4$^+$CD25$^+$ Treg cells, the numbers of which are below normal in patients with MS. The other subset of Treg cells, CD4$^+$CD25$^+$Foxp3$^-$, produces high levels of IL-10, resulting in the suppression of activated T cells. This inhibition by IL-10 is reversed by IL-10 antagonist or monoclonal antibody to IL-10.

The CD4$^+$CD25$^+$FoxP3$^+$ Treg cells recognize an epitope corresponding to 61–73 residues of the α chain of IL-2 receptors. This may be clinically relevant in finding new treatments for autoimmune diseases.

Regulatory T Cells and Antibody Production

The autoantibody production in autoimmune diseases may be attributed to the inability of Treg cells to control their synthesis. In an autoimmune model, T cells regulate the mechanisms through which B cells that were autoreactive to self-antigens do not produce autoantibodies, suggesting a role for suppressor T cells. The administration of irradiation, thymocytes, lymph nodes or spleen cells inhibits the production of autoantibodies, which is attributed to the suppressor T cells.

Mechanisms of Induction of Treg Cells

Treg cells are induced by low doses of oral antigen, but high doses result in anergy. CD4$^+$ and CD8$^+$ Treg cells can be produced in autoimmune murine models when self-antigen is administered by the oral route. These antigen-specific Treg cells are of the TH$_3$ subgroup. Antigen-specific Treg cells can also be induced when nonobese diabetic mice are administered human insulin orally or by aerosol. This causes these animals to be hyporesponsive to human insulin when an immunostimulatory route is used. CD8$^+$ Treg cells are produced in kidney grafts in rats after oral exposure to alloantigen, and transfer of these cells to naïve animals will prolong graft survival. Allogeneic cardiac graft survival is prolonged following intratracheal delivery of allogeneic peptides. This is mediated via production of IL-4- and IL-10-secreting Treg cells.

The antigen recognition by TCR involves activation of multiple signal pathways involving additional ligands. CD40 ligand (CD154), expressed on CD4$^+$ T cells, is important in initiating costimulatory signals after it binds to CD40 on APCs. After activation of CD40, other molecules including CD80 and CD86 are upregulated resulting in the proliferation of T cells and generation of an immune response. CD80 and CD86 also serve as receptors for TCRs, CD28 and CTLA-4. CD28 is an activator of the immune response via IL-2 secretion and induction of T-cell proliferation, whereas CTLA-4 inhibits T-cell responses. Consequently, CTLA-4 is involved in inducing immune tolerance by inhibiting the signals that are responsible for T-cell activation in response to an antigen and result in the induction of Treg cells.

Alloantigen-specific Treg cells can also be generated by a number of other ligands. Administration of anti-CD40LmAb to antagonize CD40–CD40L pathway results in the generation of alloreactive T-cell responses by cell–cell interaction. The antagonism of downstream signal transduction such as inhibition of NF-κB results in the generation of Treg cells. Agonist-like signals can also be used to generate Treg cells. For example, LFA-3 or CD58-mediated engagement of CD2 on naïve CD4$^+$

T cells results in the differentiation of Treg cells that are HLA-specific. Similarly, the induction of inducible costimulatory molecule on T cells and notch during antigen presentation will result in the generation of Treg cells. But the most potent positive signals for their generation are provided by cytokines, specifically IL-10 and TGF-β. IL-10 downregulates the expression of CD40, CD80 and CD86, probably resulting in inhibition of generation of CD4$^+$ and CD8$^+$ cells. This environment is optimal for the induction of Treg cells. Other proteins and soluble peptides could also produce antigen-specific Treg cells.

Antigen Specificity of Regulatory T Cells and Mechanisms of Suppression

Antigen-specific Treg cells function through antigen presentation, activation and recognition of target cells. The antigen presentation is achieved in the context of MHC molecules and in association with costimulatory and regulatory signals. After induction, coming in contact with the same antigen renders functional activity in Treg cells, which is followed by antigen-specific recognition of target cells by Treg cells.

Treg cell-induced suppression is mediated by several different mechanisms. The downregulation of CD40, CD80 and CD86 molecules results in a lack of T-cell activity when inhibited by CD8$^+$CD28$^-$ Treg cells. Another mechanism is mediated via cytokines, IL-10 and TGF-β, secreted by antigen-activated Treg cells. IL-10 downregulates CD80 and CD86 molecules via activation of JAK–STAT pathways and inhibits NF-κB activation. As a result, T-cell activation and IL-12 production are affected. The effects of TGF-β are mediated via Smad complex. Another proposed mechanism includes the killing of the effector CD8$^+$ T cells, which kill the graft by Treg cells that involve Fas–Fas ligand pathway.

Foxp3 Expression and Regulatory T-Cell Activity

The function of Treg cells is controlled by Foxp3 gene, an X-chromosome-linked factor, in a binary function, resulting in the maintenance of immune tolerance. The immune suppressive activities of T cells are regulated by Foxp3 gene. As a consequence, most attention has focused on equating abnormalities in Treg cells with immunological diseases. The effector function of Treg cells is as important as their numbers in regulating the immune response. For example, in diabetic NOD mice, there are lower levels of Foxp3 in Treg cells of intraislet as opposed to other peripheral lymphoid organs, but the frequency of Treg cells expressing this gene in different parts of the body is not different. The differences in the level of expression in diabetic NOD mice are not found in other mouse strains that are not susceptible to diabetes. One of the regulatory mechanisms may be the degree of gene switching, which determines expression levels of Foxp3. Consequently, immune disease may be a result of decreased Foxp3 expression. It seems that decreased Foxp3 expression causes defects in the function of Treg cells, and their differentiation into effector cells results in an augmented immune response that produces a loss of tolerance and probable development of the autoimmune disease.

Toll-Like Receptors and Regulatory T Cells

Innate and acquired immune responses are induced and regulated by the TLR. MyD88, a protein associated with TLR-mediated signal transduction pathway in dendritic cells, is involved in the suppression of Treg cell activity, resulting in an augmented immune response. Toll-like receptor signaling is required for the maturation of dendritic cells, and mature dendritic cells are potent inhibitors of Treg cell function. However, mature dendritic cells induce Treg cell expansion in association with TLR, IL-1 and IL-6. Small doses of IL-2 are required to maintain the suppressive function of Treg cells, which is inhibited by high doses of IL-2. The mechanism of IL-2-mediated loss of Treg cell activity is not known and is not mediated via Foxp3 gene. Furthermore, IL-6 and the strength of TCR signal help overcome the suppressive effects of Treg cells on effector cells. The suppression of Treg cells by TLR is attributed to TLR-2, which can recognize bacterial lipoproteins, and the removal of the TLR-2 influence results in the reestablishment of the suppressive abilities of Treg cells. TLR-2 is expressed on $CD4^+$ and $CD8^+$ T cells and can activate TCR-primed T cells as well as memory T cells. Treg cells and effector T cells are distinctly regulated by TLR-2-dependent signal transduction. Although T-cell function is not affected by TLR-2 signaling alone, the proliferation of TCR-primed Treg cells is strongly augmented by the agonists of TLR-2, which makes the Treg cells temporarily inactive. During infection, bacterial lipoproteins also increase the proliferation and IL-2 production from the TCR-triggered effector cells. This IL-2 increases the proliferation of both the effector and Treg cells; however, Treg cells are not able to suppress the effector cell function, which is a unique mechanism by which TLR regulates the function of Treg cells. After the bacterial infection is under control and the pathogens have been eliminated, Treg cells regain their suppressive function and IL-6 plays a role in this process. Consequently, this avoids the development of autoimmune disease that may result from the unregulated activity of effector T cells.

CTLA-4 and Regulatory T Cells

CTLA-4 is constitutively expressed on Treg cells; its deficiency is associated with fatal autoimmune proliferative disease, and inhibition of CTLA-4 by specific antibodies results in the development of autoimmune disease. Its polymorphism has a role in the development of autoimmune diseases including diabetes, Addison's disease and thyroid disease. Regulatory cells are involved in the development of disease resulting from antagonisms of CTLA-4, which may be the result of depletion of Treg cells by antibodies to CTLA-4. The activation of $CD4^+CD25^-$ T cells as a result of blockade of CTLA-4 receptors may block the suppressive effects of Treg cells on CD4$^+$CD25$^-$ T cells. The precise mechanism through which CTLA-4 is involved in the suppressive effects of Treg cells has not been determined.

Regulatory T Cells and Disease States

Allergic Disease

The role of Treg cells in the prevention of allergic disease and asthma is of considerable interest as the prevalence of this disease continues to rise. On the basis of the ability of Treg cells to prevent sensitization to allergens, they could be potentially used for the treatment of the allergic disease, and the prevention and regulation of TH$_2$-mediated responses may be possible. Mouse CD4$^+$CD25$^+$ T cells, after preactivation with differentiated TH$_2$ cells, inhibit TH$_2$ cytokine production and suppress TH$_2$ cell differentiation from naïve CD4$^+$ T cells without the requirement of cytokines. CD25$^+$ cells also inhibit IgE production in transgenic mice with monoclonal populations of T and B cells, and Tr1 cells inhibit TH$_2$ sensitization and IgE production provided that their adoptive transfer is before sensitization.

In humans, undesired TH$_2$ responses to environmental allergens are prevented by Treg cells, and the allergic disease may result from inadequate suppression of unwanted TH$_2$ responses by both naturally occurring Treg cells and Tr1 cells. An overall defect in the regulatory ability of Treg cells is not present in atopic individuals but a diminished suppressive ability of CD4$^+$CD25$^+$ T cells is observed in atopic individuals compared with their nonatopic counterparts. Treg cells from both asthmatic and nonasthmatic individuals have the same ability to suppress anti-CD3$^-$ and anti-CD28$^-$-stimulated cytokine production. Nonetheless, T-cell activation by allergens is suppressed by Treg cells, and atopic individuals may be deficient in these regulatory mechanisms. Nonatopic individuals possess higher numbers of allergen-specific IL-10-producing CD4$^+$ cells compared to atopic individuals. In addition, allergen-activated TH$_2$ cells can be inhibited by IL-10-producing T cells, and this can be reversed by either anti-IL-10 or TGF-β.

The mechanisms that alter the intricate balance between regulatory and suppressive responses after allergen exposure have not been elucidated, although a number of possibilities that may lead to atopy exist. According to the hygiene hypothesis, microorganisms alter APCs, which may result in the production of Treg cells when exposed to allergens. As a consequence, it is feasible that regulation by Treg cells is restricted by LPS-induced activation of TLRs, specifically TLR4. IL-10-producing Treg cells are also involved in the suppression of allergic responses as a result of prior exposure to allergens and mycobacterial antigens. In young children, exposure to cat antigens protects them from later development of allergies to cats due to a dominant IL-10 response, resulting in modified TH$_2$ responses where it seems that both Treg and Tr1 cells are responsible for developing tolerance.

Corticosteroids remain the main hallmark for the treatment of allergic disease/asthma despite their adverse side effects. Although their administration via inhalers and nebulizers has alleviated some concerns because of their less detrimental side effects, the development of drugs that will induce Treg cells will provide an attractive alternative. Corticosteroids, in addition to inhibiting TH_2 cytokines, stimulate Tr1 cells and increase the effector function of Treg cells by increasing IL-10 production. Inhaled corticosteroids also increase Foxp3 expression in asthmatics. Glucocorticoid-resistant asthma patients have impaired Tr1 cells, and they do not show improvement after treatment with corticosteroids. These observations provide a rationale for the development of a new class of drugs that may selectively increase the effector function of naturally occurring, as well as IL-10-induced Treg cells, and enhance Foxp3 expression. The concept is further strengthened by the observation that immunotherapy for allergic disease results in the induction of IL-10 production from Treg cells, and these patients have increased numbers of IL-10-producing Treg cells after they have received injections of allergen extracts.

Lastly, under naturally induced circumstances of tolerance such as bee keepers receiving multiple bee stings, or children who are no longer allergic to cow's milk, there is an increase in IL-10-producing Treg cells. The evidence is convincing that Treg and Tr1 cells regulate responses to allergens in nonatopic individuals, and their function may be impaired in atopy, specifically after prolonged antigen exposure. This suggests the need for the development of either new corticosteroid-like drugs that target only these mechanisms or alternate novel immunotherapeutic regimens.

Autoimmune Diseases

Human autoimmune diseases are complex genetic disorders strongly associated with the MHC on the chromosomes. Defects in the function of $CD4^+CD25^+$ Treg cells are associated with various autoimmune diseases including MS, type 1 diabetes, psoriasis and myasthenia gravis. Patients with MS have a significant decrease in the effector function of Treg cells, despite no differences in their frequency compared to normal controls. Patients with rheumatoid arthritis and juvenile idiopathic arthritis also have altered $CD4^+CD25^+$ Treg cells. Adult patients with rheumatoid arthritis exhibit high numbers of Treg cells in synovial fluid as opposed to the peripheral blood. Patients with juvenile arthritis disease have similar increases of Treg cells in the synovial fluid. These cells also contain CD27 marker with higher expression of Foxp3. Treg cells function differentially in various autoimmune diseases; for example, in human autoimmune poly-glandular syndromes and MS, the function of Treg cells is decreased, whereas Treg cells isolated from patients with autoimmune arthritic diseases exhibit augmented effector function.

Infections

Infections pose a challenge to the immune system that requires measured proinflammatory anti-infectious agent response without being detrimental to self. This intricate balance is subject to control by Treg cells, and a role of Treg cells in chronic viral and bacterial infections has been suggested. An increase in peripheral Treg cells is observed in patients with hepatitis B and C infections. Furthermore, Treg cells prevent antiviral response, since T-cell responses to hepatitis C virus

Table 9.4 Treg Cells and Viral Infections

Treg Type	Impairment of Antiviral Response
Treg cells	HCV, HIV, HSV
Tr1 cells	EBV, HCV, MLV
CD8$^+$ Treg cells	HCV, HIV
NKT cells	HIV

(HCV), hepatitis B virus (HBV), HIV, antigens and cytomegalovirus are induced after removal of Treg cells from peripheral blood of patients with viral infection (Table 9.4). Treg cells have a protective effect in HIV infection, where a decrease in Treg cells results in immune hyperactivity in HIV-inffected patients. A strong HIV-specific Treg cell activity is associated with lower levels of virus in plasma and higher CD4$^+$:CD8$^+$ ratios in HIV-infected patients. Consequently, intact Treg cell activity is desirable in patients infected with HIV.

Following HIV infection, Treg cells control the levels of activation of the immune response to avoid immune system exhaustion as well as tissue damage due to a robust immune response. However, this causes the dysfunction of immune response specifically due to inhibition of the generation of HIV-specific effector cells. Furthermore, the role of Treg cells changes during the different stages of HIV infection, as HIV infection causes alterations in the frequencies of Treg cells in the peripheral blood. As the disease progresses, there is a decrease in the frequency of Treg cells in the peripheral blood despite elevated CD25^{bright} expression on CD4$^+$ T cells, which may be attributed to their expression of CD4 and other chemokine receptors that are targets for HIV.

Chronic infections such as HIV result in immunosuppression as a result of induction of CD4$^+$ Treg cells. Increased risk of Kaposi's sarcoma, non-Hodgkin's lymphoma and liver cancer is associated with long-term infections such as HIV. Immunosuppression in HIV-infected individuals occurs before the development of AIDS, which proceeds before the depletion of CD4$^+$ T cells, and the induction of Treg cells may play a role in this process. The mechanisms are IL-10-independent and include the involvement of TGF-β secreted via signaling through cell–cell interaction involving CTLA-4.

CD8$^+$ T-cell response plays an important role in the viral replication of HBV, resulting in liver damage. These CD8$^+$ T cells are virus-specific, and a defect is found in these cells in patients with chronic HBV infection as opposed to recovered individuals. This may be due to a high antigen dose deletion and a lack of help from CD4$^+$ T cells or may be a result of Treg cells. For example, in herpes simplex virus-infected mice, the clonal expansion and effector function of virus-specific CD8$^+$ T cells is augmented by Treg cells, suggesting that Treg cells can be modulated in the periphery after viral infection. Another example is the regulation of HCV-specific T cells by Treg cells in patients with HCV infection. The circulating CD4$^+$CD25$^+$ Treg cells in patients with HBV infection suppress the activation of HBV-specific CD8$^+$ T cells. This may also serve as a feedback to avoid excessive pathogenic responses in chronic HBV infection and also helps to avoid complete clearance of the viral antigen in patients who have resolved HBV infection.

Cancer

Immune dysfunction and poor tumor-specific immune responses are observed in cancer patients with enhanced Treg cell activity. Furthermore, many different types of tumors possess high frequency of Treg cells that inhibit various immune functions including T-cell proliferation, cytokine production and cytotoxic activity. Tr1 cells also participate in a poor anticancer response. The infiltrating lymphocytes in Hodgkin's lymphoma contain both Treg and Tr1 cells that suppress various immune functions.

Transplantation

Treg cells play an important role in suppressing alloantigen-specific immune response following an organ graft. The maintenance of tolerance by various protocols in allograft transplantation is mediated via the induction of Treg cells in otherwise alloresponsive T cells. The responding T cells have a suppressive effect if naïve T cells are repetitively stimulated with immature allogeneic dendritic cells. The maximum T-cell response after allograft is dependent on the maturation stage of dendritic cells in the grafted tissue. Immature dendritic cells that result in suboptimal T-cell responses and limited costimulatory signals and cytokine production are ideal for the induction of Treg cells in responding T cells. In addition to the developmental stage, the particular subset of dendritic cells is also important in inducing Treg cells. Treg cells that are naturally occurring play an important role in the generation of induced Treg cells, resulting in a Tr1 suppressor phenotype from the graft-killing effector T cells. Furthermore, TGF-β can convert nonregulatory T cells to CD4$^+$CD25$^+$ suppressor cells. The tolerant grafts contain TGF-β and induce Treg cells in the graft, and there is also a presence of CD8$^+$ Treg cells in allografts. Patients undergoing allogeneic bone marrow transplantation and who do not exhibit graft-versus-host disease have T cells that produce IL-10 and IFN-γ, but low levels of IL-2.

Future Direction and Conclusion

Recent understanding of Treg cells provides a number of unique opportunities for their use in therapeutic intervention. The expectation is that the therapeutic application of Treg cells will result in reestablishing tolerance, the breakdown of which has resulted in the development of the disease. The treatment of individuals with antigen-specific Treg cells will be an interesting approach. Treg cells will have several advantages including a long half-life, ability to condition other cell types, induction of nonregulatory cells to secrete IL-10 and suppressive effects on the costimulatory activity of APCs. However, so far the therapeutic use of Treg cells has been limited due to their low frequency in circulating blood, and this will require industrial cultures for the propagation of clinical therapy. Initially, instead of commercial cultures of CD4$^+$CD25$^+$ T cells, large cultures of Tr1-like cells have been employed, which produce very large amounts of IL-10, but low levels of IL-2 and IL-4. Nonetheless, it has not yet been possible to use these cells therapeutically since their culture conditions stimulate TH$_2$ responses. This drawback can be overcome by propagating them in the presence of dexamethasone and vitamin D3, which results in the production of a T-cell line that secretes large amounts of IL-10. Although they also produced suppression independent of IL-10, these Tr1-like cells did not express FoxP3 gene as was expected from a cell that has a distinct lineage from CD4$^+$CD25$^+$ T cells.

Recent development of culture techniques that have allowed the ability of continuous culture of CD4$^+$CD25$^+$ cells and the generation of these Treg cells from CD25$^-$ cells after FoxP3 gene is transduced by using a retroviral vector has permitted the production of these cells on a large scale. A number of protocols have been published utilizing different stimuli for the culture of Treg cells on a large scale, and the efficacy of in vivo models has been established. The use of CD86^{hi} dendritic cells as APCs compared with whole splenocytes is far superior and efficient in expanding ovalbumin-specific Treg cells. Furthermore, mature dendritic cells are a better stimulator than immature dendritic cells, and IL-2 or IL-15 secretion from dendritic cells is not required for their successful function. Before these cells are used in clinical trials, safety and efficacy concerns need to be addressed. The challenge will be to isolate disease-specific Treg cells and expand them in large numbers. Another uncertainty is whether the results obtained in animal models will translate to humans and obviously ethical questions will also be raised by some.

Bibliography

Allez M, Mayer L. 2006. Regulatory T cells, peace keepers in the gut. Inflamm Bowel Dis. 10: 666–676.
Baecher-Allan C, Hafler DA. 2004. Suppressor T cells in human disease. J Exp Med. 200:273–276.
Baecher-Allan C, Hafler DA. 2006. Human regulatory T cells and their role in autoimmune disease. Immunol Rev. 212:203–216.
Barnett Z, Larussa SH, Hogam M, Wei S, et al. 2004. Bone marrow is a reservoir for CD4$^+$CD25$^+$ regulatory T cells that traffic through CXCL12/CXCR4 signals. Cancer Res. 64:8451–8455.
Battaglia M, Gregori S, Bacchetta R, Roncarolo MG. 2006. Tr 1 cells: From discovery to their clinical application. Semin Immunol. 18:120–127.
Beissert S, Schwarz A, Schwarz T. 2006. Regulatory T cells. J Invest Dermatol. 126:15–24.
Belkaid Y, Rouse BJ. 2005. Natural regulatory T cells in infectious disease. Nat Immunol. 6: 353–360.
Beyer M, Schultze JL. 2007. CD 4$^+$CD25$^+$high Foxp3$^+$ regulatory T cells in peripheral blood are primarily of effector memory phenotype. J Clin Oncol. 25:2628–2630.
Bluestone JA, Abbas AK. 2003. Natural versus adaptive regulatory T cells. Nat Rev Immunol. 3:253–257.
Bluestone JA. 2004. Therapeutic vaccination using CD4$^+$CD25$^+$ antigen-specific regulatory T cells. Proc Natl Acad Sci USA. 101:14622–14626.
Brimnes J, Allez M, Dotan I, Mayer L. 2005. Defects in CD8$^+$ regulatory T cells in the Lamina Propria of patients with inflammatory bowel disease. J Immunol. 174:5814–5822.

Buckner JH, Ziegler SF. 2004. Regulating the immune system: The induction f regulatory T cells in periphery. Arthritis Res Ther. 6:215–222.

Cao D, Vollenhoven RV, Klareskog L, Trollmo C, Malstrom V. 2004. CD 25$^+$CD4$^+$ regulatory T cells are enriched in inflamed joints of patients with chronic rheumatic disease. Arthritis Res Ther. 6:R335–346.

Chatenoud L, Salomon B, Bluestone JA. 2001. Suppressor T cells – they're back and critical for regulation of autoimmunity! Immunol. Rev. 182:149–163.

Chen Z, Benoist C, Mathis D. 2005. How defects in central tolerance impinge on a deficiency in regulatory T cells. Proc Natl Acad Sci USA. 102:14735–14740.

Cohen IR, Quintana FJ, Mimran A. 2004. Tregs in T cell vaccination: Exploring the regulation of regulation. J Clin Inv. 114:1227–1232.

Dejaco C, Duftner C, Grubeck-Loebenstein B, Schrimer M. 2006. Imbalance of regulatory T cells in human autoimmune diseases. Immunology. 117:289–300.

Dieckmann D, Bruett CH, Ploettner H, Lutz MB, Schuler G. 2002. Human CD4$^+$CD25$^+$ regulatory, contact-dependent T cells induce interleukin 10 producing, contact-independent type 1-regulatory T cells. J Exp Med. 196:247–253.

Fehervari Z, Sakaguchi S. 2004. CD 4$^+$ Tregs and immune control. J Clin Inv. 114:1209–1217.

Ferretti G, Felici A, Pino MS, Cognetti F. 2006. Does CTLA4 influence the suppressive effects of CD25$^+$CD4$^+$ regulatory T cells? J Clin Oncol. 24:5469–5470.

Fields RC, Bharat A, Mohanakumar T. 2006. The role of regulatory T-cells in transplantation tolerance. ASHI Quarterly. (first quarter):10–15.

Franzeee O, Kennedy PTF, Gehring AJ, Gotto GM., et al. 2005. Modulation of the CD8+ T cell response by CD4$^+$CD25$^+$ regulatory T cells in patients with Hepatitis B virus infection. J Virol. 79:3322–3328.

Gershon RK, Kondo K. 1970. Cell interactions in the induction of tolerance: The role of thymic lymphocytes. Immunology. 18:723–737.

Gershon RK. 1975. A disquisition on suppressor T cells. Transplant Rev. 26:170–185.

Godfrey DI, Kronenberg M. 2004. Going both ways: Immune regulation via CD1d-dependent NKT cells. J Clin Inv. 114:1379–1388.

Gould HJ, Sutton BJ, et al. 2003. The biology of IgE and the basis of the allergic disease. Ann Rev Immunol. 21:579–628.

Green DR, Flood PM, Gershon RK. 1983. Immuno -regulatory T cell pathways. Ann Rev Immunol. 1:439–463.

Groux H. 2003. Type 1 T-regulatory cells: Their role in the control of immune response. Transplantation. 75:85–125.

Hawrylowicz CM, O'Garra A. 2005. Potential role of interleukin-10 secreting regulatory T cells in allergy and asthma. Nat Rev Immunol. 5:271–283.

Hong J, Zang YC, Nie H, Zhong J. 2006. CD 4$^+$ regulatory T cell responses induced by T cell vaccination in patients with multiple sclerosis, PNAS. 103:5024–5029.

Hori S, Nomura T, Sakaguchi S. 2003. Control of regulatory T cell development by the transcription factor Foxp3. Science. 299:1057–1061.

Ito T, Wang Y-H, Duramad O, Hanabuchi S, et al. 2006. OX 40 ligand shuts down IL-10 producing regulatory T cells. PNAS. 103:13188–13141.

Iwashiro M, Messer RJ, Peterson KE, Stronnes IM, et al. 2001. Immunosuppression of CD4$^+$ regulatory T cells induced by chronic retroviral infection. PNAS. 98:9226–9230.

Janeway Jr. C. 1988. Do suppressor T cells exist? A reply. Scand J Immunol. 27:621–623.

Jerne NK. 1974. Towards a network theory of the immune system. Ann Immunol. 125:373–389.

Joosten SA, Meigjaarden KE, Savage N, Boer TD, et al. 2007. Identification of a human CD8$^+$ regulatory T cells subset that mediates suppression through the chemokine CC chemokine ligand 4. PNAS, 104:8029–8034.

Karlsson MR, Rugtveit J, Brandtzaeg P. 2004. Allergen -responsive CD4$^+$CD25$^+$ regulatory T cells in children who have outgrown cow's milk allergy. J Exp Med 12:1679–1688.

Kim JM, Rudensky A. 2006. The role of the transcription factor Foxp3 in the development of regulatory T cells. Immunol Rev. 212:86–98.

Knoechel B, Lohr J, Kahn E, Bluestone JA, Abbas AK. 2005. Sequential development of interleukin 2-depdent effector and regulatory T cells in response to endogenous systemic antigen. J Exp Med. 202:1375–1386.

Kretschmer K, Apostolou I, Jaechel E, Khazaie K., et al. 2006. Making regulatory T cells with defined antigen specificity: Role in autoimmunity and cancer. Immunol Rev. 212:163–169.

Latour RP, Dujardin HC, Mishellany F, Defranoux OB, et al. 2006. Ontogeny , function, and peripheral homeostasis of regulatory T cells in the absence of interleukin-7. Blood. 108: 2300–2306.

Liu H, Komai-Koma M, Xu D, Liew FY. 2005, Toll-like receptor2 signaling modulates the function of CD4+CD25+ regulatory T cells. PNAS. 103:7048–7053.

Lohr J, Knowchel B, Abbas AK. 2006. Regulatory T cells in periphery. Immunol Rev. 212: 149–162.

Lu L, Wermeck MBF, Cantor H. 2006. The immunoregulatory effects of Qα-1. Immunol Rev. 212:51–59.

Masteller EL, Tang Q, Bluestone JA. 2006. Antigen -specific regulatory T cells-ex vivo expansion and therapeutic potential. Semin Immunol. 18:103–110.

Nishizuka Y, Sakakura T. 1969. Thymus and reproduction: Sex linked dysgenesia of the gonad after neonatal thymectomy in mice. Science. 166:753–755.

O'Garra A, Vieira P. 2004. Regulatory T cells and mechanisms of immune system control. Nat Med. 10:801–805.

O'Garra A, Vieira P, Goldfeld A. 2004. IL -10 producing and naturally occurring CD4+ Tregs: Limiting collateral damage. J Clin Inv. 114:1372–1378.

Picca CC, Larkin III J, Boesteanu A, Lerman MA, Rankin AL, Caton AJ. 2006. Role of TCR specificity in CD4+CD25+ regulatory T cell selection. Immunol Rev. 212:74–85.

Qizhi T, Bluestone JA. 2006. Regulatory T cell physiology and application to treat autoimmunity. Immunol Rev. 212:217–237.

Reibke R, Garbi N, Hammerling GH, et al. 2006. CD 8+ regulatory T cells generated by neonatal recognition of peripheral self antigen. PNAS. 103:15142–15147.

Roncarlo MG, Gregori S, Battaglia M, Bacchetta R, et al. 2006. Interleukin -10 secreting type 1 regulatory T cells in rodents and humans. Immunol Rev. 212:28–50.

Rouse BT, Sarangi PP, Suvas S. 2006. Regulatory T cells in virus infections. Immunol Rev. 212:272–286.

Sakaguchi S. 2004. Naturally arising CD4+ regulatory T cells for immunologic self tolerance and negative control of immune response. Ann Rev Immunol. 22:531–562.

Sakaguchi S. 2005. Naturally arising Foxp3-expressing Cd25+CD4+ regulatory T cells in immunological tolerance to self and nonself. Nat Immunol. 6:345–352.

Sakaguchi S. 2006. Regulatory T cells: Meden Agan, Immunological reviews, 212:1–5.

Sakaguchi S, Ono M, Setoguchi R, Yagi H, et al. 2006. Foxp 3+CD25+CD4+ natural regulatory T cells in dominant self-tolerance and autoimmune disease. Immunol Rev. 212:8–27.

Sansom DM, Walker LS. 2006. The role of CD28 and cytotoxic T-lymphocyte antigen-4 (CTLA-4) in regulatory T cell biology. Immunol Rev. 212:131–148.

Schultze JL. 2006. Regulatory T cells: Timing is everything. Blood. 107:857.

Shevach EM. 2000. Regulatory T cells in autoimmunity. Ann Rev Immunol. 18:423–449.

Shevach EM. 2002. CD 4+CD25+ suppressor T cells: More questions than answers. Nat Rev Immunol. 2:389–400.

Shevack EM. 2004. Fatal attraction: Tumors beckon regulatory T cells. Nat Med. 10:900–901.

Shevach EM, DiPaolo RA, Anderson J, Zhao DM. 2006. The lifestyle of naturally occurring CD4+CD25+Foxp3+ regulatory T cells. Immunol Rev. 212:60–73.

Snoeck V, Peters I, Cox E. 2006. The IgA system. Vet Res. 37:455–467.

Sutmuller RPM, den Brok HMGM, Kramer M, Bennink EJ, et al. 2006. Toll -like receptors control expansion and function of regulatory T cells. J Clin Invest. 116:485–494.

Taams LS, Palmer DB, Akbar AN, Robinson DS, et al. 2006. Regulatory T cells in human disease and their potential for therapeutic manipulation. Immunology. 118:1–9.

Tarbell KV. Yamazaki S, Steinman RM. 2006. The interaction of dendritic cells with antigen specific, regulatory T cells that suppress autoimmunity. Semin Immunol. 18:93–102.

Thomas D, Zaccone P, Cooke A. 2005. The role of regulatory T cell defects in Type I diabetes and the potential of these cells for therapy. Rev Diabet Stud. 2:9–18.

Umetsu DT, DeKruyff RH. 2006. The regulation of allergy and asthma. Immunol Rev. 212: 238–255.

Vigouroux S, Yvon E, Biagi E, Brenner MK. 2004. Antigen -induced regulatory T cells. Blood. 104:26–33.

Walker LS, Chodos A, Eggena M, Dooms H, Abbas AK. 2003. Antigen -dependent proliferation of CD4$^+$CD25$^+$ regulatory T cells in vivo. J Exp Med. 198:249–258.

Walsh PT, Taylor DK, Turka LA. 2004. Tregs and transplantation tolerance. J Clin Inv. 114: 1398–1403.

Walsh PT, Buckler JL, Zhang J, Gelman AE, et al. 2006. PTEN inhibits IL-2 receptor mediated expansion of CD4$^+$CD25$^+$ Tregs. J Clin Inv. 116:2521–2531.

Wan YY, Flavell RA. 2006. The roles for cytokines in the generation and maintenance of regulatory T cells. Immunol Rev. 212:114–130.

Wau Y, Flavell RA. 2007. Regulatory T cell functions are subverted and converted owing to atten-uated FoxP-3 expression. Nature. 445:766–770.

You S, Thieblemont N, Alayankian MA, Bach J, Chatenoud L. 2006. Transforming growth factor-β and T-cell-mediated immunoregulation in the control of autoimmune diabetes. Immunol Rev. 212:185–202.

Zorn E, Nelson EA, Mohseni M, Porchery F, et al. 2006. IL -2 regulates FOXP3 expression in human CD4$^+$CD25$^+$ regulatory T cells through a STAT-dependent mechanism and induces the expression of these cells in vivo. Blood. 108:1571–1579.

Chapter 10
Gene Therapy

Introduction

Although considered a modern therapeutic procedure, the theory of gene therapy goes back many decades. Historically, gene therapy has always been a part of science fiction with hope that its clinical applications will be realized. In the literature, these proposals go back many decades, and many have not yet been achieved. In ancient cultures, breeding plants and animals for certain goals was a way of life. Selective breeding practices were established before the development of modern genetics. The transfer of genes within DNA was first reported by Avery, Macleod and McCarthy in 1944, and a few years later, it was observed that viruses possess the ability to transfer genes. On the basis of the ability of viruses to transfer genes, Tatum in 1966 hypothesized the role of viruses in gene therapy and the possibility of isolating or synthesizing genes and inserting them into the organs with defective genes. Aposhian in 1969 suggested the use of genes in place of drugs to treat diseases and the initial gene therapy animal models included ducks and rats. The first gene therapy experiment in humans was performed by Rogers in 1969 to treat arginase deficiency, using Shope papilloma virus, but this was not successful. A significant advance in realizing this goal was achieved when Michael Wigler and Richard Axel in 1977 transferred a thymidine kinase gene into mammalian cells. Cline performed a gene therapy experiment in humans in 1980 to treat thalassemia, but his ethics was questioned and this resulted in his reprimand. In 1983, retroviral vectors were used to transfer a functional gene into cells followed by the development of helper-free retroviral packaging cell line. This resulted in the development of a number of gene transfer systems based on viruses. The foundation laid by this important work led to the first approved clinical trial of gene therapy in 1990 to treat ADA-deficient severe combined immunodeficiency. In this chapter, the use of various techniques applied in treating various defects by gene therapy will be described.

The concept of gene therapy was originally based on treating diseases that were the result of a single gene defect. A defective gene causes abnormal protein synthesis or no synthesis at all, resulting in the etiology and pathogenesis of an inherited disease. The clinical manifestation could be mild or may be severe with a broad spectrum of abnormalities dependent on the importance of the protein in the

M.M. Khan, *Immunopharmacology*, DOI: 10.1007/978-0-387-77976-8_10,
© Springer Science+Business Media, LLC 2008

physiological processes. Providing the missing protein to the patient has always been a challenge due to concerns regarding the stability of the protein once it is administered. However, inserting a healthy gene that could synthesize the missing protein in the cell can overcome the challenge and help ameliorate the clinical symptoms of the disease. This scheme is plausible if the defective gene is expressed in only a few tissues. The genes that are expressed in all cells result in severe abnormalities and generally the pregnancy does not proceed. Today gene therapy is limited to addition of a corrected copy of a gene to the somatic cells that are either accessible stem cells or terminally differentiated, post-mitotic, long-lived cells. Among various types of stem cells, stem cells obtained from the hematopoietic system and the skin are optimally suited for gene therapy. Other potential targets with good suitability include neuronal cells and photoreceptor cells in the retina. Despite some moderate progress in the development of these techniques, there remain many obstacles such as poor gene transfer, mutagenesis when viral vector systems are used and immunogenicity of the vector and the transgene.

Altered Genes and Diseases of Inherited Disorders

Each individual often carries some defective genes, without any symptoms or disease. Almost 3000 diseases that are caused by an abnormal gene impacting about 10% of the population starting from birth to any stage of life are known. The presence of a defective gene does not assure an abnormality in most individuals due to the presence of a second copy of the same (but healthy) gene with the exception of genes on the Y chromosome. However, in the case of a dominant gene, the disease will be manifested although the other copy of the gene is normal and will require gene therapy. Under ideal circumstances, the insertion of a healthy gene should correct the problem; however, even calculating the amount of gene product that will provide the required levels of the lacking function protein is very tedious for some diseases. The strategy for gene therapy today is not only for the genetic disorders but also for the acquired disorders and as mediators of behavior, such as alcoholism and promiscuity, where these techniques are being used, at least experimentally, to treat disorders such as cardiovascular disease, cancer and metabolic diseases.

Transgene is the genetic information that a cell or a tissue receives via a vector that is used to carry this genetic information, and the type of transgene defines its expected therapeutic effect. There could be multiple reasons as to why gene therapy is being performed; one is to compensate for a gene mutation where a healthy gene is provided for a missing or a mutated gene. Examples of clinical trials have included the addition of a cystic fibrosis transmembrane conductance regulator gene in the lung epithelium and the restoration of p53 gene in cancer cells. The other type of gene therapy is purely therapeutic where the introduction of transgene causes the amelioration of disease pathology such as in autoimmune diseases. Purely therapeutic gene therapy is also used for ischemic heart disease by introducing the VEGF gene that expresses the growth factor responsible for the development of blood vessels. An additional form of gene therapy uses transgenes which are "suicide genes" that kill cancer cells after they are expressed in these cells. Examples of these genes include thymidine kinase, VSV-TK, deoxycytidine K and cytosine deaminase, proapoptic genes and enzyme prodrug combinations, such as HSV-TK and acyclovir, VSV-TK and Ara M, deoxycytidine K and Ara-C, and cytosine deaminase and 5-fluorocytidine. The suicide genes convert the prodrug to a toxic substance that kills the cancer cells, and the side effects could be controlled by limiting the doses of the prodrug. Lastly, gene transfer-mediated vaccination, which could potentially be used for both infectious and noninfectious diseases, is another approach for gene therapy. An example for noninfectious diseases would be to produce an immune response against cancer cells with the induction of tumori-cidal cytokines, tumor-infiltrating lymphocytes and induced expression of lympho-cyte costimulatory molecules, and the use of gene therapy instead of vaccination for infectious diseases may have therapeutic implications for AIDS.

Vectors for Gene Therapy

The potential gene therapy vectors come in many different forms to deliver the genes, all of which have advantages and disadvantages. The priorities in vector selection include its ability to deliver maximum levels of the gene, optimal expres-sion of the gene for a desired period and its ability to enter the target cell or tissue while sparing unintended tissues. A listing of different gene transfer vectors used for gene therapy is provided in Table 10.1.

Adenoviruses

Adenoviral vectors are considered the most versatile and easily manipulated viral vectors for delivering genes, and over 50 identified serotypes of human adenovirus are known belonging to six species. They infect the respiratory tract in normal popu-lations, resulting in upper respiratory infection, but their effects are self-limited.

Adenoviruses contain double-stranded DNA, are not enveloped and do not require the division of the host cells for their proliferation. They are ideal as a vector

Table 10.1 Gene Transfer Vectors

Viral	Nonviral
Adenovirus	Lipoplexes
Adeno-associated virus	Polyplex
Retrovirus	Oligonucleotides
Lentivirus	Antisense RNA
Herpes simplex virus-1	SIRAN
Vaccinia virus	RNA/DNA chimera
Amplicon-based	Aptamers
	Ribozyme
	Naked DNA
	Lipoplexes
	Cationic lipids
	Molecular conjugates

for gene therapy because they can penetrate various tissues including respiratory epithelium, cardiac and skeletal muscle, vascular endothelium, hepatocytes, peripheral and central nervous system and a number of different tumors. The ability to infect various tissues is dependent on the presence of coxsackie and adenovirus receptor (CAR) and alpha 5 integrins. This could be a potential problem since certain tissues and forms of cancer that could benefit from gene therapy are not rich in expressing CAR required for optimal infection with the adenovirus. The modification of proteins present on the capsid of adenovirus that interact with CAR results in changing the specificity of adenoviruses for the target tissue—allowing targeted delivery and avoiding entry into nontarget tissue. Infectivity could be increased by replacing existing knob proteins and adding different knob proteins from different adenovirus serotypes, other motifs or single-chain antibodies. This allows targeting the tissues that may not previously contain abundant receptors for targeting and provides an opportunity for tissue-targeted delivery.

Adenoviruses are ubiquitous as most adults have been exposed to them, and their rearrangement is not observed at a high rate resulting in the stability of inserted genes after several cycles of viral replication. They contain 36 kilobases of double-stranded DNA. After their entry into the host cell E1a and E1b genes, which are the genes from the early region, are transcribed. There is an expression of E_1, E_2, E_3 and E_4 genes as the virus replicates, and these genes regulate transcription, which leads to genome expression. Most of the viral transcription is regulated by the late promoter after DNA replication is initiated. There are two types of elements Cis and Trans, which perform separate viral functions. The virus itself, with the help of Cis genes, is responsible for the replication and condensation of DNA. The Trans genes can be removed or helped by other inserted genes. The foreign DNA can be placed in one of the three regions, which include E_1, E_3 and a small portion of E_4. Immunogenicity of the adenovirus can be reduced by removing various E_1, E_2 and E_4 genes, resulting in lower toxicity to the host. However, this viral redesign does not reduce MHC class II-dependent T helper responses.

Adenoviral vectors have been modified to create vectors that do not contain all viral protein-coding DNA sequences. This is a helper-dependent vector system in which all the viral genes required for replication are present, but its packaging domain has a deficiency resulting in its inability to form a virion. Its combination, a second vector, possesses therapeutic genes as well as normal packaging recognition signal, which allows its genome to form a virion and be released. This vector system is also called "gutless" adenoviral vectors. These viruses can be delivered by a wide variety of modes of administration including intravenous, intraperitoneal, intrabiliary, intracranial, intrathecal and/or intravesicular injections.

Adenoviruses are used for the treatment of cystic fibrosis and cancer by gene therapy. For the treatment of cystic fibrosis, aerosolization has been used to deliver the gene for cystic fibrosis transmembrane conductance epithelial cell surface. However, many hurdles still exist in this therapeutic procedure.

Various protocols have been used to treat cancer by gene therapy using adenoviruses, which include gene-based immunotherapy, prodrug therapy and gene replacement approaches. These procedures are often employed in combination with standard chemotherapy. One example of gene-based immunotherapy to treat cancer is the use of adenoviruses to augment normal B-cell and T-cell responses against tumor-associated antigens. Another example involves transducing autologous tumor cells to make them produce GM-CSF, which has been used to treat malignant melanoma. A different approach that has been experimentally used to treat brain tumors utilizes prodrug therapy in which gene therapy using adenoviruses is based on delivering genes for enzymes into tumor cells, which then convert prodrugs to cytotoxic chemotherapeutic agents. Other cancer gene therapy protocols are based on the observations that tumor cells are more responsive to conventional chemotherapy if they express wild-type p53 molecules that are associated with apoptosis, and interference with this function results in resistance to chemotherapy by tumor cells. Gene therapy could be employed to remedy defects in p53 by reestablishing apoptosis-inducing regulators. Alternatively, genes that induce apoptosis such as TNF-α, caspase-8 and FAS-ligand can be added.

Adeno-Associated Virus

Adeno-associated virus is so named because it is often found in cells that are also infected with adenovirus. This nonenveloped, nonautonomous parvovirus has been well studied for gene therapy, has single-stranded DNA and can integrate into the genome at chromosome 19 of nondividing cells. Adeno-associated virus naturally expresses transgenes for an extended period as opposed to adenovirus and has multiple serotypes that enable the virus to infect a wide variety of tissues. The virus is not associated with human disease, and the infected host does not generate either antibodies or an inflammatory response against it. Adeno-associated virus is attractive for gene therapy because it lacks pathogenicity, can infect nondividing cells, stably integrates into host cell genome at a specific site, and its ability to integrate only at chromosome 19 makes it superior to other viruses, which by random insertion and mutation may cause cancer. Its disadvantages are that it has a limited transgene capacity and two vectors are needed simultaneously to appropriately express the transgene. The virus uses integrins as well as heparan sulfate proteoglycans as a primary receptor and fibroblast growth factor 1 as a coreceptor 1 for infecting the host cell.

Adeno-associated virus transgene has shown some encouraging results as a delivery system for cystic fibrosis where AAV serotype 2 is used in clinical trials since this serotype infects lung epithelium. Other serotypes being tested for gene therapy include AAV_1, AAV_5, AAV_6 and AAV_8. AAV_1 and AAV_5 are efficient in delivering genes to vascular endothelial cells, AAV_6 has been more effective than AAV_2 in infecting airway epithelial cells and AAV_8 is very effective in transducing hepatocytes.

Retroviruses

Retroviruses have advantages over adenovirus and AAV vectors due to their ability to integrate the transgene DNA into the host and their potential of long-term expression, which make them a preferable candidate for gene therapy specifically for curative treatments. They are not immunogenic, and the host lacks preexisting antibodies or T cells against them. Despite these advantages, an undesirable feature is their ability to integrate into host cells permanently. Furthermore, their application for gene therapy could be limited only to dividing cells. In SCID trials where retroviruses were used as vectors for gene therapy, the children developed leukemia, which was attributed to proliferation of T cells as a result of mutation.

Retroviruses were first used for gene therapy in 1989. They contain two copies of RNA genome and are packaged in an envelope that is like a cell membrane. The envelope fuses with the host cell membrane in some retroviruses, whereas it is endocytosed in others. The positive-strand RNA genome is transcribed to cDNA by the reverse transcriptase contained in the viral envelope, and cDNA then migrates into the nucleus followed by integration into the host's cell DNA. This process is mediated by viral integrase, which is also present in the envelope of the retrovirus. The viral mRNA is transcribed by the integrated provirus, which is subsequently processed and translated into viral proteins. The integrated provirus synthesizes a positive-strand RNA genome, which is then packaged with other proteins, and particles are released from the host cell via a process called budding.

A retroviral genome is composed of 5' and 3' long terminal repeats (LTRs), a group-specific antigen gene (gag), reverse transcriptase (pol) and envelope protein (env). Furthermore, the psi sequence and a cis-acting element are critical for packaging. Without a psi sequence, RNA cannot be packaged.

The most commonly used retroviruses for gene therapy are the murine leukemia viruses. The proviral form of the virus is used to construct retroviral vectors. A standard protocol involves the removal of gag, env and pol genes, which are responsible for the replication of the virus. Furthermore, this deletion makes way for the space where the therapeutic gene will be placed. Retroviruses can accommodate up to 8 kilobases of therapeutic genes. This retrovirus will not synthesize any viral proteins, resulting in protection from any form of immune response that may result if viral antigens were produced. Once the genes encoding viral structural proteins are deleted, special viral packages or packaging cell lines require insertion of the deleted genes; however, this is done on a different chromosome; thus, these genes will not be able to package with the vector. This system allows production of the gene-deleted vector at a large scale, which can be used to insert the therapeutic genes and then used for gene therapy. The recombinant proviral DNA is inserted into the package system to produce the recombinant vector. Many different techniques have been used to insert the recombinant gene into the packaging cell line.

The recombinant vector can now be administered in the patient by either ex vivo transduction or directly injecting the virus into the patient's target tissue. Alternatively, the retrovirus-producing cell can also be injected. However, the most commonly used technique in clinical trials has been the ex vivo transduction of the patient's cells, which allows quantification of gene transfer and targeting of the specific cell population. The transduction efficiency is further improved by employing this technique since it allows the use of a high ratio of viral particles to target cells.

As opposed to the tissues that are easily accessible, others present a challenge for delivery, such as brain tumors. Retrovirus provides a unique opportunity to treat cancer because of its ability to transduce only the dividing cells. In experimental models, T cells have been used to successfully deliver retrovirus to metastatic tumor deposits in the lung and liver. T cells have also been engineered to produce retroviral vectors using sirolimus induction. Sirolimus has been used because it mediates heterodimer formation between the FK-binding protein and lipid kinase FRAP, and by fusing FKBP domains to DNA-binding domain and FRAP to a transcriptional activation domain, assembly of an active transcription factor and expression of a target gene can be made dependent on the presence of sirolimus.

The disadvantages of retroviruses include the limited number of cells that are transduced by virus due to dilute retrovirus preparations and their ability to transduce only the dividing cells. To overcome these problems, retrovirus-producing cell lines are directly administered into the tumors. Although there have been no reports from the clinical trials, the mutagenicity after integration of the retrovirus into the host genome has been of concern.

Lentiviruses

Lentiviruses are a subclass of retroviruses and can integrate into both dividing and nondividing cells. Despite the rapid evolution of these viruses, many elements in their genome have been conserved over time. The gag and pol genes are conserved, but the env gene is much more variable. After the virus enters the cell, its genome, which is RNA, undergoes transcription to DNA by reverse transcriptase and a viral integrase randomly incorporates it into the host genome. As a result, a provirus is formed and will propagate with the proliferation of the host cell. The provirus may interfere with the normal functioning of the host cells and could even induce oncogenes. For gene therapy, lentivirus vector is modified to remove the genes needed for replication. Lentivirus is produced by transfecting several plasmids into a packaging cell line, which includes packaging plasmids encoding the virion proteins and genetic material that the vector will deliver. The single-stranded RNA viral genome is used to package the genome into the virion.

Lentiviruses can be very effective vectors for gene therapy since they can change the expression of genes in target cells for up to 6 months. They are useful for nondividing and terminally differentiated cells including muscle cells, hepatocytes, neurons, macrophages, retinal photoreceptors and hematopoietic stem cells. However, lentiviruses cannot enter quiescent cells in which reverse transcription is blocked.

Herpes Simplex Virus-1 Vector

Herpes simplex virus-1 (HSV-1) vector is a human neurotropic vector that possesses double-stranded DNA, replicates in the nucleus of the infected cells and can infect both dividing and nondividing cells. It could also live in a nonintegrated state. The virus has received attention as a vector for gene transfer in the nervous system since HSV-1 infects neurons and may enter a lytic life cycle or live as an intranuclear episome, which is a latent state. The latent virus does not replicate despite the presence of active neuron-specific promoters, and an immune response is not produced against the latently infected neurons.

Viral genome can be inserted with a large DNA sequence by homologous recombination, and the recombinant virus, which is defective in replication, can be plaque purified by using transcomplementing cells. However, this viral vector has several disadvantages; it is difficult to obtain preparations that are completely defective in replication, the modified vector generates immune response and antibodies specific for HSV-1 are present.

Vaccinia Vectors

Vaccinia virus, a member of the poxvirus family, has been used for smallpox vaccination and for other infectious agents. It is a large enveloped virus with DNA as genetic material, a large genome of 130–200 kilobases and can infect a wide variety of cell types including both dividing and nondividing cells. As nonintegrated genome, its gene expression is for a short period. The recombinant virus is produced by homologous recombination, and it requires a viral vector DNA and a packaging cell line, as its large genome allows the insertion of large genes. An immune response is produced against almost 200 antigens of the virus; although the adverse reactions are not that common, its use in gene therapy has to be limited to individuals who have not previously received smallpox vaccination. This system has the potential to allow for vaccination against diseases that are not currently treatable including HIV infection. The main concern with this vector in addition to its immunogenicity is its replication, specifically in immunodeficient patients. Vaccinia vector has been used for clinical trials for cancer where it is expected to directly stimulate the immune response to destroy cancer cells. A summary of the characteristics of the above described viral vectors used for gene therapy is depicted in Table 10.2.

Amplicon-Based Vectors

HSV amplicons are plasmids that contain HSV replication origin and packaging sequence. These vectors lack parts of the HSV genome that is required for replication and depend upon naturally occurring defective interfering particles (DI vectors), which are infectious agents formed during HSV propagation. Plasmid DNA containing HSV packaging signal can be packaged into DI particles. After the insertion of the gene into an amplicon, it is transfected into a cell line that is appropriate for HSV proliferation, which is followed by the infection of the cells with helper HSV virus. The proliferation of the virus results in the production of

Table 10.2 Some Viral Vectors Used for Gene Therapy

Vector	Packaging Capacity	Titer	Integration	Expression	Tropism	Transduction Efficiency
Adenovirus	8 kb	10^{11}	No	Transient	Broad	High
Adeno-associated virus	<5 kb	10^9	Yes (?)	Transient	Broad except for hematopoietic cells	High
Retrovirus	8 kb	10^7–10^{10}	Yes	Variable	Only dividing cells	Low
HSV-1	40–150 kb	10^8	No	Transient except neurons	Strong for neurons	High
Vaccinia	35 kb maximum	N/A	No	Transient	Broad	High

DI vectors and helper viruses. However, the use of amplicon-based vectors for gene therapy has been restricted because of the inability to purify the DI vectors that are not contaminated with helper virus. Nonetheless, HSV amplicon-based factors allow genetic transfer of multiple transgene copies in the absence of viral genes. Consequently, due to their relative flexibility, as opposed to other viral vector systems, they are used in clinical trials for cancer gene therapy as part of oncolytic virotherapy and immunotherapy. The advantages and disadvantages of various viral vectors used in gene therapy are described in Table 10.3.

Table 10.3 Advantages and Disadvantages of Various Viral Vectors

Vector	Genetic Material	Advantages	Disadvantages
Adenovirus	dsDNA	Excellent transduction of most tissues	Potent inflammatory response against the virus
Adeno-associated virus	ssDNA	No toxicity. No inflammatory response	Limited packaging capacity
Retrovirus	RNA	Transduction efficiency is high. Sustained expression of vector after integrating into host genome. Host does not express vector proteins	Dividing cells are required for infectivity. Random integration. Oncogenesis may be induced after integration
HSV-1	dsDNA	Large packaging capacity. Strong tropism for neurons	Transgene expression is transient except neurons. Inflammatory response against the virus
Vaccinia	dsDNA	Allows insertion of large genes. Minimum adverse reactions. Anticancer effects of immunogenicity	Immunogenic. Limited to non-smallpox vaccinated or immunodeficient subjects

Nonviral Gene Therapy Vectors

The procedures for nonviral vectors in gene delivery include physical methods such as naked DNA delivery by ultrasound or electroporation or endocytosis-mediated mechanisms. These protocols provide an alternative to the viral vectors for gene delivery, but the methods are not as efficient as viral vectors because of the method of entry and DNA degradation in the endosomes or lysosomes. However, limited success has been reported by the use of liposomes for delivering genes to treat cancer. In these experiments, adenovirus mu protein or protamine sulfate was used for protecting DNA degradation. One of the drawbacks of cationic liposomes is their limited ability in delivering transgene to the target cells and the induction of its expression.

Other vehicles under development for nonviral gene delivery are polymers that are positively charged and filled with large negatively charged DNA to facilitate endocytosis. The targeting is further improved by the incorporation of ligands into polymers. In some studies, folate and transferrin have been used for receptor-mediated endocytosis. A recent advance in designing a nonviral transgene delivery system has been the development of biodegradable polymers. These polymers slowly release transgene around the cells as they degrade, which increases the transfection efficiency and also provides extended expression of the gene. This efficiency has been further improved by the use of polymers like PEG against which an immune response is not produced.

Various strategies have been developed to overcome difficulties associated with gene delivery when using nonviral vectors. Some of these examples include internalization of the carrier, polynucleotide degradation in the extracellular space, dissociation of the polynucleotide from the carrier, entry of the polynucleotide into the nucleus and intracellular trafficking from endosome to lysosome. The focus has been the development of multifunctional nonviral vectors that could penetrate different barriers. The major parts of such a system are molecules, such as protamine and polyethylamine, that not only can condense polynucleotides but also can protect against nucleases when injected into the blood DNA–cationic carrier complex aggregate. However, this positive charge could be protected by PEGylation of the cationic carrier that enhances blood circulation and impedes aggregation. The PEGylated nanocarriers home in richly vasculated tissue, and if ligands are then attached to the distal end of the PEG to induce receptor-mediator endocytosis following the cell surface recognition, the cellular uptake is significantly augmented. The encapsulated polynucleotide is released in the cell following endosome rupture caused by endosomolytic activity contained in the components of the carrier.

Although most of the studies using nonviral vectors are being performed in animal models, O'Malley and colleagues have performed IL-2 gene therapy by using nonviral vector in patients with advanced head and neck cancer. In their phase I clinical trial, cationic liposomes prepared with CMV-IL-2 plasmid were injected into the tumor. All patients completed the study but most died due to the progression of the disease. One patient exhibited a decrease in the burden of the tumor.

RNA Interference Gene Therapy

RNA interference is a technique where a strand of RNA in a cell destroys another RNA strand and the second strand of RNA is responsible for relaying protein-coded messages from a gene, and its destruction means that the gene's message can no longer be carried out. The rationale for using RNAi-based gene therapy is to knock out genes that are overexpressed in diseases such as macular degeneration or cancer. For the treatment of viral infections, RNAi will disable the gene required for the survival of the virus, and the expectation is that this technique will shut off genes for viruses such as HIV or hepatitis and consequently the disease will not result, due to the inability of these viruses to replicate. The drawbacks of this technique have been the production of immune response and inadvertent switching off of the wrong genes, resulting in unexpected toxic effects. At present three human RNAi gene therapy trials are in progress, one for RSV pneumonia and two for macular degeneration. Others are working on using this form of gene therapy for viral hepatitis.

Gene Therapy Using RNA Aptamers

Aptamers are small structured single-stranded RNAs or DNAs that have been produced by using a technique called systemic evolution of ligands by exponential enrichment (SELEX). Aptamers recognize and bind their cognate ligands utilizing complementary three-dimensional structures. SELEX consists of selection and amplification phases that reduce the library of nucleic acid, and the final product is made up of one or more sequences of nucleic acids purified from complex randomized sequences with affinity for a specific target. After binding, aptamers become encapsulated by the ligand and the target is incorporated into the intrinsic structure of the aptamer. Aptamers are distinct as they are three-dimensional globular structures, as opposed to ribozymes and antisense oligonucleotides that are linear and act at mRNA levels to alter protein expression. They are specific molecules with high affinity for the target that act like antagonists by binding to the functionally relevant part of the protein. The stability and bioavailability of aptamers can be augmented by simple chemical modification. They do not produce an immune response and lack any toxic effects. Their clinical use was initially limited due to poor availability, limited cellular delivery, rapid clearance and sensitivity to nucleases, but many of these issues have been resolved with new preparations of aptamers.

The first aptamer drug, Macugen, was approved for the treatment of wet age-related macular degeneration (AMD); it is an antagonist of VEGF and binds to the heparin-binding domain. Additional aptamers under clinical developmental stages include transcriptional decoys and thrombin-specific aptamers. Transcriptional decoys are made up of small RNAs or double-stranded DNAs and contain a specific transcription factor's consensus binding sequence. The inhibition of gene expression is achieved after a decoy binds to its transcription factor's site, resulting in the prevention of the binding of the protein to the target gene's promoter region.

Several transcriptional decoys are under development for cardiovascular diseases, inflammation, arthritis, liver disease and dermatitis.

Other aptamers under clinical trials are for coagulation factors including thrombin, Factor VIIA, Factor IXA and Factor XII. Additional aptamers are being tested for kidney and lung cancers. A distinct advantage of aptamers is the availability of antidotes for them, which are short complementary sequences to the aptamers that serve as their antagonists and could be used to avert aptamer toxicity.

For the long-term gene therapy against HIV, focus has been placed on combining RNA-based therapies including aptamers (decoy RNA), siRNA and ribozyme, as their combination helps overcome the shortcomings associated with each form of therapy. Furthermore, distinct phases of the HIV cell cycle can be targeted by using specific combination of RNA-based therapies. An example is a lentiviral-based construct that contains transgenes for both CCR5 ribozyme and U16TAR decoy. CCR5 ribozyme is a downregulator of HIV-1 coreceptor and U16TAR decoy is an inhibitor of Tat-activation transcription in HIV. The primary $CD34^+$-derived monocytes possessing this vector exhibit strong resistance against HIV. Another example is a vector combining three forms of RNA-based therapies which include a U6 transcribed nucleolar-localizing TAR RNA decoy, a U6 Pol III promoter-driven short hairpin RNA and a VA1-derived Pol III cassette and is effective in inhibiting the proliferation of HIV.

Treatment of Diseases by Gene Therapy

Cancer

Gene therapy is a broad field and for cancer it can be divided into three distinct forms of treatment: (a) immunotherapy, (b) oncolytic virotherapy and (c) gene transfer. The concept of immunotherapy to treat cancer, where an augmented host response against the tumor cells is intended, is not a new one. A major current focus of this approach is to develop recombinant cancer vaccines that are employed to treat and not prevent the disease. This is being done by the administration of the immunostimulatory genes, which after entering the tumor cells produce proteins resulting in the production of humoral and cellular response against the tumor cells.

Another type of protocol is used to alter the immune response of the patient, resulting in sensitization against cancer cells. In this form of gene therapy, a tumor antigen or a stimulatory gene is added to the peripheral blood lymphocytes or bone marrow cells of the patient, which are then primed to produce an immune response against the cancer cells. Initial trials have been performed for cancer treatment by using first-generation vaccines, but have provided mixed results. Nevertheless, these trials have allowed a better understanding for developing the next generation of cancer vaccines. A number of next-generation vaccines are under trial for various types of cancer. A success story so far has been the treatment of non-small lung cell cancer, which responds very poorly to the conventional treatment. GVAX, a vaccine produced by modifying autologous tumor cells to express GM-CSF, has been used alone and in combination with cyclophosphamide to treat lung cancer, and the results have been promising. Other clinical trials are under way with this form of gene therapy to treat other types of cancers including prostate cancer. In other studies, the immunostimulatory genes are directly inserted into the tumor cells as has been the case for clinical trials for malignant melanoma patients. In this specific case, genes for IL-24 have been inserted into the tumor cells and adenovirus has been used as the vector. Other vaccines that are undergoing clinical trials include TRICOM, PANVAC-VF and PROSTVAC. In TRICOM vaccines, a cancer antigen is added to vaccinia vector, in which genes for ICAM-1, B7-1 and LFA-3 have also been included. PANVAC vaccine is undergoing clinical trials for pancreatic cancer in which a modified vaccinia virus has been used as a vector that was supplemented with genes for immunostimulation, CEA and MUC-1. This vaccine is injected subcutaneously, and booster vaccine of a fowlpox virus modified similarly to the vaccinia virus is subsequently administered.

Another form of gene therapy to treat cancer involves oncolytic virotherapy. This involves the use of oncolytic vectors, which are virus-designed to home and kill the tumor cells without harming the normal cells in the body. The cancer cells are killed by cell lysis as a result of the production of cytotoxic proteins or due to the propagation of the virus itself. The viruses that have been used to produce oncolytic vectors include adenovirus, vaccinia, reovirus, HSV-1 and Newcastle disease virus.

The ongoing clinical trials include the use of adenovirus and herpes virus vectors. One example of adenoviral vector is ONYX-015, which lacks E1B protein, required for replication with a normal p53 pathway and RNA export during viral replication. It has been used to treat squamous cell carcinoma of the head and neck and has also been tested as a preventive treatment for oral precancerous tissue. The concept behind using this vector is that ONYX-015 will proliferate in p53 pathway-deficient tumor cells and kill them.

Two other vectors that have been used clinically are G207 and NV1020, both of which are modified HSV-1. They have been modified such that they could successfully replicate only in tumor cells. G207 cannot proliferate in nondividing cells and has high affinity for neurons. The mutations in NV1020 include insertion of thymidine kinase gene controlled by $\alpha4$ promoter, deletions in several components of the genome and deletion in the thymidine kinase region. These vectors are being tested for cancer gene therapy in treating malignant glioma (G207) and for colorectal and liver cancer (NV1020). They utilize two mechanisms to kill cancer cells: (a) lytic portion of life cycle directly causes killing and (b) thymidine kinase produced by the virus makes tumor cells susceptible to ganciclovir.

The third type of gene therapy for cancer treatment is gene insertion. These genes include suicide genes, cellular stasis genes and antiangiogenesis genes. The gene delivery systems used are both viral vectors and nonviral systems such as DNA transfer, oligodendroma DNA coatings and electroporation. The choice depends on the need for specificity or period of desired expression of the gene. The real challenge in this form of therapy is in avoiding the incorporation of these genes into the unintended target cells. The early experiments encountered the problem of gene silencing, where either the inserted gene was not expressed or its expression was very limited. The treatment agents employed in this form of gene therapy include TNFerade, Rexin-G and adenovirus delivering the HSVtk gene. TNFerade is a modified adenovirus-based vector that delivers the gene for TNF-α and is under the control of a promoter that is induced by radiation, which necessitates receiving radiation for the patient after this vector is injected. The combination of radiation and TNF-α kills the tumor cells, and this form of gene therapy has been tested for the treatment of melanoma, rectal cancer, pancreatic cancer and esophageal cancer. Rexin-G is a retroviral-based vector that delivers a gene that interferes with the function of cyclin G1 gene. The gene after entering into tumor cells disrupts the cell cycle and has been tested for pancreatic and colon cancers that have spread to the liver.

Cell-targeted suicide is another form of gene therapy, which is achieved by delivering a gene. Insertion of HSV-thymidine kinase gene via adenoviral vector into malignant cells in conjunction with the systemic administration of ganciclovir has gone through clinical trials to treat cancer. Ganciclovir is not toxic before it is metabolized by the enzyme. Consequently, the treatment would affect only the cells that have received the HSV-thymidine kinase gene. This treatment has been tested for patients with glioblastoma.

Other approaches involve the delivery of p53 gene into tumor cells via adenoviral vector. A mutation in p53 gene results in some types of cancer, and providing a healthy copy of the gene may stop proliferation of the tumor cells and also cause apoptosis. INGN201 has been used as a carrier of p53 gene for gene therapy in patients with squamous cell carcinoma of head and neck, glioma, bladder cancer, ovarian cancer and prostate cancer.

Cystic Fibrosis

Gene therapy has presented itself as an attractive option for treating cystic fibrosis over the decades. Most of the morbidity and mortality in patients with cystic fibrosis results from pulmonary dysfunction, and this is the most common genetic disorder among Caucasians. The lung is not suited for *ex vivo* gene transfer methods, and the removal and regrafting of airway cells is not practical. The organization of the cystic fibrosis transmembrane conductance regulator gene and its mutations causing the pathophysiological symptoms are now better understood. However, clinical success has been difficult to achieve despite human clinical trials using both the viral and nonviral vectors for gene delivery. The major problem has been the inability of the vector to reach the target cell in cystic fibrosis patients. The airway epithelium has proven to be difficult to penetrate due to its physical and immunological barriers, and vector transduction has not been optimal. This has been attributed to the low expression or absence of cellular receptors and coreceptors that are required for viral binding and entry into the target cell.

An effective gene therapy for patients with cystic fibrosis will require the delivery of cDNA encoding the cystic fibrosis transmembrane conductance regulator protein to the nucleus of the epithelial cells lining the bronchial trees within the lungs. Furthermore, an effective treatment could be achieved only if the transgene is expressed for the rest of the life of the patient. So far, while it has been possible to deliver the transgene to the nucleus of the target cells, the optimal expression of the gene has not been more than 30–40 days in clinical trials. The problem of limited expression has been further complicated by the ability of the host's immune response to reduce the effectiveness of viral vectors (adenovirus and adeno-associated virus) when administered as repeated doses. To overcome the problem of repeated dosing, nonviral vectors such as cationic liposomes have also been clinically tested, but they are less effective than viral vectors. Additional ligands are being used, which may facilitate endosomal escape or contain a nuclear localization signal. This may enhance gene delivery by cationic liposomes. The retroviral vectors are not suitable for cystic fibrosis because they require cell division, and the target cells in the airway have a very low replication rate.

Vasculature

The gene therapy for vasculature involves delivery of the therapeutic gene(s) to many different types of vascular cells including myocardium, endothelium, vascular smooth muscle and/or tissues involved in regulating the lipid levels. At least four different protocols have been used for this form of gene therapy, which include ex vivo gene therapy, cell-based genetic alteration, delivering the gene to the target locally in vivo and system transgene delivery. Ex vivo gene therapy can be used if therapeutic gene delivery to a tissue is possible safely and effectively. An example is the gene therapy of vein graft failure, which is done when a vein can be reached during a coronary artery bypass surgery, and this procedure has distinct advantages since there is a minimum immune response to the virus due to transduction with a limited amount of vector, and local delivery prevents systemic transgene expression.

The development of local delivery devices is essential in targeting vascular tissues for gene therapy since most of the tissues are inaccessible when the vector is administered systematically. Catheters that can be used under X-ray fluoroscopy guidance for gene delivery have been devised. This provides a contrast medium for arterial wall gene transfer where only the lumen of the vessel can be seen. It is expected that this process in combination with magnetic resonance imaging will demonstrate the atherosclerotic lesions and gene interaction. The devices under consideration for local gene therapy are stents, channel balloon, microporous coated stents, nipple catheters, microspheres, double balloon and hydrogel-coated catheters.

Cardiovascular Diseases

Ischemia

Gene therapy provides an alternative to amputation and heart transplant for the treatment of ischemia. The focus of gene therapy for ischemia has been on genes for angiogenic growth factor and fibroblast growth factor where the goal is to deliver the genes for these growth factors to the site of ischemia. Angiogenic growth factor is involved in the angiogenesis of endothelial cells. In clinical trials of patients with peripheral arterial disease, the intramuscular administration of angiogenic growth factor gene using adenovirus as the vector has demonstrated an increased endothelial cell function and lowering of extremity flow reserve. In another study, the administration of this gene resulted in the improvement of angina symptoms.

Atherosclerosis

Gene therapy for atherosclerosis is challenging since the disease is complex and involves the interaction of both genetic and environmental factors. Gene transfer is also difficult to achieve in the vasculature specifically when the administration involves atherosclerotic lesions as targets. A number of diverse genes have been considered for the treatment of atherosclerosis. Some examples are genes for LDL and VLDL receptors, hepatic lipases and lipoprotein, apoB mRNA editing enzyme, apolipoprotein A-I and lecithin–cholesterol acyltransferase. LDL receptor efficiency, a determinant of atherosclerosis, is a major genetic abnormality. The transfer of LDL or VLDL receptor gene overcomes this LDL receptor efficiency. Apolipoprotein E (ApoE), which is a protein present in circulation, has pleiotropic atheroprotective properties and has drawn serious consideration for the treatment of cardiovascular disease and hypercholesterolemia.

Hypertension

Hypertension is a major risk factor for strokes, atherosclerosis and peripheral vascular disease. The hyperactive renin–angiotensin system has been established as a contributing factor for primary hypertension. The first ever clinical trial of cell-mediated gene therapy for the treatment of pulmonary hypertension started on November 7, 2006. The clinical trial called PHACeT (pulmonary hypertension: assessment of cell therapy) uses engineered stem cell-like "endothelial progenitor cells," which are obtained from the blood of the patient. Since patients with pulmonary hypertension exhibit deficiencies in the production of nitric oxide, a DNA vector carrying a gene for endothelial nitric oxide synthase (eNOS) has been added and is delivered to the endothelial progenitor cells. This enzyme is involved in the production of nitric oxide, which is a vasodilator, and also plays an important role in the repair and regeneration of blood vessels. These engineered cells carrying the transgene are injected into the lung. This trial is expected to be completed in 2009.

Atrial natriuretic peptide (ANP) is involved in lowering the blood pressure by a number of mechanisms, which include relaxation of the blood vessel's smooth muscle cells, increasing the diameter of the blood vessel and altering the effects of vasoconstrictive agents. ANP is also involved in affecting the elimination of sodium and inhibiting the sympathetic nervous system. In an animal model, the ANP gene was added by using a modified adenoviral vector that was also connected to a gene regulatory system turned on by the drug mifepristone. The results of the study suggest that the experiments were successful in returning the blood pressure to the normal levels in this group. Additional targets for lowering blood pressure by gene therapy include endothelin and kallikrein.

Thrombosis

This endothelial cell dysfunction results in clot formation, and the anticlotting genes are associated with antiplatelet activity. The tissue plasminogen activator (TPA) gene is a suitable target to treat thrombosis; however, the continued expression of the TPA gene will be required since TPA has a short half-life. Additional antithrombosis gene products include antistatin, tissue factor pathway inhibitor, thrombomodulin and hirudin. The antithrombotic gene therapy could be useful in multiple clinical conditions including peripheral artery angioplasty, percutaneous transluminal coronary angioplasty, intravascular stenting and coronary artery bypass graft. Due to a delay in gene expression, it seems more appropriate that gene therapy for thrombosis is used for the prevention of chronic arterial narrowing and reocclusion.

Familial Hypercholesterolemia

Patients with familial hypercholesterolemia have an inherited disorder and possess deficient LDL receptors, resulting in the inability to process cholesterol correctly. These patients develop elevated levels of plasma cholesterol, which cause arteriosclerosis at an early age due to high levels of artery-clogging fat resulting in heart attacks and strokes. The disease provides an intricate gene therapy approach where the concept is to modify the liver so that it is able to express LDL receptors. The procedure involves ex vivo transgene therapy in which there is a partial surgical removal of the liver followed by insertion of corrected copies of the deficient gene and transplantation of the liver segment back to the patient.

On the basis of the experiments on gene delivery, only the adenovirus (among viral vectors) exhibits sufficient expression to produce a therapeutic response in some animal models, but a severe immune response remains a major hurdle. The other alternative has been the cationic liposome complex intravenous gene delivery, which has exhibited gene expression in vascular endothelial cells and monocytes/macrophages, and immune response is not a problem with these vectors. However, the efficiency of gene delivery is low, which limits their use. Some of these gene delivery efficiency problems have been overcome by using ligand-facilitated transfer of cationic liposome DNA.

Cell-based genetic modification has been used to treat familial hypercholesterolemia. This form of gene therapy involves collection of the target cells from the patient, insertion of the desired genes and regrafting of the modified autologous cells. Specifically, hepatocytes are harvested and transduced with retrovirus-expressing genes for LDL receptors and subsequently reimplanted into the patient but the procedure has shown only a limited success in correcting the problem, and the success of statins has further diminished the enthusiasm for the procedure.

Human Immunodeficiency Virus

HIV infection is reaching up to 40 million people (new revised numbers) worldwide and more than 1 million people in the United States. The mortality figures are about 3 million people per year worldwide and approximately 14,500 per year in the United States. The development of new drugs and the awareness about the infection have not slowed down the infection rate of the virus and the resulting death rate worldwide. The treatment regimens used reduce viral loads and enhance life expectancy. The cost of the drugs and dependency on a strict compliance make their usefulness very limited, specifically in the developing countries where the infection is spreading at an alarming rate. Furthermore, the spread of drug-resistant strains has added to the concerns about the disease.

Gene therapy has been under serious consideration as an antiviral treatment. One approach has been the RNA-mediated inhibition of HIV, which provides a superior alternative to antiretroviral agents that do not carry the risk of adverse side effects associated with the available reverse transcriptase and protease inhibitors. In addition to the antisense RNA therapy, RNAi and ribozyme-based gene therapy are also under investigation. These treatments are responsible for posttranscriptional gene silencing or splicing of viral RNA and inhibit HIV transcripts or HIV receptors. The recognition of the cognate RNA sequence is required for these treatments to succeed and the sequence-specific breakdown of the HIV genome could be achieved by injecting the cell with short interfering RNA. The entry and proliferation of HIV into the $CD4^+$ cells can be prevented by targeting CD4 receptors and the gag gene by using RNAi. However, for these treatments to succeed, long-term expression of siRNA is essential. Long-term expression of siRNA will enable the sustained production of anti-HIV genes.

Since HIV has a very high mutation rate, the effectiveness of both siRNA and ribozyme sequence treatments has limitations, and reduces the impact of their treatment as HIV mutates and evades siRNA due to its short length. Ribozyme recognition sequences are also small, and any mutation in its cleavage site will render resistance. Antisense RNA therapy overcomes these limitations because long antisense sequences targeting HIV force the virus to mutate at a much higher rate to escape inhibition of replication caused by antisense RNA. In a phase I clinical trial, a lentiviral vector expressing a gene against HIV envelope has been tested. The clinical trial employed lentiviral vector VRX496 containing a long antisense sequence to HIV envelope while retaining the full LTRs of HIV and without its self-inactivation. It was anticipated that using long antisense sequence can target multiple sites of HIV, which may limit the ability of the virus to form resistant mutants. For this trial, peripheral blood lymphocytes were obtained by apheresis, $CD4^+$ T cells were purified and transduced with the VRX496 vector, and were then infused into the patient. All patients in the study tolerated a single IV infusion of gene-modified autologous CD4 T cells. The key observations from the study were that these patients with late-stage HIV infection retained modified $CD4^+$ T cells over the long term with minimum immunogenicity. However, a complete safety profile of this gene vector therapy has not been established at this time, since there is a latency period of 3 years for adverse effects. A total of five patients were enrolled in this study; four of the five patients exhibited an increase in the cellular response to HIV, three patients had improved T-cell memory responses and one subject produced a very strong antiviral response.

Hematopoietic Stem Cell as a Target for Gene Therapy

Various inherited and acquired disorders could potentially be corrected by the introduction of transgenes into bone marrow stem cells. Hematopoietic stem cells are self-renewable and are able to differentiate and proliferate. They are readily accessible and can be easily administered back to the patient by autologous transplantation. This form of treatment would benefit a number of disease states including thalassemias, sickle-cell disease, various lymphocyte disorders, chronic granulocyte diseases, chemotherapy-induced myelosuppression and AIDS. Lentiviral vectors could be efficient tools for hematopoietic stem cell gene delivery since the stem cells are quiescent and consequently are difficult targets for vectors that require dividing cells. However, gene therapy of hematopoietic stem cells is still limited by their quiescent nature, low frequency of the target cells, poor grafting ability of gene-transduced stem cells and inability of a growth advantage for genetically modified cells.

Epilepsy

Gene therapy is a potential viable alternative for the treatment of epileptic patients who do not respond well to the conventional antiepileptic drugs. With the use of transgene viral vectors, various gene targets, specifically the inhibitory and excitatory neurotransmitters, have been the focus of interest. In animal models, success has been reported when neuropeptide genes including neuropeptide Y and galanin have been transduced in specific areas of the brain. The intracerebral applications of the transgene were performed with an emphasis on preventing the seizures by using recombinant adeno-associated viral vectors, and the results were a reduction in seizures, delay in fully kindled seizures and the production of neuroprotection. The concerns that need to be addressed before human clinical trials include the production of an immune response, specifically the synthesis of antibodies to the viral vector, promoter silencing, stability of the transgene and loss of transduced cells. Only a single experiment has been reported for gene therapy to treat human epilepsy in which adeno-associated viral vector containing a lacZ marker gene was used for direct gene transfer into human epileptogenic hippocampal tissue using brain slices. However, the technology has not been used for clinical practice, and the expression of the transgene in neurons will be critical for the success of such a protocol. Additional targets for these studies include genes for adenosine, anti-NMDA, cholecystokinin, GLUT-1, neurotrophins and calbindin.

Challenges Associated with Successful Gene Therapy

The idea of gene therapy was born to treat diseases that are caused by defects in single genes; clinical trials have been conducted for four diseases—ornithine transcarbamylase deficiency, hemophilia resulting from Factor IX deficiency, chronic granulomatous disease and severe combined immunodeficiency (SCID)—with mixed results. Although no sustained clinical benefits were observed after gene therapy in patients with ornithine transcarbamylase deficiency and Factor IX deficiency hemophilia, the development of a functional immune response was reported in patients who received gene therapy for SCID. The clinical trials for the treatment of SCID were performed in France, Italy and Britain with the report of one death. The patients receiving gene therapy (MDS1-EVI1) for chronic granulomatous disease developed functional neutrophils and clonal myeloproliferation.

These clinical trials have pointed to a number of adverse effects, problems with in vivo administration of viral vectors and the severe concerns of insertional oncogenesis. Several problems associated with in vivo administration of viral vectors have been identified, which includes the following:

(1) An immune response is produced against the viral vectors that make subsequent administration of viral vectors less feasible, specifically when a sustained expression of the gene is a major issue.
(2) The viral vector infects nontarget cells also resulting in oncogenesis or severe inflammatory responses.
(3) Although germline transduction has not yet been reported, its risks continue to be a major concern for the success of gene therapy protocols.

In all the patients developing cancer, the genetic defect was treated by inserting a therapeutic gene into a modified retrovirus. This vector was then infected into bone marrow stem cells isolated from each patient and subsequently administered back into the patient. The transgene encoded the common γ chain of the IL-2 receptor (γ_c), which is defective in patients with SCID. Unfortunately, these vectors also activated a cancer-promoting gene, which was due to the transduction of the retroviral vector close to the promoter of LM02, which is a proto-oncogene. The development of leukemia has been a major concern for the future of gene therapy both in Europe and the United States due to reactions from regulatory authorities, but despite the risk of leukemia, gene therapy is a superior alternative to mismatched bone marrow transplantation for SCID patients. Nevertheless, all gene transfer techniques that lead to the integration of DNA into chromosomes may cause mutagenesis due to inactivation of tumor suppressor genes or activation of a proto-oncogene. There has also been concern that patients with chronic granulomatous disease undergoing gene therapy may develop leukemia and myelodysplasia. Nevertheless, so far, the studies have not provided sufficient data to link gene therapy with oncogenesis in human clinical trials. Four possible ways have been recommended to decrease the risk of oncogenesis:

(1) Improvement in vector design;
(2) Buffer the genome from the effect of viral integration;

(3) Controlling transgene integration; and
(4) Production of genetically modified stem cells in vitro and infusion of these nononcogenic cells into the patient.

Conclusion

Although significant progress has been made over the past several years in developing new technologies for gene therapy, the outcomes have not matched the expectations, and the "Gelsinger Case" in Philadelphia continues to raise a red flag. The identification of therapeutically suitable genes, development of efficient viral and nonvector systems, and efficacy and safety of the clinical trials are essential to attain the high expectations from this form of treatment.

Bibliography

Alba R, Bosch A, Chillou M. 2005. Gutless adenovirus: Last generation adenovirus for gene therapy. Gene Ther. 12:518–527.
Alemany R. 2007. Cancer selective adenoviruses. Mol Aspects Med. 28:42–58.
Amalfitano A, Parks RJ. 2002. Separating fact from fiction: Assessing the potential of modified adenovirus vectors for use in human gene therapy. Curr Gene Ther. 2:111–133.
Anderson WF, Fletcher JC. 1980. Gene therapy in human beings: When is it ethical to bring? NEJM. 303:1293–1297.
Anderson WF. 1984. Prospects for human gene therapy. Science. 226:401–409.
Anderson WF. 1992. Human gene therapy. Science. 256:808–813.
Anson DS, Smith GJ, Parsons DW. 2006. Gene therapy for cystic fibrosis airway disease – Is clinical success imminent? Curr Gene Ther. 6:161–179.
Baum C, Dullmann J, Li Z, Fehse B, et al. 2003. Side effects of retroviral gene transfer into hematopoietic stem cells. Blood. 101:2099–2114.
Blaese RM, Culver KW, Miller AD, Carter CS, et al. 1995. T lymphocyte directed gene therapy for ADA deficiency SCID: Initial trial results after four years. Science. 270:475–480.
Boucher RC. 1999. Status of gene therapy for cystic fibrosis lung disease. J Clin Inv. 103:441–445.
Brand TC, Tolcher AW. 2006. Management of high risk metastatic prostate cancer: The case for novel therapies. J Urol. 176:576–580.
Brunetti-Pierri N, Ng P. 2006. Progress towards the application of helper-dependent adenoviral vectors for liver and lung gene therapy. Curr Opin Mol Ther. 8:446–454.
Caplen NJ. 2003. RNA1 as a gene therapy approach. Expert Opin Biol Ther 3:575–586.
Cavazzana-Calvo M, Thrasher A, Mavilio F. 2004. The future of gene therapy. Nature. 427: 779–781.
Cavazzana-Calvo M, Fischer A. 2007. Gene therapy for severe combined immunodeficiency: Are we there yet? J Clin Inv. 117:1456–1465.
Crittenden M, Gough M, Chester J, Kottle T, et al. 2003. Pharmacologically regulated production of targeted retrovirus from T cells for systemic antitumor gene therapy. Cancer Res. 63: 3173–3180.
Culver KW. 1996. Gene Therapy. A Primer for Physicians, Mary Ann Liebert Inc., New York.
DiPaola RS, Plante M, Kaufman H, Petrylak DP, et al. 2006. A phase I trial of pox PSA vaccines (PROSTVAC-VF) with B7-1, ICAM-1 and LFA-3 co-stimulatory molecules (TRICOM) in patients with prostate cancer. J Transl Med. 4:1–1.

Djordjevic M. 2007. SELEX experiments: New prospects, applications and data analysis in inferring regulatory pathways. Biomol Eng. 24:179–189.

Dodge JA. 1998. Gene therapy for cystic fibrosis: What message for the recipient? Thorax. 53: 157–158.

Eck SL, Wilson JM. 1996. Gene-based therapy. In: Goodman and Gilman's The Pharmacological Basis of Therapeutics. Hardman JG, Limbird LE, et al., Eds. McGraw Hill, New York, pp. 77–1022.

Editorial, 1997. Supervising gene therapy openly. Lancet. 350:79.

Einfeld DA, Roelvink PW. 2002. Advances towards targetable adenovirus targetable adenovirus vectors for gene therapy. Curr Opin Mol Ther. 4:444–451.

Epstein AL, Marconi P, Argnani R, Manservigi R. 2005. HSV-1 derived recombinant and amplicon vectors for gene transfer and gene therapy. Curr Gene Ther. 5:445–448.

Flotte TR, Carter BJ. 1997. In vivo gene therapy associated with adeno-associated virus vectors for cystic fibrosis. Adv. Pharmacol. 40:85–101.

Freidmann T. 1989. Progress toward human gene therapy. Science. 244:1275–1281.

Fuch M. 2006. Gene therapy: An ethical profile of a new medical territory. J Gene Med. 8: 1358–1362.

Galanis E, Vile R, Russell SJ. 2001. Delivery systems intended for in vivo gene therapy of cancer: Targeting and replication competent viral vectors. Crit Rev Oncol Hematol. 38:177–192.

Galanis E, Okuno SH, Nascimento AG, Lewis BD, et al. 2005. Phase I–II trial of ONYX-015 in combination with MAP chemotherapy in patients with advanced sarcomas. Gene Ther. 12:437–445.

Gardlik R, Palffy R, Hodosy J, Lukacs J, et al. 2005. Vectors and delivery systems in gene therapy. Med Sci Monitor. 11:RA110–121.

Gordon EM, Lopez FF, Cornelio GH, Lorenzo CC, et al. 2006. Pathotropic nanoparticles for cancer gene therapy Rexin-G IV: Three year clinical experience. Int J Oncol. 29:1053–1064.

Gordon EM, Chan MT, Geraldino N, Lopez FF, et al. 2007. Le morte du tumour: Histological features of tumor destruction in chemo-resistant cancers following intravenous infusions of pathotropic nanoparticles bearing therapeutic genes. Int J Oncol. 30:1297–1307.

Gutermann A, Mayer E, von-Dehn-Rothfelser K, Breidenstein C, et al. 2006. Efficacy of oncolytic herpes virus NV1020 can be enhanced in combination with chemotherapeutics in colon carcinoma cells. Hum Gene Ther. 17:1241–1253.

Harris JD, Evans V, Owen JS. 2006. ApoE gene therapy to treat hyperlipidemia and atherosclerosis. Curr Opin Mol Ther. 8:275–287.

Hege KM, Jooss K, Pardoll D. 2006. GM-CSF gene modified cancer cell immunotherapies: Of mice and men. Int Rev Immunol. 25:321–352.

Hermonat PL, Mehta JL. 2004. Potential of gene therapy for myocardial ischemia. Curr Opin Cariol. 19:517–523.

Joyner A, Keller G, Phillips RA, Bernstein A. 1983. Retrovirus transfer of a bacterial gene into mouse hematopoietic progenitor cells. Nature. 305:556–558.

Kaiser J. 2005. Gene therapy. Putting the fingers on gene repair. Science. 310:1894–1896.

Kanai R, Tomita H, Hirose Y, Ohba S, et al. 2007. Augmented therapeutic efficacy of an oncolytic herpes simplex virus type 1 mutant expressing ICP345 under the transcriptional control of musashi 1 promoter in the treatment of malignant glioma. Hum Gene Ther. 18:63–73.

Kanerva A, Hemminki A. 2005. Adenoviruses for treatment of cancer. Ann Med. 37:33–43.

Kaplan JM. 2005. Adenovirus based cancer gene therapy. Curr Gene Ther. 5:595–605.

Katz B, Goldbaum M. 2006. Macugen (pegaptanib sodium), a novel ocular therapeutic that targets vascular endothelial growth factor (VEGF). Int Opthamol Clin. 46:141–154.

Kaufman HL, Cohen S, Cheung K, DeRaffele G, et al. Local delivery of vaccinia virus expressing multiple costimulatory molecules for the treatment of established tumors. Hum Gene Ther. 17:239–244.

Kay MA, Liu D, Hoogerbrugge G. 1997. Gene therapy. Proc Nat Acad Sci. USA. 94:12744–12746.

Kay MA, Glorioso JC, Naldini L. 2001. Viral vectors for gene therapy: The art of turning infectious agents into vehicles of therapeutics. Nat Med. 7:33–40.

Kemeny N, Brown K, Covey A, Kim T, et al. 2006. Phase I. Open-label, dose-escalating study of a genetically engineered herpes simplex virus, NV1020, in subjects with metastatic colorectal carcinoma to the liver. Hum Gene Ther. 17:1214–1224.

Kim HJ, Jang SY, Park JI, Byun J, eet al. 2004. Vascular endothelial growth factor-induced angiogenic gene therapy in patients with peripheral artery disease. Exp Mol Med. 36:336–344.

Kolb M, Martin G, Medina M, Ask K, et al. 2006. Gene therapy for pulmonary diseases. Chest. 130:879–884.

Kootstra NA, Verma IM. 2003. Gene therapy with viral vectors. Ann Rev Pharmacol Toxicol. 43:413–439.

Kudo-Saito C, Wansley EK, Gruys ME, Wiltrout R, et al. 2007. Combination therapy of an orthotropic renal cell carcinoma model using intratumoral vector-mediated costimulation of systemic interleukin-2. Clin Canc Res. 13:1936–1946.

Kyte JA, Mu L, Aamadal S, Kvalheim G, et al. 2006. Phase I/II trial of melanoma therapy with dendritic cells transfected with tumor-RNA. Cancer Gene Ther. 13:905–918.

Levine BL, Humeau LM, Boyer J, Rebello T, et al. 2006. Gene transfer in humans using a conditionally replicating lentiviral vector. PNAS. 103:17372–17377.

Levy B, Panicalli D, Marshall J. 2004. TRICOM: Enhanced vaccine as anticancer therapy. Exp Rev Vaccines. 3:397–402.

Li D, Jiang W, Bishop S, Ralston R, et al. 1999. Combination surgery and nonviral Interleukin-2 gene therapy for head and neck cancer. Clin Cancer Res. 5:1551–1556.

Li D, Shugert E, Guo M, Bishop JS, et al. 2001. Combination non viral interleukin-2 and interleukin-12 gene therapy for head and neck squamous cell carcinoma. 127:1319–1324.

Li M, Li H, Rossi JJ. 2006. RNAi in combination with ribozyme and TAR decoy for treatment of HIV infection in hematopoietic cell gene therapy. Ann NY Acad Sci. 1082:172–179.

Li SD, Huang L. 2006. Gene therapy progress and prospects. Gene Ther. 13:1313–1319.

Liu L, Liu H, Visner G, Fletcher BS. 2006. Sleeping beauty-mediated eNOS gene therapy attenuates monocrotaline-induced pulmonary hypertension in rats. FASEB J. 20: 2594–2596.

Lou E, Marshall J, Aklilu M, Cole D, et al. 2006. A phase II study of active immunotherapy with PANVAC or autologous, cultured dendritic cells infected with PANVAC after complete resection of hepatic metastasis of colorectal carcinoma. Clin Colorectal Cancer 5: 368–371.

Lu X, Yu Q, Binder GK, Chen Z, et al. 2004. Antisense-mediated inhibition of human immunodeficiency virus replication by use of an HIV type I based vector results in severely attenuated mutants incapable of developing resistance. J Virol. 78:7079–7088.

Lundstrom K, boulikas T. 2003. Viral and nonviral vectors in gene therapy: Technology development and clinical trials. Technol Canc Res Treat. 2:471–486.

Luque F, Oya R, Macias D, Saniger L. 2005. Gene therapy for HIV infection: Are lethal genes a valuable tool? Cell Mol Biol. 51:93–101.

Madan RA, Arlen PM, Gulley JL. 2007. PANVAC-VF: Poxviral-based vaccine therapy targeting CEA and MUC1 in carcinoma. Expert Opin Biol Ther 7:543–544.

Mann R, Mulligan RC, Baltimore D. 1983. Construction of a retrovirus packaging mutant and its use to produce helper-free defective retrovirus. Cell. 33:153–159.

Marathe JG, Wooley DP. 2007. Is gene therapy a good therapeutic approach for HIV positive patients? Genet Vaccines Ther 5:5–5.

McConnell MJ, Imperiale MJ. 2004. Biology of adenovirus and its use as a vector for gene therapy. Hum Gene Ther. 15:1022–1033.

McLaughlin JM, McCarty TM, Cunningham C, Clark V, et al. 2005. TNFerade, an adenovector carrying the transgene for human tumor necrosis factor alpha, for patients with advanced solid tumors: Surgical experience and long term followup. Ann Surg Oncol. 12: 825–830.

Middleton PG, Alton EW. 1998. Gene therapy for cystic fibrosis: Which postman, which box? Thorax. 53:197–199.

Miller AD, Jolly DJ, Friedmann T, Verma IM. 1983. A transmissible retrovirus expressing human hypoxanthine phosphoribosyltransferase (HPRT): Gene transfer into cells obtained from human deficient in HPRT. Proc Nat Acad Sci USA. 80:4709–4713.

Miller AD. 1992. Human gene therapy come of age. Nature. 357:455–460.

Neff T, Beard BC, Kiem HP. 2006. Survival of the fittest in vivo selection and stem cell gene therapy. Blood. 107:1751–1760.

Nemunaitis J. 2005. Vaccines in cancer: GVAX, a GM-CSF gene vaccine. Exp Rev Vaccines. 4:259–274.

Nemunaitis J, Murray N. 2006. Immune-modulating vaccines in non-small cell lung cancer. J Thorac Oncol. 1:756–761.

Nemunaitis J, Nemunaitis J. 2007. A review of vaccine clinical trials for non-small cell lung cancer. Exp Opin Biol Ther. 7:89–102.

Niidome T, Huang L. 2002. Gene therapy progress and prospects: Nonviral vectors. Gene Ther. 9:1647–1652.

Noe F, Nissinen J, Pitkanen A, Gobbi M., et al. 2007. Gene therapy in epilepsy, the focus on HPY. Peptides. 28:377–383

Osterman JV, Waddell A, Aposhian HV. 1970. DNA and gene therapy: Uncoating of polyoma pseudovirus in mouse embryo cells. Proc Nat Acad Sci USA. 67:37–40.

Palmer DH, Young LS, Mautner V. 2006. Cancer gene therapy: Clinical trials. Trends Biotechnol. 24:76–82.

Petrulio CA, Kaufman HL. 2006. Development of the PANVAC-VF vaccine for pancreatic cancer. Exp Rev Vaccines 5:9–19.

Philpott NJ, Thrasher AJ. 2007. Use of nonintegrating lentiviral vectors for gene therapy. Hum Gene Ther. 18:483–489.

Pickles RJ. 2004. Physical and biological barriers to viral vector-mediated delivery of genes to the airway epithelium. Proc Am Thorac Soc. 1:302–308.

Ponnazhagam S, Curiel DT, Shaw DR, Alvarez RD, et al. 2001. Adeno-associated virus for cancer gene therapy. Canc Res. 61:6313–6321.

Poreteus MH, Connelly JP, Pruett SM. 2006. A look to future directions in gene therapy research. PLOS Genet. 2:133–2141.

Prchal JT. 2003. Delivery on demand – A new era of gene therapy? NEJM. 348:1282–1283.

Preuss MA, Curiel DT. 2007. Gene therapy: Science fiction or reality? South Med J. 100:101–104.

Puck JM, Malech HL. 2006. Gene therapy for immune disorders: Good news tampered by bad news. J All Clin Immunol. 117:865–869.

Puddu GM, Cravero E, Ferrari E, Muscari A, et al. 2007. Gene based therapy for hypertension – Do preclinical data suggest a promising future? Cardiology. 108:40–47.

Puhlmann M, Brown CK, Gnant M, Huang J, et al. 2000. Vaccinia as a vector for tumor directed gene therapy: Biodistribution of a thymidine kinase-deleted mutant. Cancer Gene Ther. 7: 66–73.

Qian SH, Channon K, Neplioueva V, Wang Q, et al. 2001. Improved adenoviral vector for vascular gene therapy. Circ Res. 88:911–917.

Que NS, Sallenger BA. 2007. Gene therapy progress and prospects; RNA aptamers. Gene Ther. 14:283–291.

Que-Gewirth NS, Sullenger BA. 2007. Gene therapy progress and prospects: RNA aptamers. Gene Ther. 14:283–291.

Rainov NG, Ren H. 2003. Clinical trials with retrovirus mediated gene therapy – What have we learned? J Neurooncol. 65:227–236.

Raizada MK, Der-Sarkissian S. 2006. Potential of gene therapy strategy for the treatment of hypertension. Hypertension. 47:6–9.

Ramqvist T, Andreasson K, Dalianis T. 2007. Vaccination, immune and gene therapy based on virus-like particles against viral infection and cancer. Exp Opin Biol Ther 7:997–1007.

Rein DT, Breidenbach M, Curiel DT. 2006. Current developments in adenovirus based cancer gene therapy. Future Oncol. 2:137–143.

Romano G. 2006. The controversial role of adenoviral-derived vectors in gene therapy programs: Where do we stand? Drug News Persp. 19:99–106.

Rosenecker J, Huth S, Rudolph C. 2006. Gene therapy for cystic fibrosis lung disease: Current status and future perspectives. Curr Opin Mol Ther. 8:439–445.

Sabbioni S, Callegari E, Manservigi M, Argnani R, et al. 2007. Use of herpes simplex virus type 1-based amplicon vector for delivery of small interfering RNA. Gene Ther. 14:459–464.

Salmons B, Gunzburg WH. 1993. Targeting of retroviral vectors for gene therapy. Hum Gene Ther. 4:129–141.

Schillinger KJ, Tsai SY, Taffet GE, Reddy AK, et al. 2006. Regulatable atrial natriuretic peptide gene therapy for hypertension. Proc Nat Acad Sci USA. 27:13789–13794.

Senzer N, Mani S, Rosemurgy A, Nemunaitis J, et al. 204. TNFerade biologic, an adenovector with a radiation-inducible promoter, carrying the human tumor necrosis factor alpha gene: A phase I study in patients with solid tumors. J Clin Oncol. 22:592–601.

Shen Y, Nemunaitis J. 2006. Herpes simplex virus 1 (SHV-1) for cancer treatment. Cancer Gene Ther. 13:975–992.

Shichiri M, Tanaka A, Hirata Y. 2003. Intravenous gene therapy for familial hypercholesterolemia using ligand facilitated transfer of liposomes LDL receptor gene complex. Gene Ther. 10: 827–831.

Simon JW, Sacks N. 2006. Granulocyte-macrophage colony stimulating factor-transduced allogeneic cancer cellular immunotherapy: The GVAX vaccine for prostate cancer. Urol Oncol. 24:419–424.

Small EJ, Sacks N, Neumanitis J, Urba WJ, et al. 2007. Granulocyte macrophage colony-stimulating factor-secreting allogeneic cellular immunotherapy for hormone-refractory prostate cancer. Clin Cancer Res. 13:3883–3891.

Song S, Morgan M, Ellis T, Poirier A, et al. 1998. Sustained secretion of human-alpha-1-antitrypsin from murine muscle transduced with adeno-associated virus vectors. Proc Nat Acad Sci. 95:14384–14388.

Tan PH, Tan PL, George AJ, Chan CL. 2006. Gene therapy for transplantation for viral vectors – How much of the promise has been realized? Expert Opin Biol Ther. 6:759–772.

Tatum EL. 1966. Molecular biology, nucleic acids and the future of medicine. Persp Biol Med. 10:19–32.

Thomson L. 1994. Correcting the Code: Inventing and Genetic Cure for Human Body. Simon and Schuster, New York. pp.189–267.

Tomanin R, Scarpa M. 2004. Why do we need new gene therapy viral vectors? Characteristics, limitations and future perspectives of viral vector transduction. Curr Gene Ther. 4: 357–372.

Tsai TH, Chen SL, Xiao X, Chiang YH, et al. 2006. Gene therapy of focal cerebral ischemia using defective recombinant adeno-associated virus vectors. Front Biosci. 11:2061–2070.

Vahakangas E, Yla-Herttuala S. 2005. Gene therapy of atherosclerosis. Handbook Exp Pharm. 170:785–807.

Verma IM, Weitzman MD. 2005. Gene therapy twenty first century medicine. Ann Rev Biochem. 74:711–738.

Von-Laer D, Hasselmann S, Hasselmann K. 2006. Gene therapy for HIV infection: What does it need to make it work? J Gene Med 8:658–667.

Vorburger SA, Hunt KK. 2002. Adenoviral gene therapy. The Oncologist 7:46–59.

Watanabe S, Temin HM. 1983. Construction of a helper cell line for avian reticuloendotheliosis virus cloning vectors. Mol Cell Biol. 3:2241–2249.

Weber E, Anderson WF, Kasahara N. 2001. Recent advances in retrovirus vector-mediated gene therapy: Teaching an old vector new tricks. Curr Opin Mol Ther. 3:439–453.

White AF, Ponnazhagan S. 2006. Airway epithelium directed gene therapy for cystic fibrosis. Med Chem. 2:499–503.

Wigler M, Silverstein S, Lee LS, Pellicer A, et al. 1977. Transfer of purified herpes virus thymidine kinase gene to cultured mouse cells. Cell. 11:223–232.

Wolff JA. 1994. Gene Therapeutics: Methods and Applications of Direct Gene Transfer. Birkhauser, Boston, MA.

Wolff JA, Lederberg J. 1994. An early history of gene transfer and therapy. Hum Gene Ther. 5: 469–480.

Woods NB, Ooka A, Karlsson S. 2002. Development of gene therapy for hematopoietic stem cells using lentiviral vectors. Leukemia. 16:563–569.

Xiao W, Chirmule N, Berta SC, McCullough B, et al. 1999. Gene therapy vectors based on adeno-associated virus type 1. J Virol. 73:3994–4003.

Young LS, Mautner V. 2001. The promise and potential hazards of adenovirus gene therapy. Gut. 48:733–736.

Young LS, Searle PF, Onion D, Mautner V. 2006. Viral gene therapy strategies. J Pathol. 208: 299–318.

Index

Printed in the United States of America

LIVERPOOL
JOHN MOORES UNIVERSITY
AVRIL ROBARTS LRC
TITHEBARN STREET
LIVERPOOL L2 2ER
TEL. 0151 231 4022